Autodesk Inventor Fundamentals: Conquering the Rubicon

VOLUME 2
Chapters 12-20

Elise Moss

ISBN: 1-58503-301-4

PUBLICATIONS

Schroff Development Corporation

www.schroff.com
www.schroff-europe.com

Lesson 12
Drawing Management

Learning Objectives:

Upon completion of this lesson, the user will be familiar with:

- Creating Base Views
- Creating Orthographic Views
- Creating Auxiliary Views
- Creating Section Views
- Creating Detail Views
- Creating Sheets
- Creating Title Blocks
- Modifying Title Blocks
- Managing Views
- Managing Sheets

We do not see either the Drawing Management toolbar or the Drawing Annotation toolbar unless we are in the drawing layout environment.

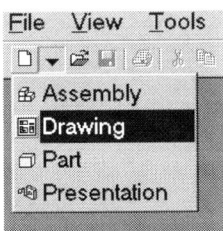

To get there, we select Drawing under the New File pull-down.

The Drawing Views Panel Toolbar

Base View

Base View creates and places a single view into a drawing, independent of any existing views. The dialog box remains open until you close it, enabling you to make changes to view setup before placing the view.

Autodesk Inventor Fundamentals

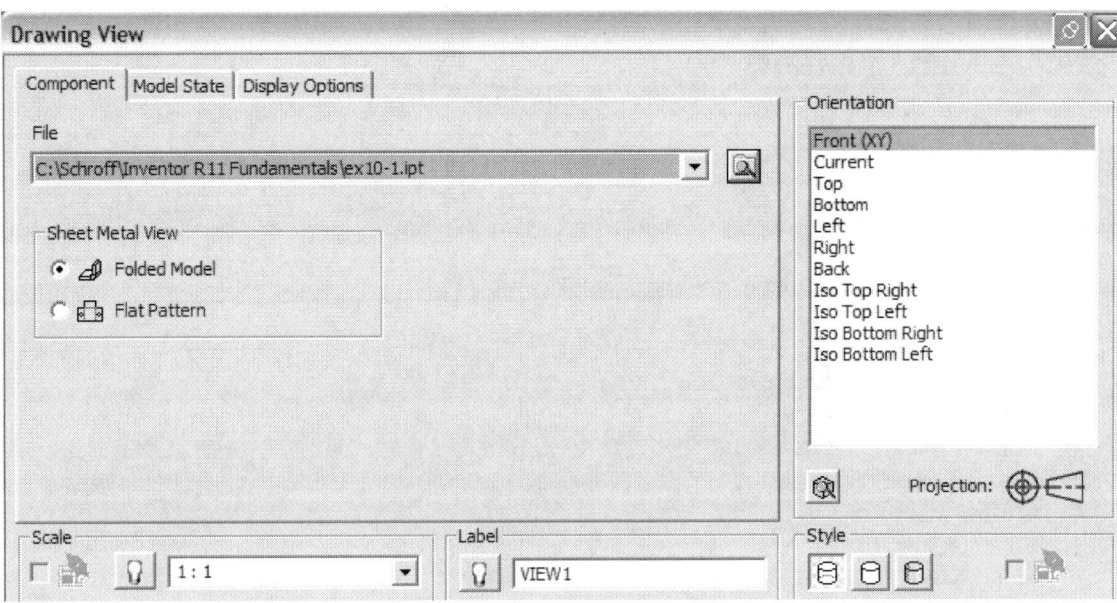

Drawing View selects the part or assembly file and sets the orientation of the view.		
File	Specifies the part, assembly, or presentation file to use for the drawing view. Specify the file name in one of the following ways. • Enter a path and file name in the box. • Click the down arrow to select from the list of open files. • Click the Explore button to browse for the file.	
Sheet Metal View	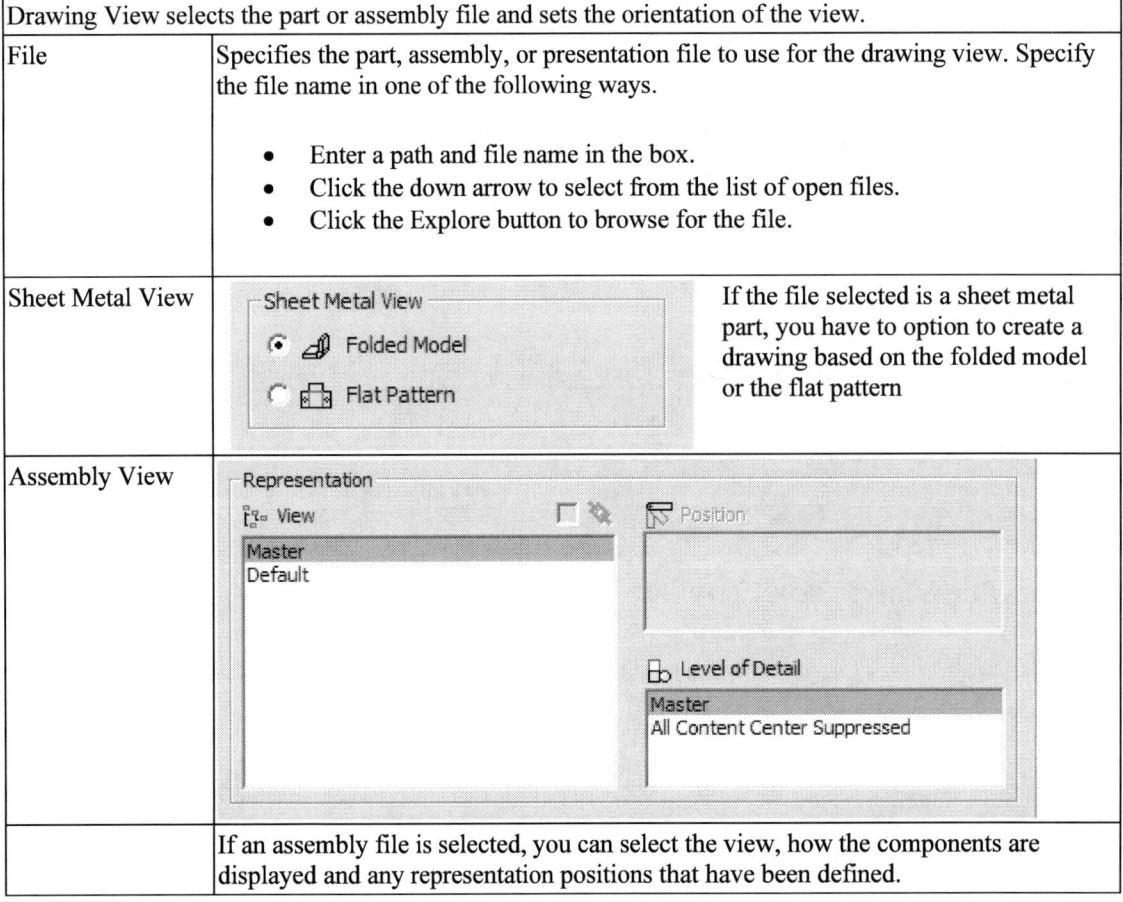	If the file selected is a sheet metal part, you have to option to create a drawing based on the folded model or the flat pattern
Assembly View		
	If an assembly file is selected, you can select the view, how the components are displayed and any representation positions that have been defined.	

Drawing Management

Orientation	Sets the view orientation. Select one of the standard views from the list. If you are creating a view from a presentation view, the last item on item on the list is the saved camera view of the presentation. You can also use the Orientation Window to set an orientation.
Display Style sets the display style for the view. To change the display style, click a button.	
	Sets the display to show hidden lines.
	Sets the display to remove hidden lines.
	Sets the display to a shaded rendering.
Scale	Sets the scale for the view. Enter the desired scale in the box or click the arrow to select from a list of commonly used scales.
Label	Displays a View label.

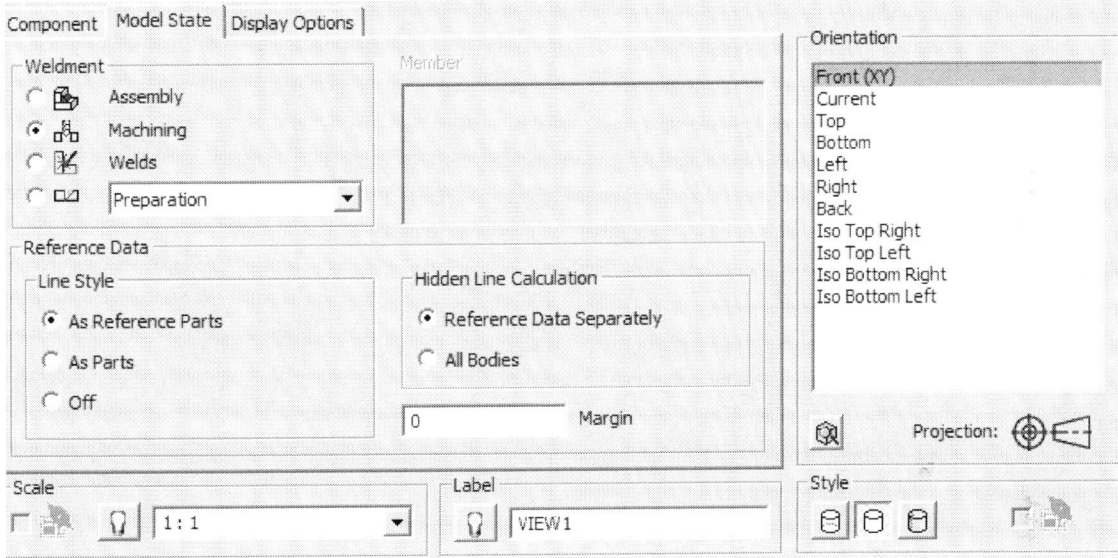

12-3

Autodesk Inventor Fundamentals

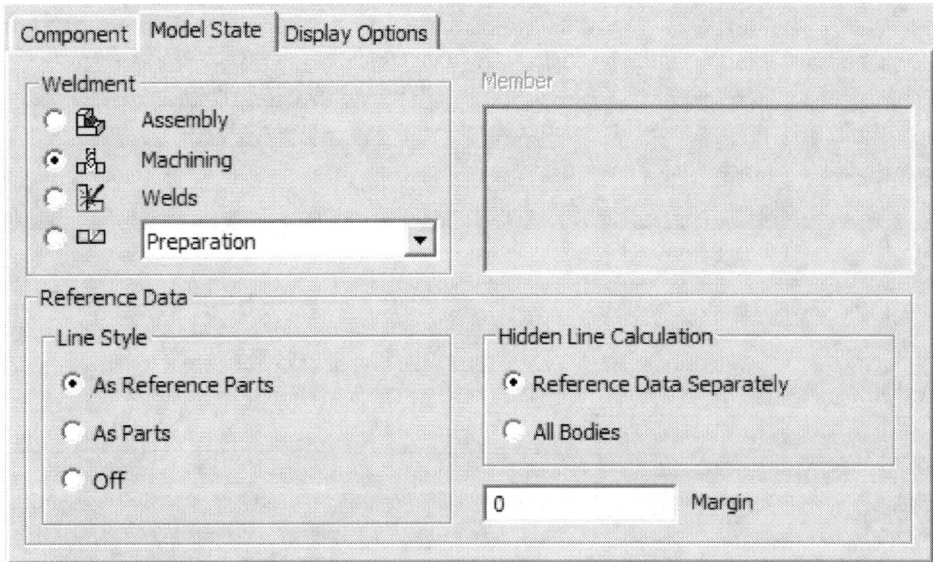

Model State tab

Weldment	
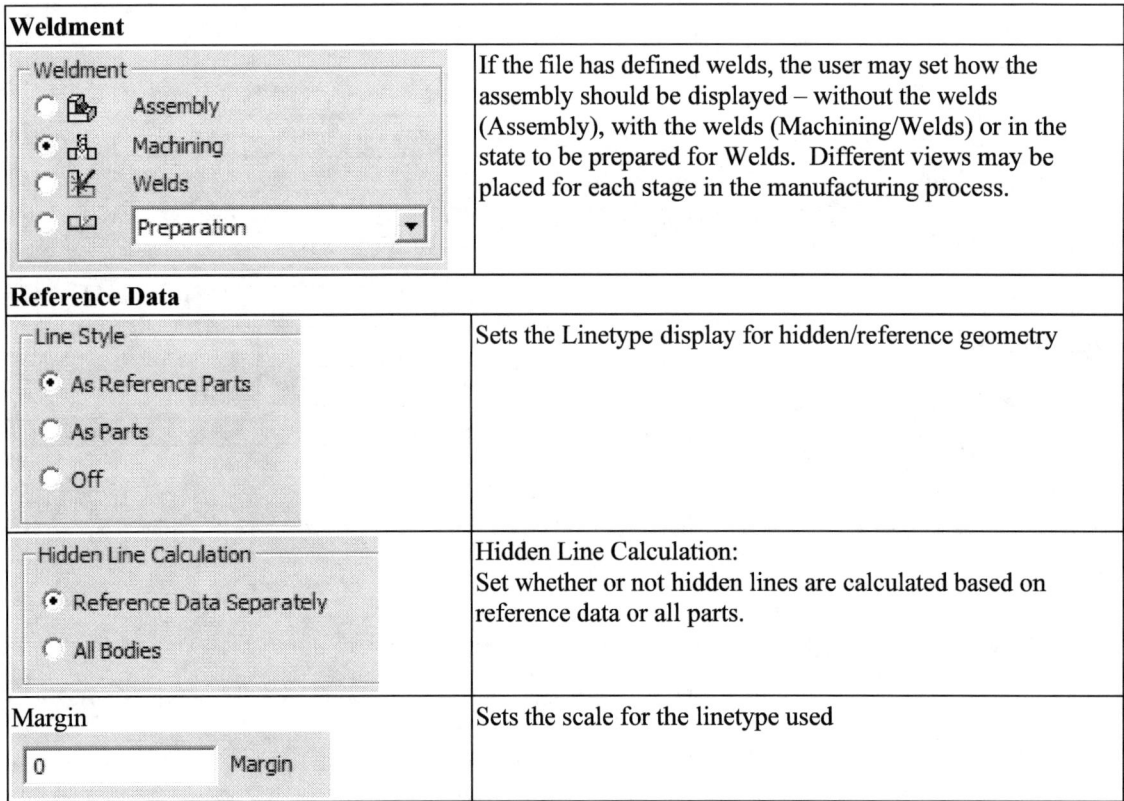	If the file has defined welds, the user may set how the assembly should be displayed – without the welds (Assembly), with the welds (Machining/Welds) or in the state to be prepared for Welds. Different views may be placed for each stage in the manufacturing process.
Reference Data	
Line Style options (As Reference Parts, As Parts, Off)	Sets the Linetype display for hidden/reference geometry
Hidden Line Calculation (Reference Data Separately, All Bodies)	Hidden Line Calculation: Set whether or not hidden lines are calculated based on reference data or all parts.
Margin	Sets the scale for the linetype used

Drawing Management

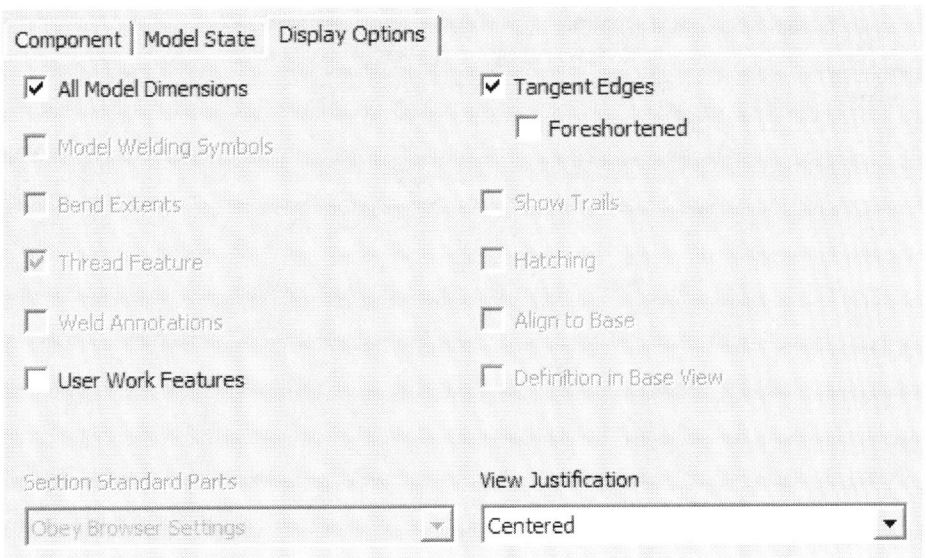

Display Options tab

Display – these options are available based on the part/assembly/presentation file selected	
Model Dimensions	Enabling this button automatically inserts model dimensions with the view. Only those dimensions that are planar to the view and have not been used in existing views on the sheet will display.
Model Weld Symbols	This feature is enabled for weldment assemblies. Toggles the display of the weldment notes
Bend Extents	This feature is enabled for sheet metal parts.
Thread feature	Toggles the display for threads
Weld Annotations	Toggles the display for weldment assemblies
Work Features	Toggle the display of work planes/axis/work points
Tangent Edges	Toggle the display of tangent edges
Show Trails	Display a dotted line in presentation files using exploded views
Hatching	Toggles the display of hatching
Align to Base	Aligns projected views to the main/parent view
Definition in Base View	Displays the projection line for the main/parent view
Section Standard Parts	Include hardware, like fasteners and washers, in section views
View Justification	Centered or Fixed.

Autodesk Inventor Fundamentals

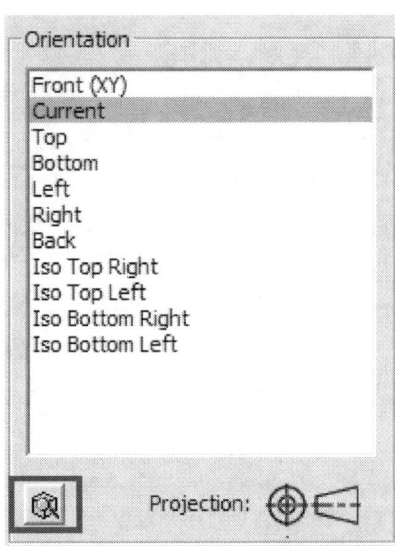

When creating a base view, if the user does not like any of the standard views set up by Inventor, he can select the **Change View** orientation button.

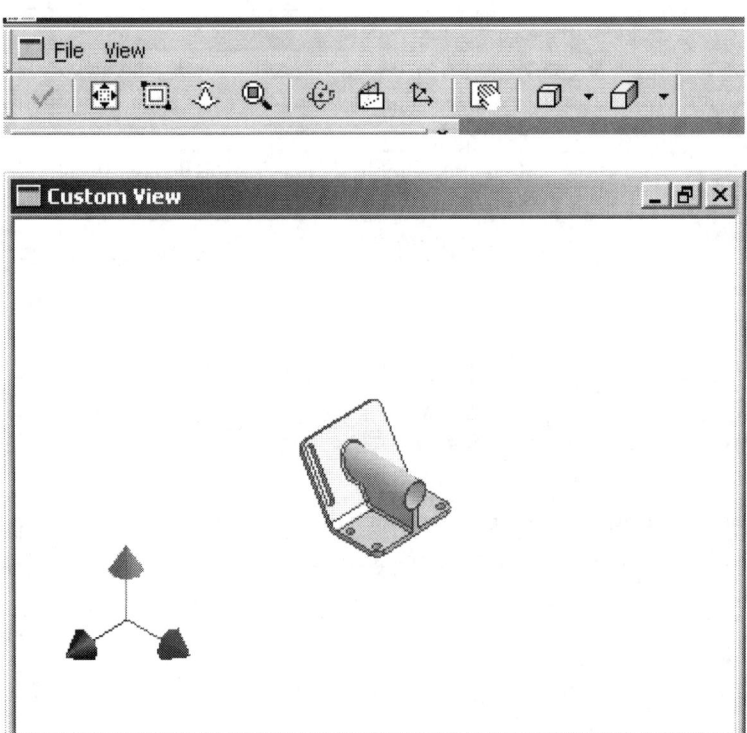

This brings up a special View window where the user can specify the orientation for the base view. Simply use the viewing tools to orient your part in the desired manner and press the check mark to set the base view.

NOTE: *In previous versions of Inventor, the view tools were located as a toolbar attached to the Custom View window. In Release 6 and on, the toolbar is located where the Standard toolbar usually is placed. This may be confusing for some users.*

Exercise 12-1:
Creating a Base View

File: Ex11-1.ipt
Estimated Time: 15 minutes

Always use default (ANSI) for dwg

1. 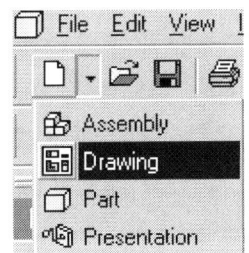 Open a new drawing file.

There is more than one way to create views in Inventor.

In the Graphics Window	New Sheet / Base View...	Right click and select 'Base View'.
From the Menu	Insert → Model Views → Base View...	Go to **Insert→Model Views→ Base View**
Drawing Management toolbar		Base View tool
In the Browser	Model / Drawing1 / Drawing Resources / Sheet... Base View...	Highlight the sheet name, right click and select 'BaseView'.

2. Select the **Base View** tool from the **Drawing Management** toolbar.

12-7

Autodesk Inventor Fundamentals

3. 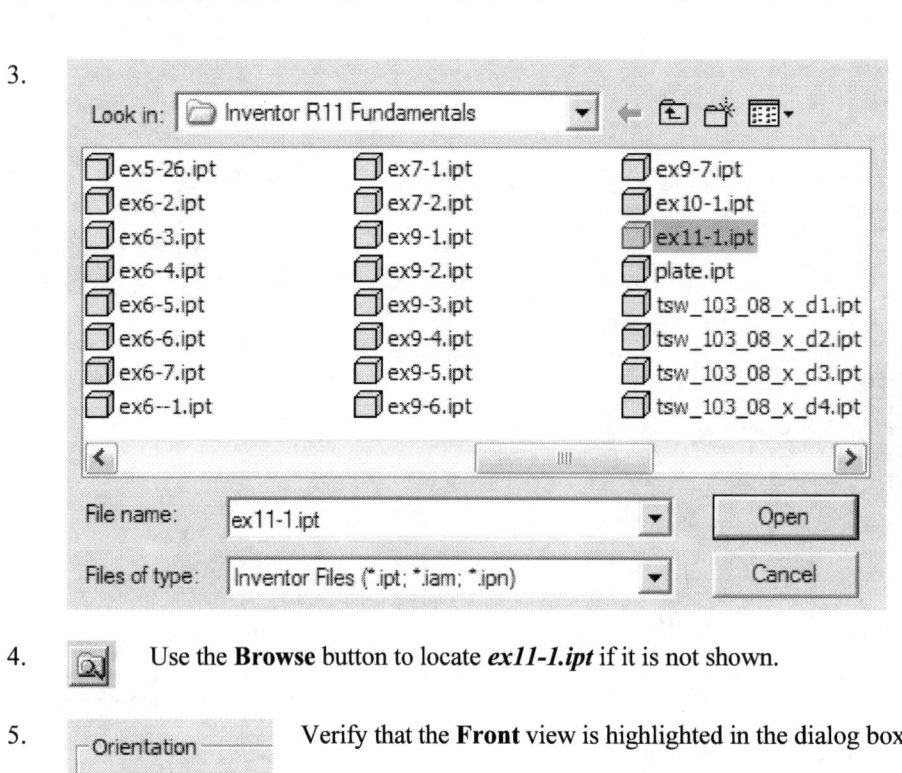 The dialog file drop-down list will show any open part, assembly or presentation files.

4. Use the **Browse** button to locate *ex11-1.ipt* if it is not shown.

5. Verify that the **Front** view is highlighted in the dialog box.

   You will also see a preview of the view at the end of your cursor.

6. Select **Hidden Line** as the **Style**.

7. Set the **Scale** to **1:1**.

   NOTE: If you want the view to be labeled with a scale, click on the light bulb to turn on the visibility.

8. Select the **Display Options** tab.

12-8

Drawing Management

9.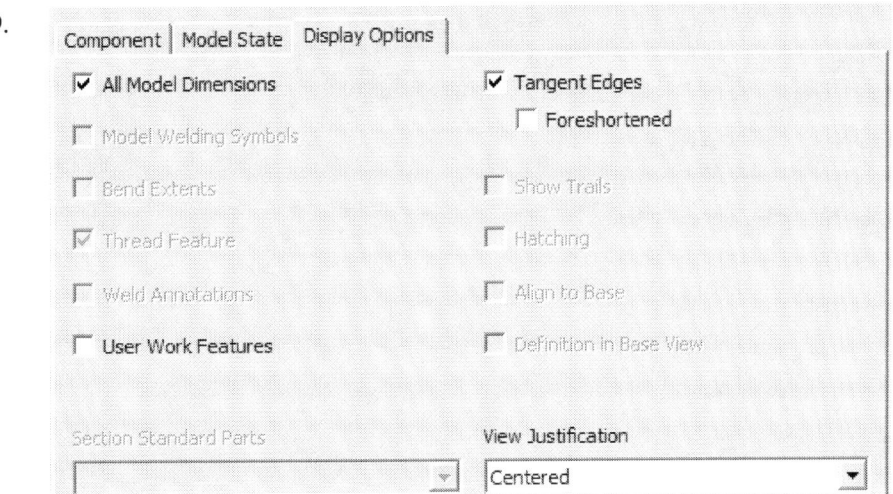

Enable **All Model Dimensions**.

Enable **Tangent Edges**.

10.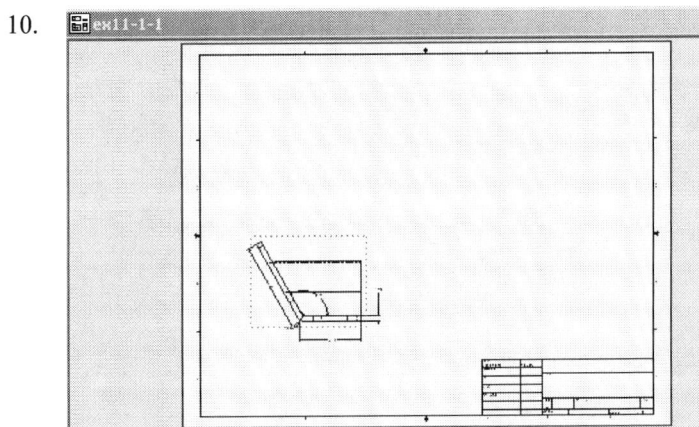

Place the view in the sheet in the lower left corner.

Note that dimensions are placed with the view.

12-9

Autodesk Inventor Fundamentals

Some people do not like the default color of the sheet, which is a beige color.

11. 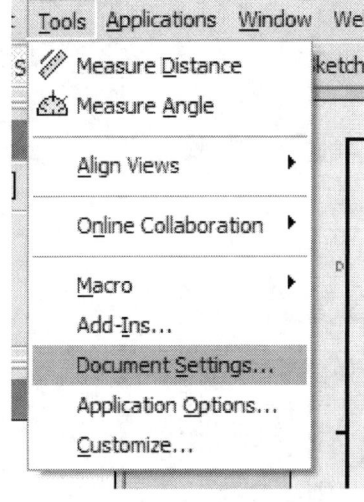 Go to **Tools→Document Settings**.

12. Select the **Sheet** tab.

Press the color rectangle for Sheet. Assign the color white to the sheet.

Press **Apply** and **OK**.

13. Note the labels; these assign the default labels that appear in your Browser and in your layout.

TIP: Once you have set up your sheet with the format you like, create a sheet format with those settings and save it to a template.

14. Save as *ex12-1.idw*.

TIP: Some of the characters in your drawing may appear as a black rectangle. This is a graphics card issue. Your plot should be OK.

12-10

Projected View

You can create a projected view with a first-angle or third-angle projection, depending on the drafting standard for the drawing. You must have a base view before you can create a projected view. Use the Projected View button on the Drawing Management toolbar.

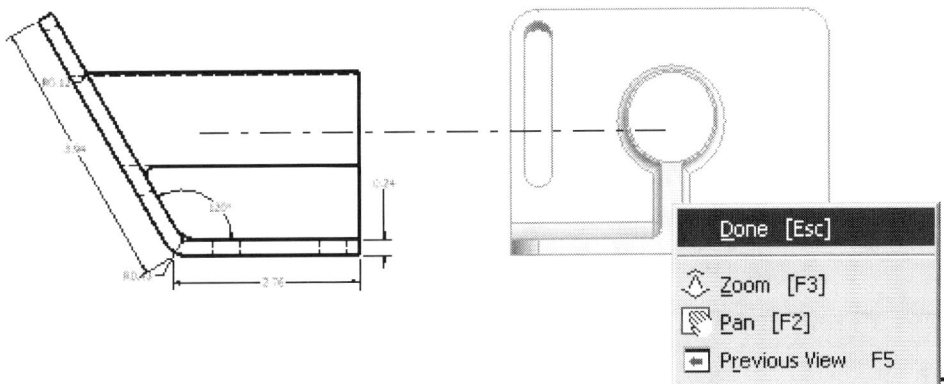

 TIP: Orthographic projections are aligned to the base view and inherit its scale and display settings. Isometric projections are not aligned to the base view. They default to the scale of the base view but do not update if you change the scale of the base view.

Exercise 12-2:
Create a Projected View

File: Ex12-1.idw
Estimated Time: 15 minutes

1. Continue working with the *ex12-1.idw* file.

There is more than one way to create the projected views:

From the Base View	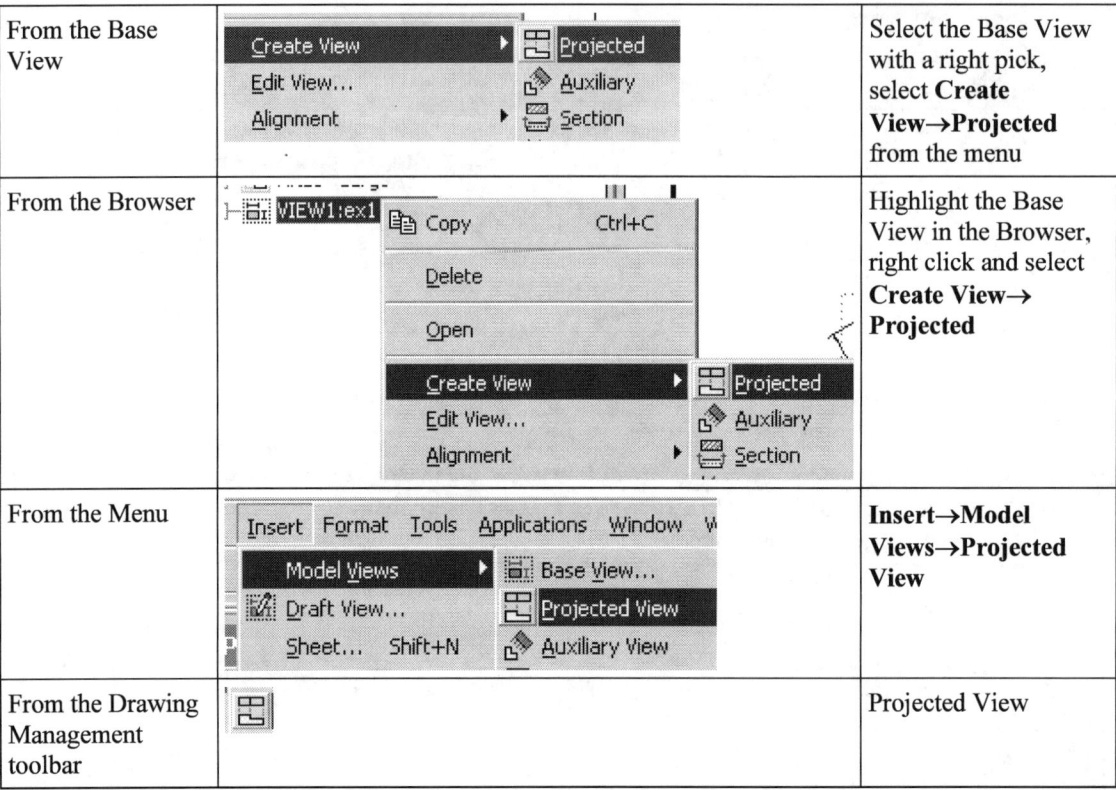	Select the Base View with a right pick, select **Create View→Projected** from the menu
From the Browser		Highlight the Base View in the Browser, right click and select **Create View→ Projected**
From the Menu		**Insert→Model Views→Projected View**
From the Drawing Management toolbar		Projected View

2. Highlight the base view by picking on it. (A red dotted rectangle should appear when the view is active.)

Right click the mouse to bring up the menu.
Select **Create View→ Projected**.

Drawing Management

3. Pick the location for the top view with the left mouse button.

4. 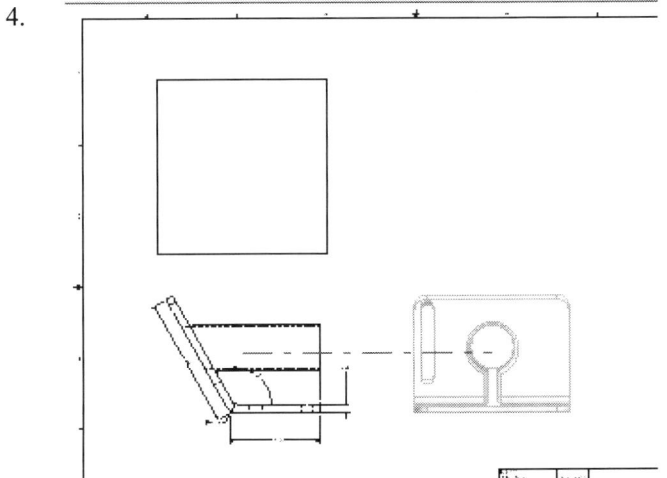 Move the mouse to the right of the base view.

    Left pick to place the right side view.

    A blank rectangle is shown to act as a placeholder for the top view, so you can keep track of which views you have already placed.

5. 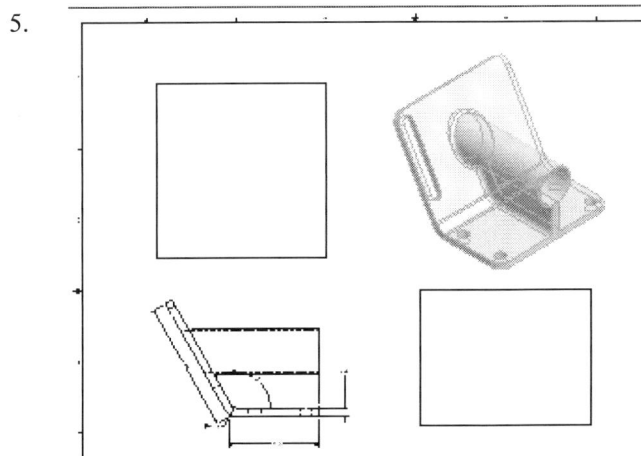 Move the mouse above the right side view and pick to place the isometric view.

12-13

6. 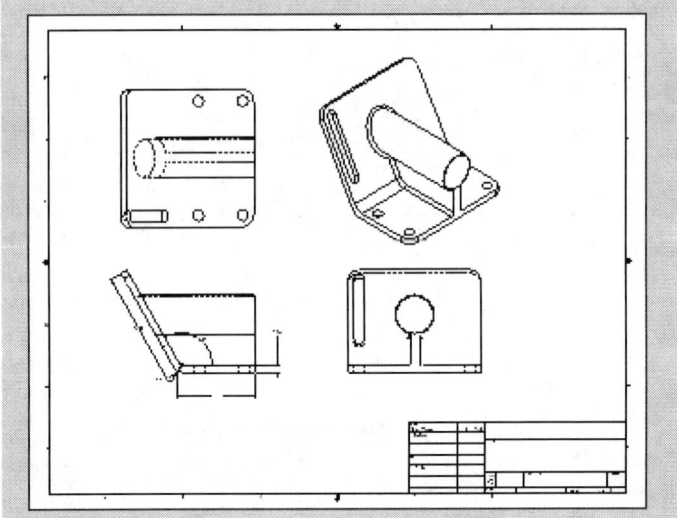 Right click the mouse and select **Create** to set the views in place and finish.

7. Save the file as *ex12-2.idw*.

Drawing Management

Auxiliary View

Places an auxiliary view by projecting from an edge or line in a base view. The resulting view is aligned to the base view.

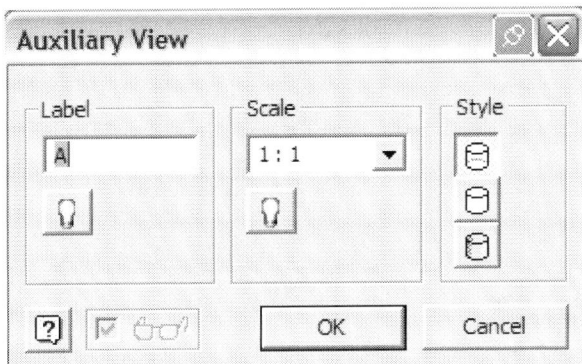

Label	Specifies the view label determined by the active drawing standard. To change the label, select the label in the box and enter the new label.
	Displays or hides the view label. Toggle to display the label or to hide the label.
Scale	Sets the scale for the auxiliary view.
	Displays or hides the view scale. Toggle to display the scale or to hide the scale.
	Sets the display to show hidden lines.
	Sets the display to remove hidden lines.
	Sets the display to a shaded rendering.

TIP: To place the view without alignment to the base view, press Ctrl as you move and place the preview.

12-15

Autodesk Inventor Fundamentals

Exercise 12-3:
Adding an Auxiliary View

File: Ex12-2.idw
Estimated Time: 15 minutes

1. Continue working with the *ex12-2.idw* file.

There is more than one way to create an auxiliary view.

From the Menu		Insert→Model Views→ Auxiliary View
From the Browser		Highlight the view you want to use as the base view for the auxiliary view. Right click and select **Create View →Auxiliary**.
From the graphics window		Highlight the view you want to use as the base view for the auxiliary view. Right click and select **Create View →Auxiliary**.
Drawing Management toolbar		Auxiliary

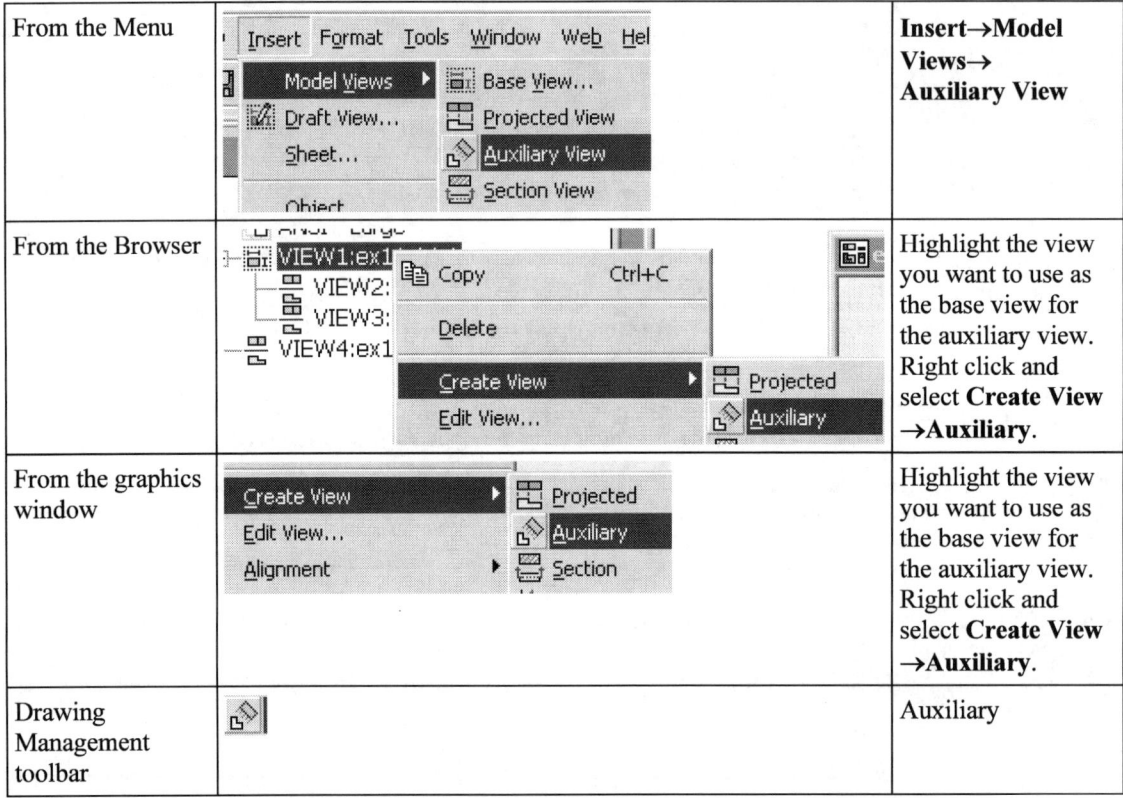

12-16

Drawing Management

2. Select the base view.
Right click and select **Create View→Auxiliary**.

3.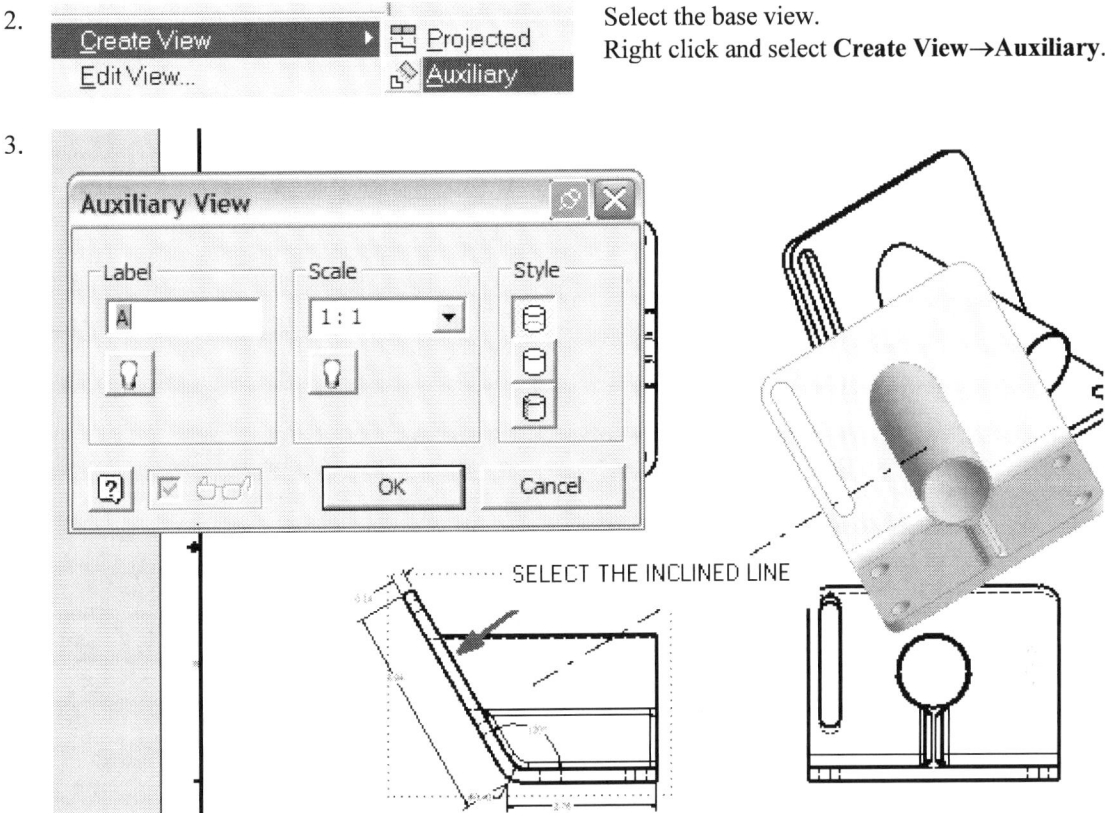

Drag and drop to place the auxiliary view.

We can move the auxiliary view to a second sheet.

You can use several methods to add an additional drawing sheet.

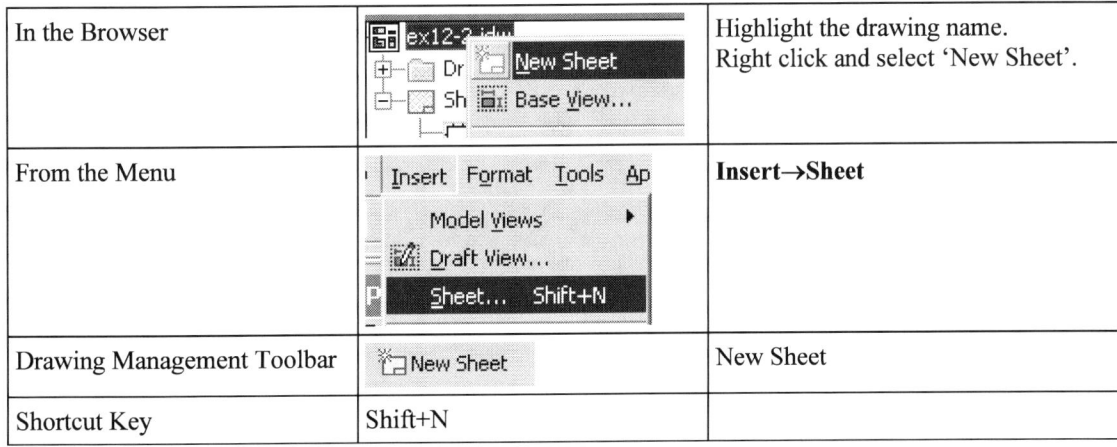

In the Browser		Highlight the drawing name. Right click and select 'New Sheet'.
From the Menu		**Insert→Sheet**
Drawing Management Toolbar	New Sheet	New Sheet
Shortcut Key	Shift+N	

4. Select the **New Sheet** tool.

12-17

Autodesk Inventor Fundamentals

5.

The New Sheet appears in the Browser and activates in the graphics window.

6.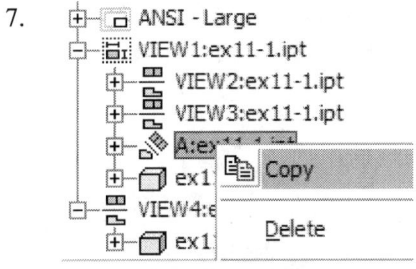

To switch back to the first sheet, highlight in the Browser. Right click and select **Activate**.

You can also activate by double left clicking on top of the paper sheet icon next to the sheet name.

7.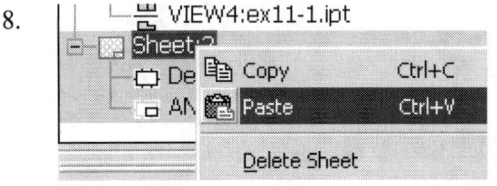

Activate the first sheet.
Highlight the Auxiliary view in the Browser.
It will also highlight in the graphics window.

Right click and select **Copy**.

8.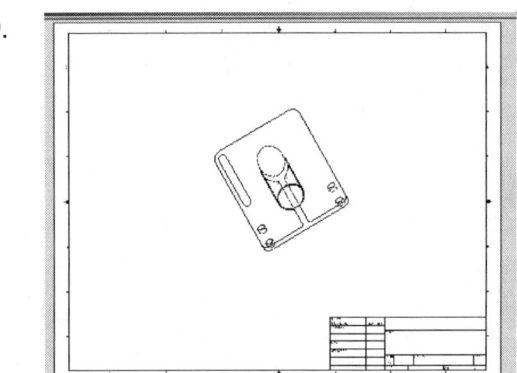

Highlight Sheet 2.

Right click and select **Paste**.

9.

The auxiliary view is copied onto the new sheet.

The view will appear in the same location on the sheet as its previous location.

12-18

Drawing Management

10. 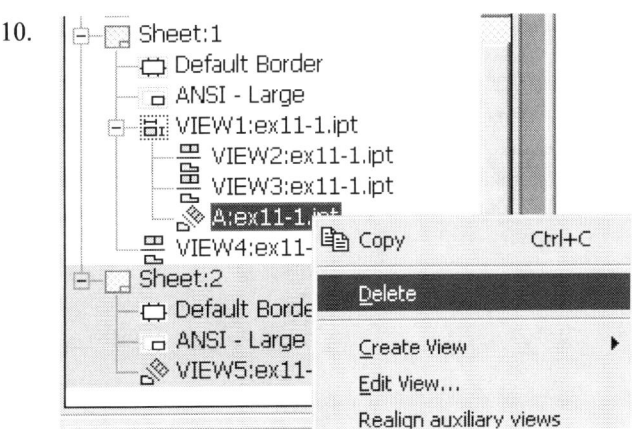 Activate Sheet 1.

    Highlight the auxiliary view on Sheet 1.

    Right click and select **Delete**.

11. 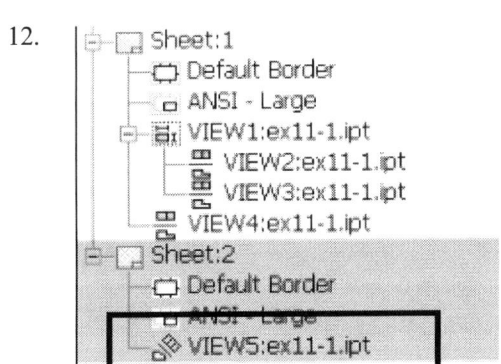 Press **OK**.

12. You still have the auxiliary view you copied to Sheet 2.

    Save the file as *ex12-3.idw*.

12-19

Section View

Creates a full, half, offset, or aligned section view from a specified base view. You can also use Section View to create a view projection line for an auxiliary or partial view. A section view is aligned to its parent view.

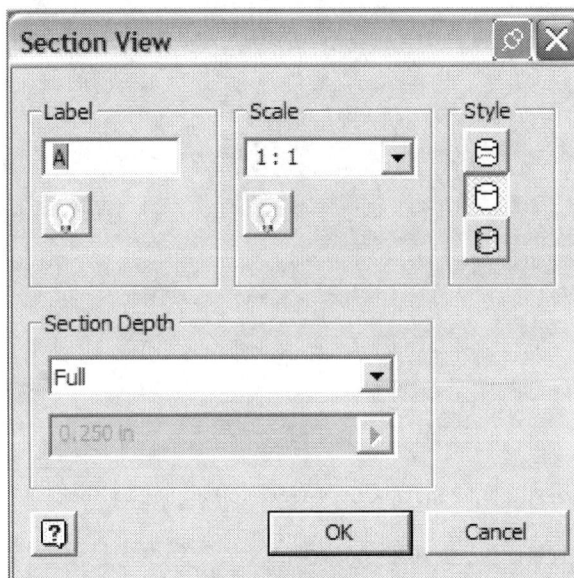

Label	Specifies the view label determined by the active drawing standard. To change the label, select the label in the box and enter the new label.
	Displays or hides the view label.
Scale	Sets the scale for the section view.
	Displays or hides the view scale.
	Sets the display to show hidden lines.
	Sets the display to remove hidden lines.
	Sets the display to a shaded rendering.
Section Depth	Can be set to Full, which goes through the entire part, or Distance.
Distance Value	If the Section Depth is set to Distance, enter the distance for the section view.

Drawing Management

There is more than one method to create a section view.

From the Browser	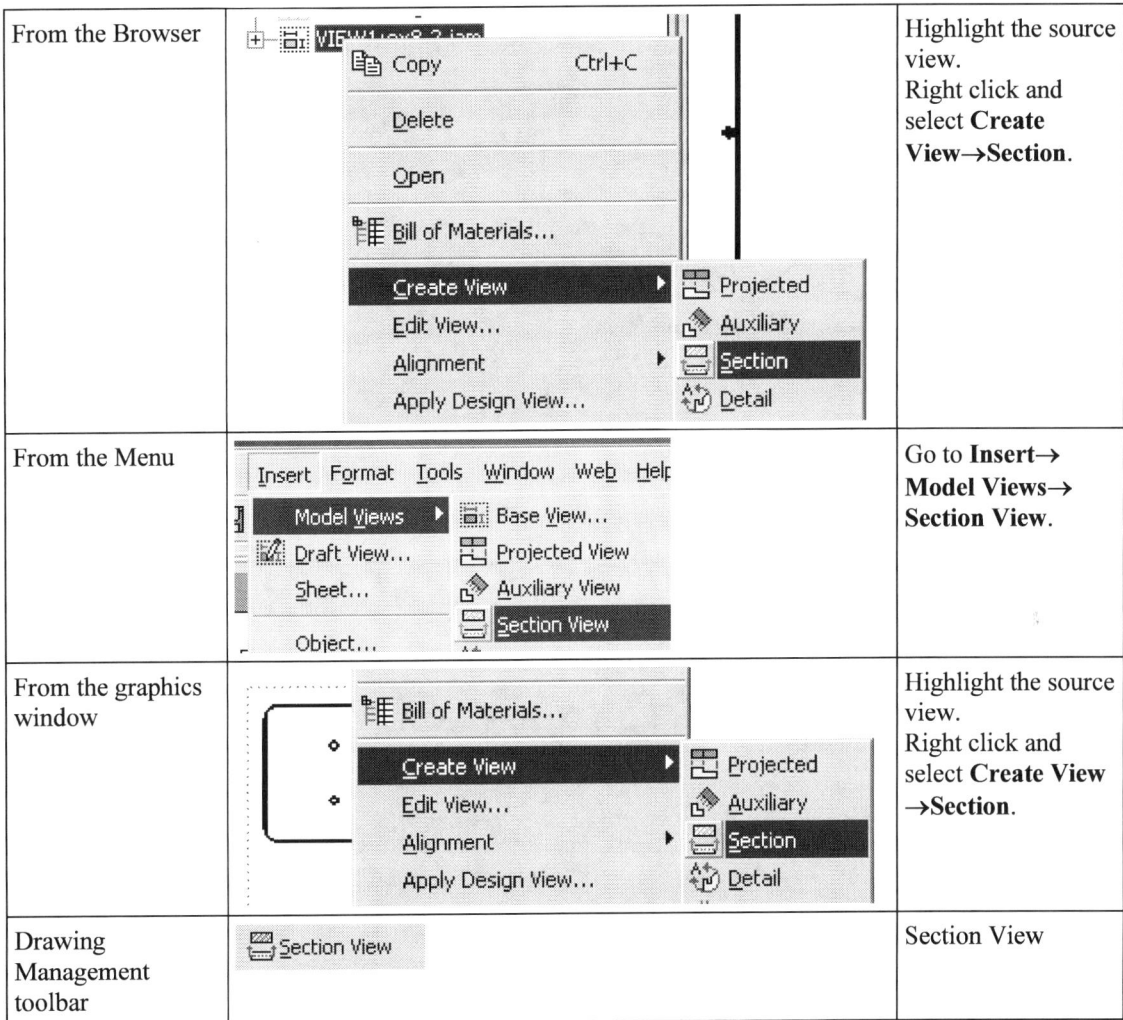	Highlight the source view. Right click and select **Create View→Section**.
From the Menu		Go to **Insert→ Model Views→ Section View**.
From the graphics window		Highlight the source view. Right click and select **Create View →Section**.
Drawing Management toolbar	Section View	Section View

12-21

Autodesk Inventor Fundamentals

Exercise 12-4:
Adding a Section View

File: New drawing using Standard
Estimated Time: 15 minutes

This exercise reviews the following:

- New Drawing
- Base View
- View Window
- Section View

1. Start a new drawing.

2. Select the **Base View** tool.

3. Select the **Browse** tool.

4. Select **'iam'** under **Files of Type**. Locate the *ex5-7.ipt* file.

   Press **Open**.

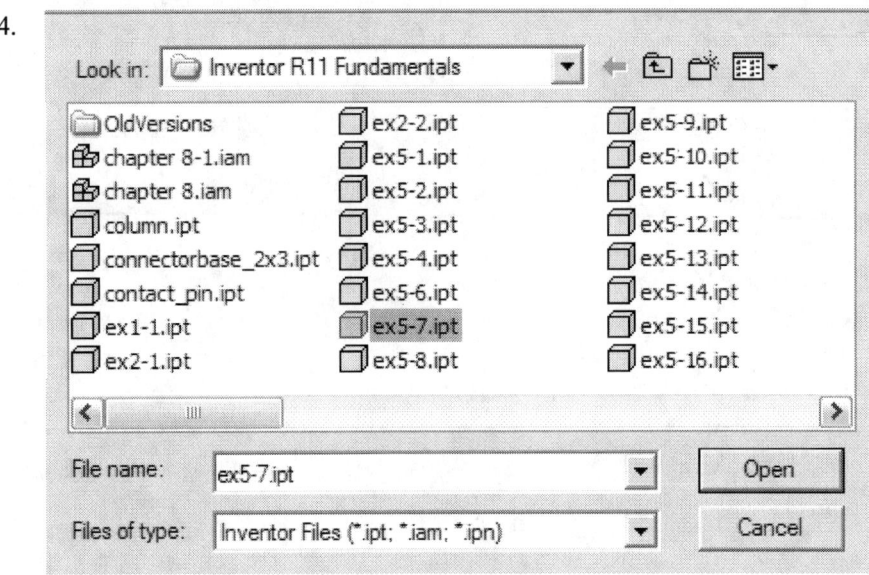

5. Set the Orientation to the **Front** view.

12-22

Drawing Management

6. 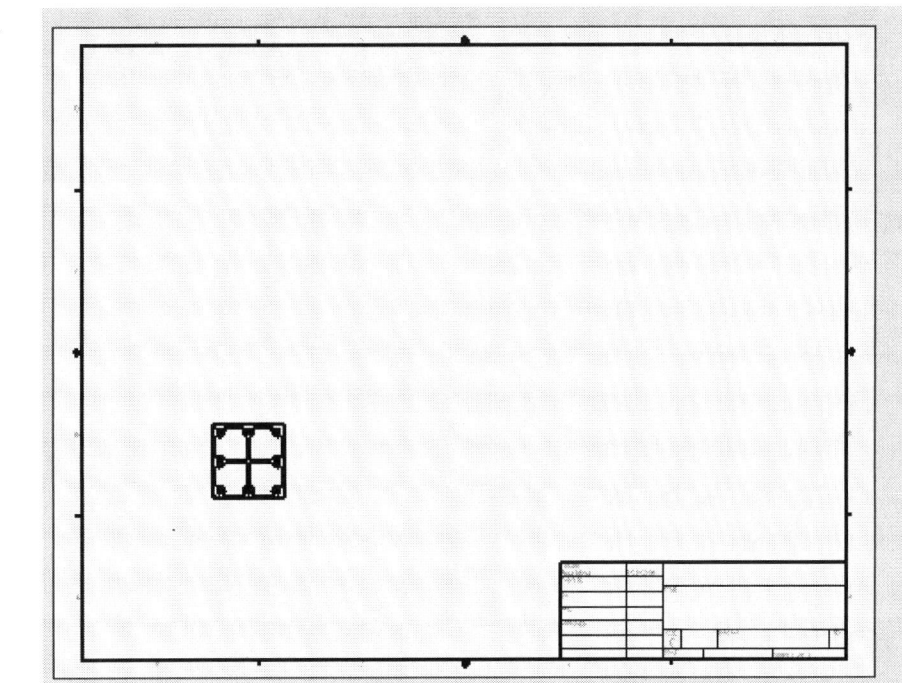   Place the base view in the sheet.

7.    Highlight the source view in the graphics window.
Right click and select **Create View→Section**.

8. 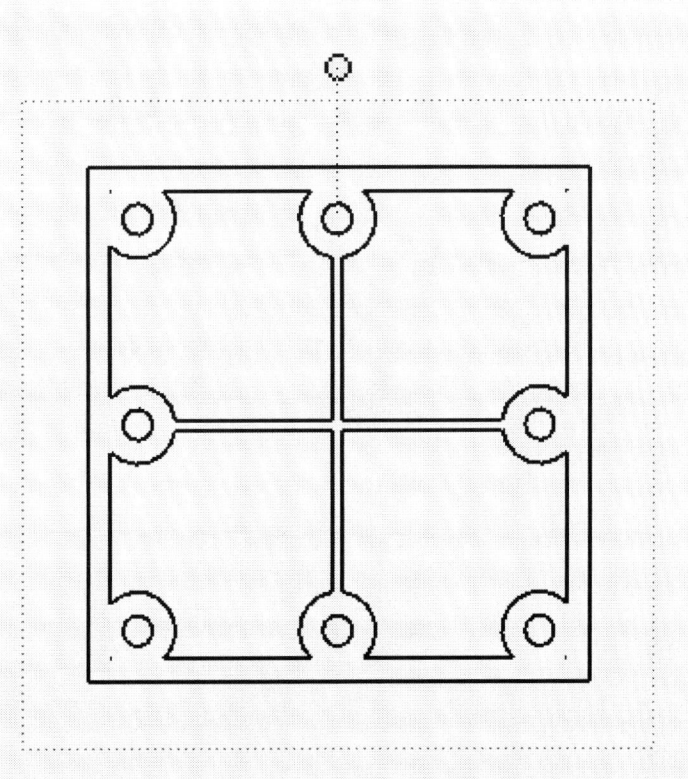   Use tracking to line up your section line with the center point of the hole of the part.

12-23

9.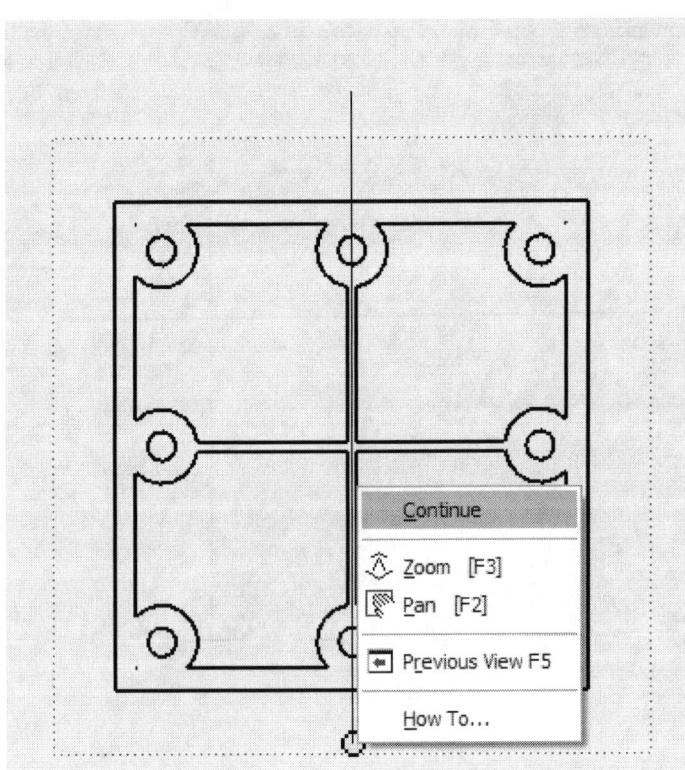

Pick a point above the plate.
Drag the line down in a straight line.
Pick a point below the plate.

If you have a problem creating a straight line, you can apply a sketch constraint; i.e. vertical/horizontal, to the line.

10. Right click and select **Continue**.

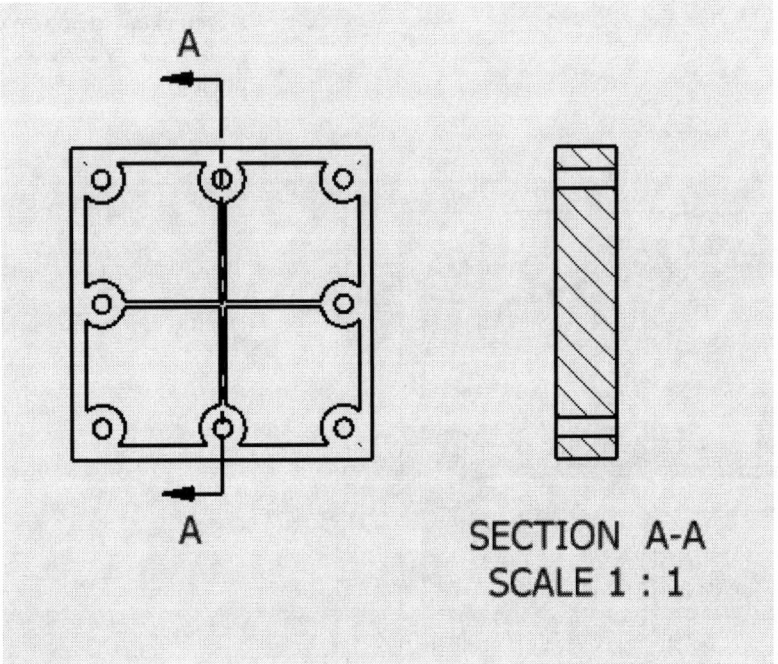

11. Drag the view to the right.
    Left click to place.

12. Save the file as *ex12-4.idw*.

Drawing Management

TIP: To get a straight section line, just draw a straight line using the object tracking. Inventor will add the shoulders, arrowheads and labels automatically.

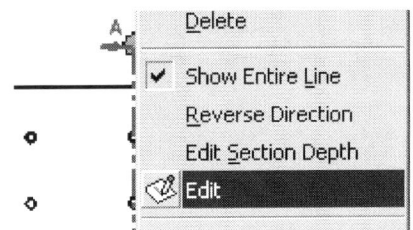

To apply a sketch constraint to the section line, select the line. Right click and select **Edit**. Place the constraint. Right click and select **Finish Sketch.**

Exercise 12-5:
Editing a Section View

File: Ex12-4.idw
Estimated Time: 15 minutes

This exercise reviews the following:

- Section View

To edit the view:

♦ Select the view, right click and select 'Edit View'
♦ Select the view in the Browser, right click and select 'Edit View'

1. Open *ex12-4.idw*.

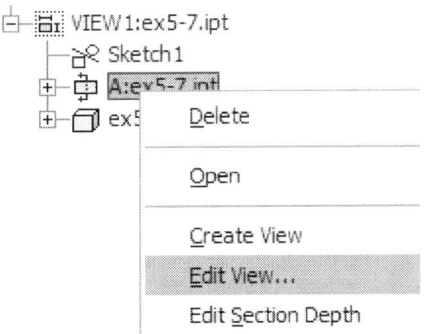

Highlight the section view.
Right click and select **Edit View**.

2. The Edit View dialog box appears.
Toggle the **Visible** button.
Press **OK**.

12-25

Autodesk Inventor Fundamentals

3. 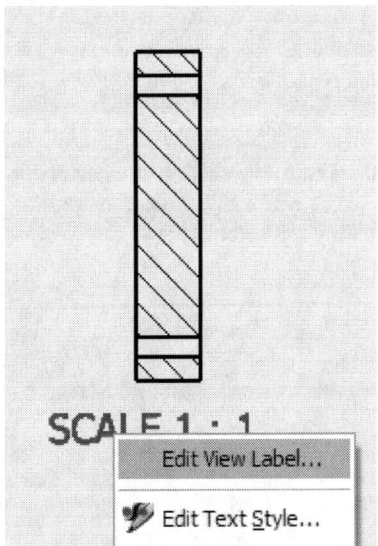 To edit the Scale Label, select, right-click and then select **Edit View Label**.

4. Add '**FOR REFERENCE**' in the edit field.

   Press **OK**.

5. Save the file as *ex12-5.idw*.

Drawing Management

Detail View

Creates and places a detail drawing view of a specified portion of a base view. The view is created without an alignment to the base view.

TIP: To set a different fence shape, right-click and select the fence shape from the menu before clicking to indicate the outer boundary. You can modify the size and location of the Detail View by using the grips.

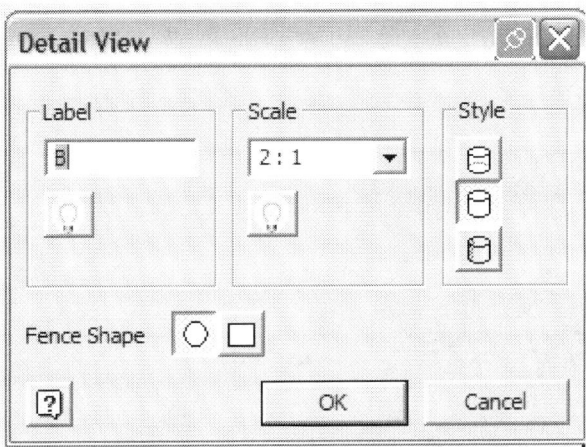

Label	Specifies the view label determined by the active drawing standard. To change the label, select the label in the box and enter the new label.
Visible	Displays or hides the view label.
Scale	Sets the scale for the section view.
Visible	Displays or hides the view scale.
Fence Shape	Sets the boundary for the detail view – either circular or rectangular in shape.
	Sets the display to show hidden lines.
	Sets the display to remove hidden lines.
	Sets the display to a shaded rendering.

12-27

Autodesk Inventor Fundamentals

There are several methods to create a Detail View.

From the Browser		Highlight the source view. Right click and select **Create View→Detail**.
From the Menu		Go to **Insert→ Model Views→ Detail View**.
From the graphics window		Highlight the source view. Right click and select **Create View→Detail**.
Drawing Management toolbar		Detail View

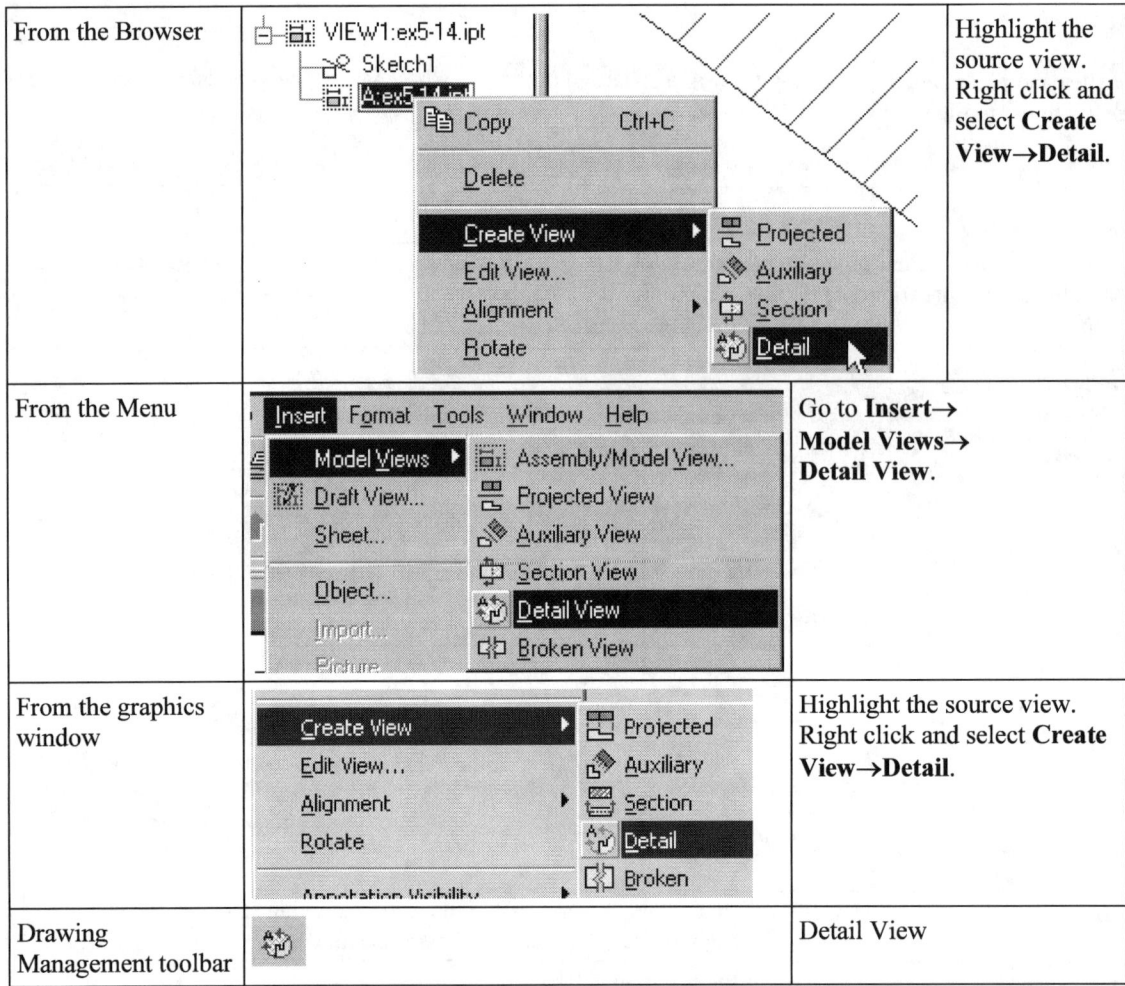

12-28

Drawing Management

Exercise 12-6:
Creating a Detail View

File: Ex12-5.idw
Estimated Time: 10 minutes

1. Open *ex12-5.idw*.

2. 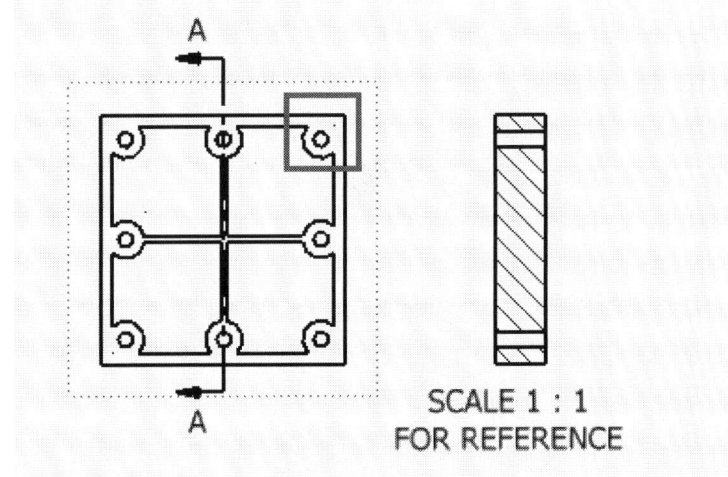 We'll create a detail view of the front view indicated.

3. Select the **Detail View** tool from the toolbar.

4. 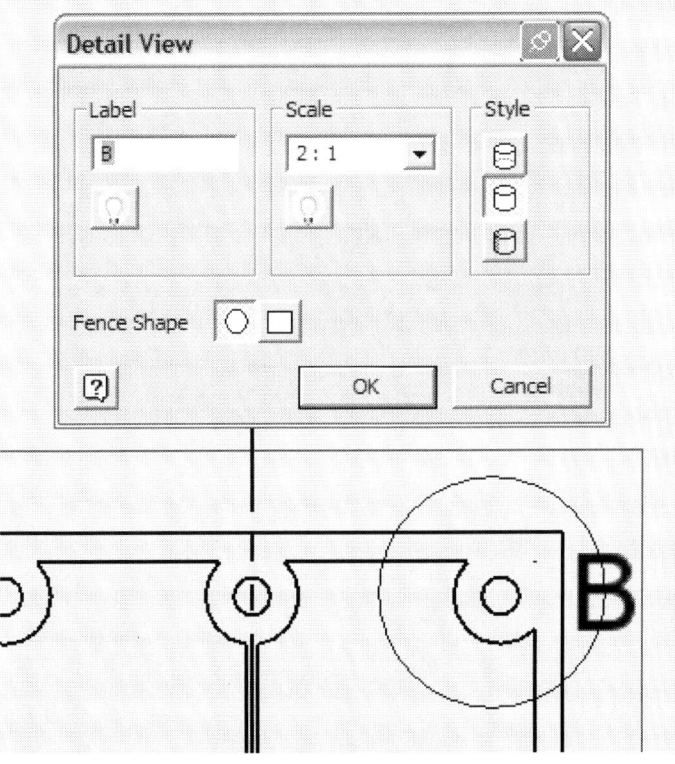 Select the **Front View**.

12-29

5. Pick the center of the hole.
   Drag to define the circle and then pick to place.

6. Place the detail view to the right of the front view and above the section view.
   The **Scale** is set to **2** in the dialog.
   Both labels are set to be **Visible**.

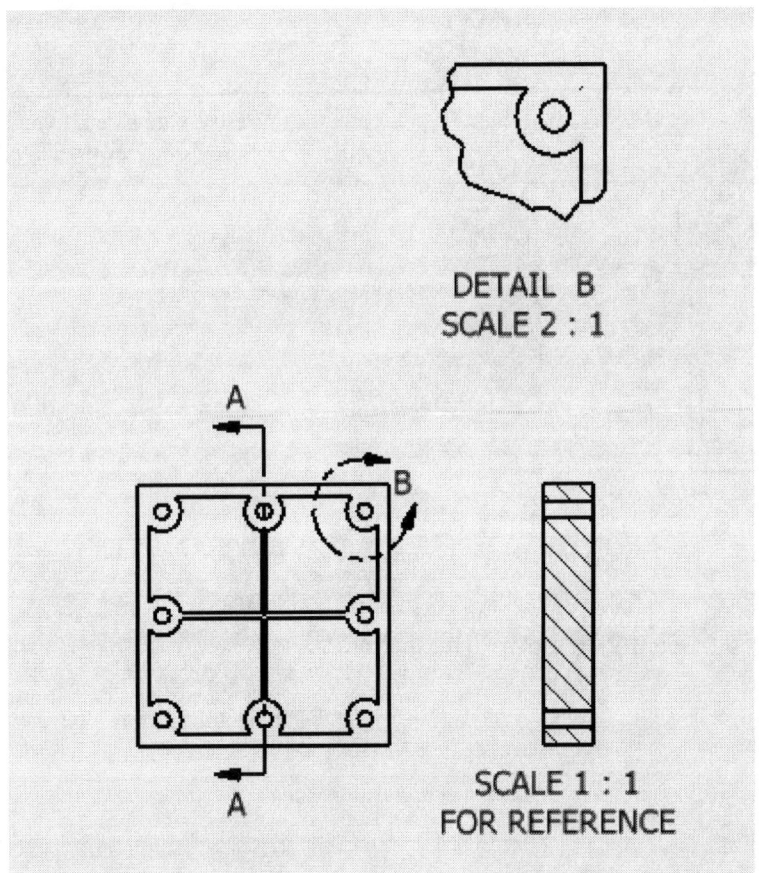

7. Save the file as *ex12-6.idw*.

Drawing Management

Broken View

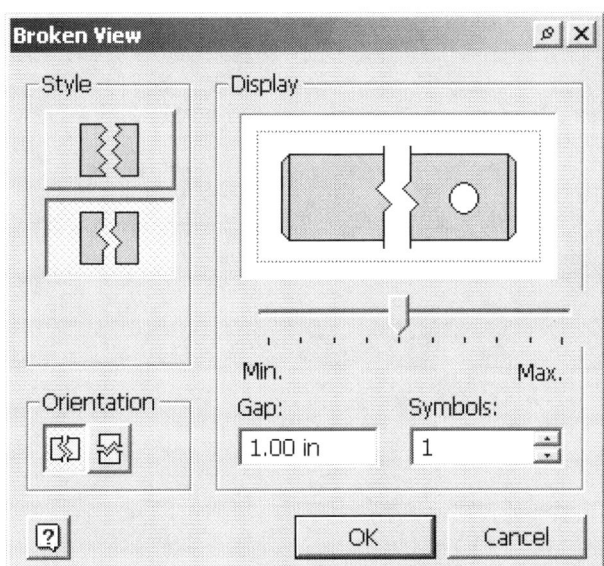

Creates a broken, foreshortened view.

STYLE: Sets the style of break to Rectangular or Structural.		
	Rectangular	Creates a broken view for non-cylindrical objects and all sectioned broken views.
	Structural	Creates a broken view using stylized break lines.
ORIENTATION: Sets break orientation to horizontal, vertical, or aligned to the view projection.		
	Sets break orientation to horizontal.	
	Sets break orientation to vertical.	
DISPLAY: Controls appearance of each break type. Works with the Style buttons. Select a Style button to activate the Display settings. Display settings preview in the Display area.		
Min._Max. slider		• With Rectangular button selected, controls quantity or pitch of break edges displayed. • With Structural button selected, controls amplitude of break line. Expressed as a percentage of the break gap.
Gap		Specifies the distance between the breaks in the broken view. Uses the units specified for the drawing.
		Specifies the number of break symbols for the selected break. Allows up to 3 symbols for each break. Available only with the Structural break.

12-31

Autodesk Inventor Fundamentals

Exercise 12-7:
Creating a Broken View

File: New
Estimated Time: 15 minutes

1. Start a new drawing file.

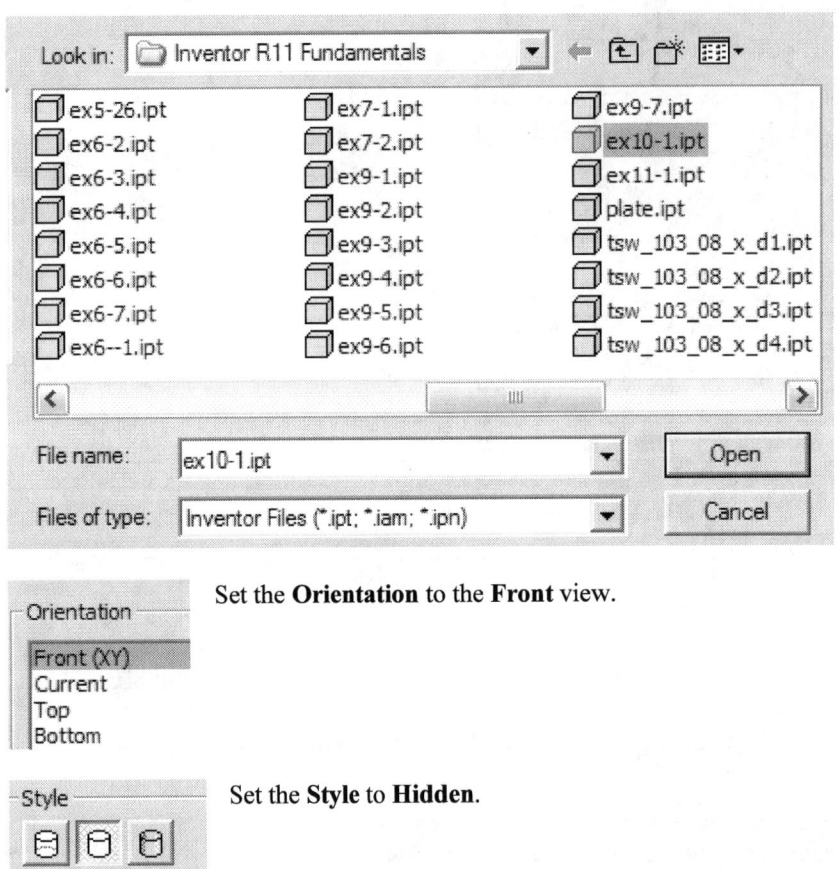

Select *ex10-1.ipt*.

Set the **Orientation** to the **Front** view.

Set the **Style** to **Hidden**.

Place the view on the sheet.

2. 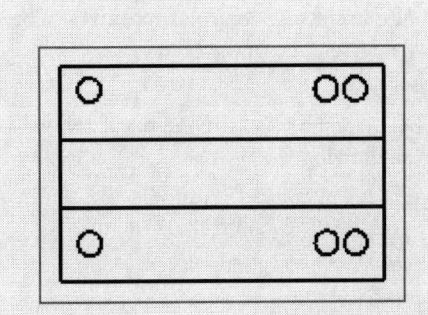 Select the **Broken View** tool.

12-32

3. 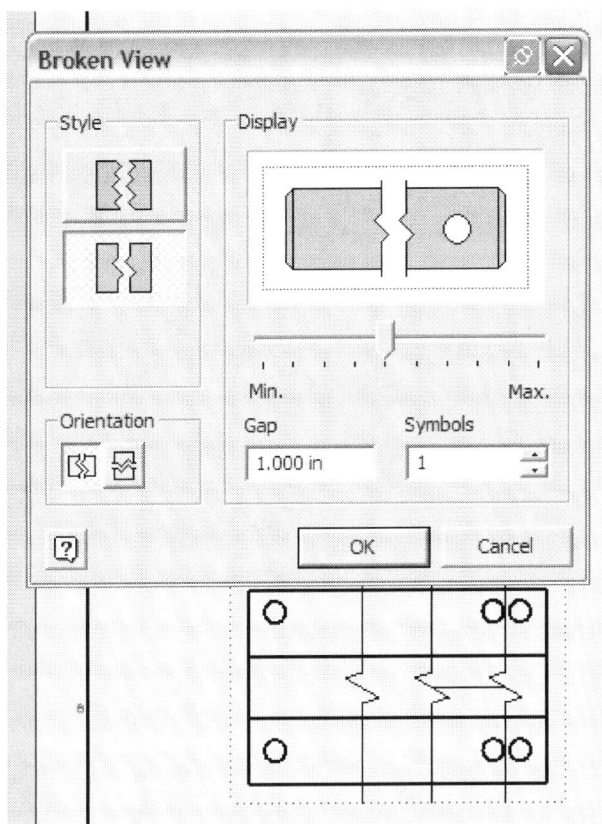 Select the base view.
Pick a point on the rod.
Drag your mouse up to establish the section you want removed.
Pick to finish.

 TIP: Break-out Views require you to place an *associated sketch* with a view. To associate a sketch with a view, select the view and highlight, then select the Sketch tool.

4. 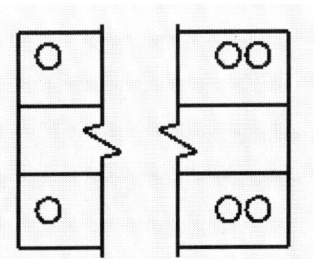 The view is updated to indicate the break.

   NOTE: If you have more than one view placed, all views will update to display the break.

5. 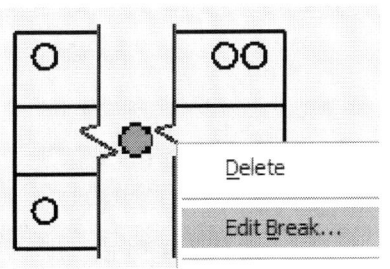 Select the break so it highlights.
Right click and select **Edit Break**.

12-33

6. 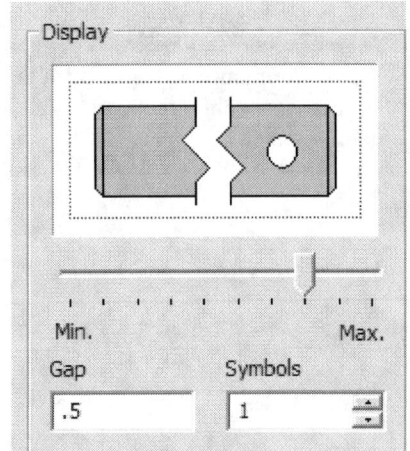 Change the **Style**.

   Modify the display using the slider.
   Set the **Gap** to **.5**.

   Press **OK**.

7. Save the file as *ex12-7.idw*.

Break-out View

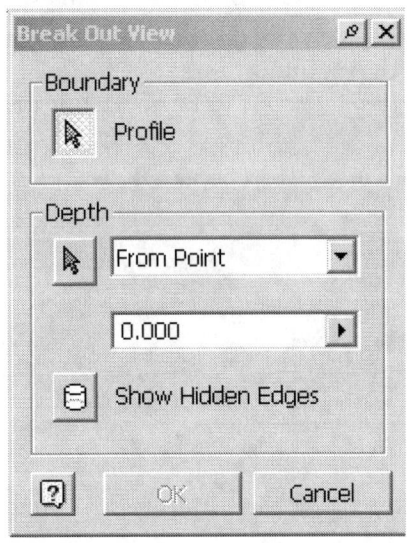 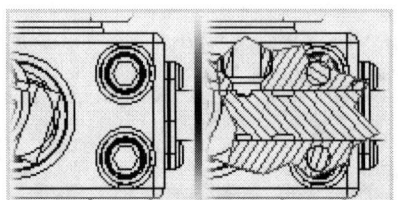

It allows you to create an exposed view of an assembly. Break-out views are commonly used to show insulation, hidden pipes or wiring.

Profile	Select closed polygon sketch to indicate the section to expose.
Depth	You can set different depths for the exposure using From Point, To Sketch, To Hole, or Through Part.
	Toggles display of Hidden Edges

Drawing Management

Exercise 12-8:
Creating a Break-out View

File: New using standard.idw (template), break-out.ipt (downloaded from publisher's website)
Estimated Time: 15 minutes

1. 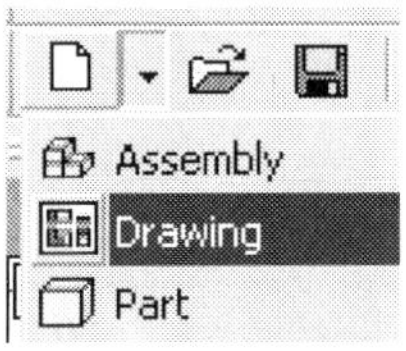 Start a new drawing.

2. 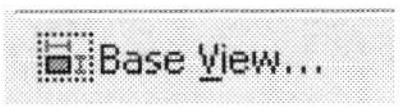 Select the **Base View** tool.

3. 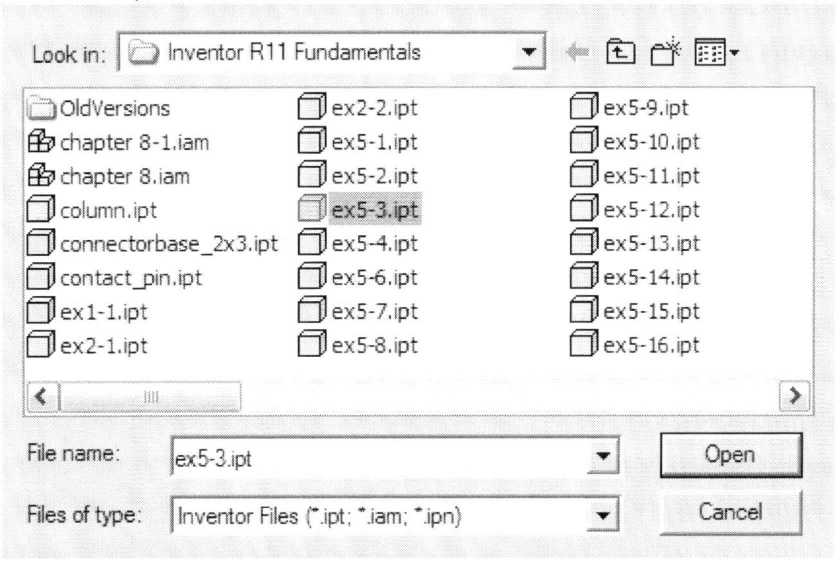 Locate *ex5-3.ipt* and press **Open**.

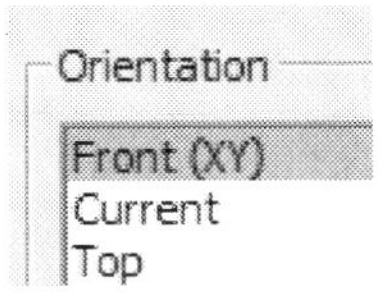

   Set the **Orientation** to **Front**.

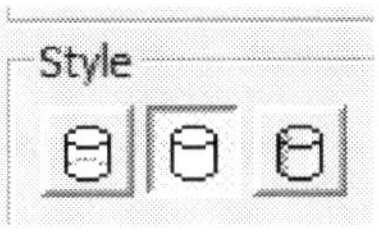

   Set the **Style** to **Hidden**.

12-35

Autodesk Inventor Fundamentals

4. 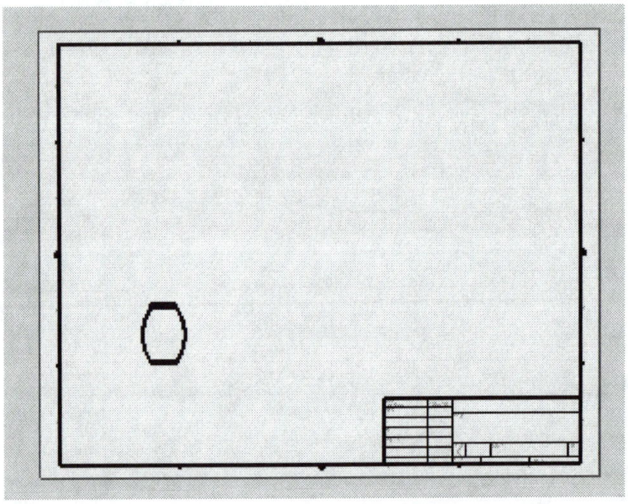 Place a left view with **Hidden Lines Removed** in the lower left of the sheet.

5. 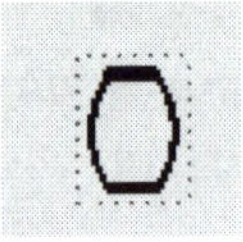 Highlight the view, so the dotted outline appears.

6.  Select the **Sketch** tool.

7. 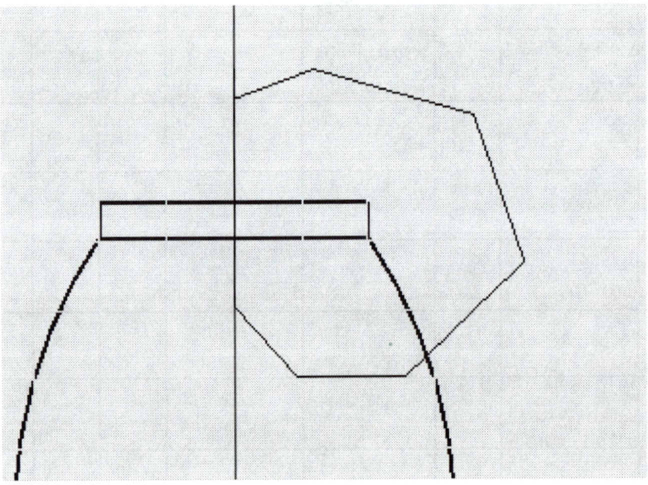 Use the **Line** tool to create an irregular closed polygon. Right click and select **Finish Sketch**.

8. 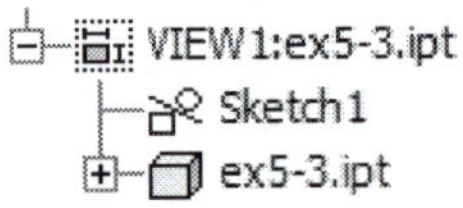 Verify that the Sketch is attached to the view in the Browser. If it is not assigned to a view, the Break Out View tool will not work.

   Verify that the sketch you created is a Closed Loop or the Break Out View tool will not work.

9.  Select the **Break Out View** tool.

12-36

Drawing Management

10.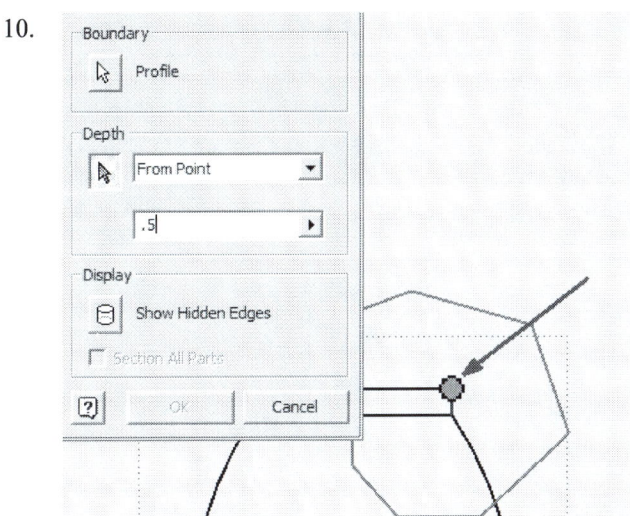

Select the view.

The tool automatically selects the sketch associated with the view.

For the point, select the top right point.

Set the **Depth** to **0.5**.

Press **OK**.

 TIP: If you get an error message stating that there is no closed profile for your break out view, edit the sketch and use **Close Loop** to create a closed profile.

11.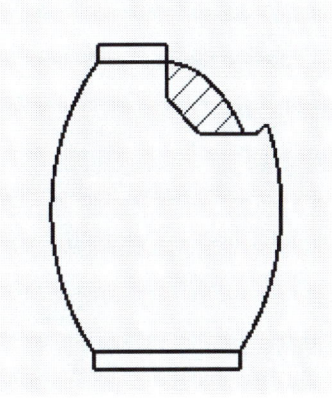

You have just created a break-out view.

12. Save the file as *ex12-8.idw*.

Things to keep in mind when creating a break-out view:

- You need to associate the sketch with the view
- If you go through the entire part, you probably will miss the details you want to show

 TIP: If you get an error message stating that there is no sketch associated with your view, highlight the view in the Browser and then activate sketch mode to create your sketch. This guarantees the sketch is associated with the highlighted view.

12-37

Overlay View

The Overlay View tool allows users to create positional views for their assemblies. Before you can place the views in your drawing, you need to define the positional views in your assembly. You can only apply positional views to assemblies that are at least three levels: a main assembly, a sub-assembly, and parts. If your assembly is a single level assembly, you can not apply overlay views.

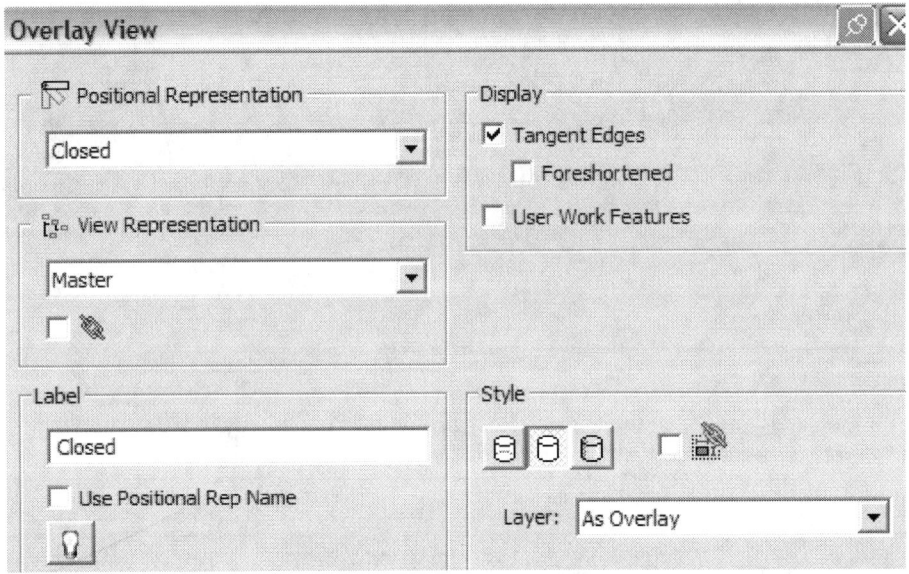

Positional Representation	Select the Positional Representation defined in the drawing to be used as the overlay
Display	
Tangent Edges	When enabled, tangent edges are displayed. This is useful for assemblies with fillets or cylindrical edges.
Foreshortened	Modifies the display of tangent edges to distinguish them from visible edges.
User Work Features	When enabled, axes, work planes, and work points are visible.
View Representation	Design Views may be used to control the visibility of specific parts, or change how parts are displayed (different colors, linetypes, etc.) Each overlay view may use a unique design view.
	When enabled, links the design view with the overlay.
Label	Assigns a name to the overlay view. If Use Positional Rep Name is enabled, the name assigned in the assembly file is used. If disabled, the user may enter their own name.
	When enabled, displays the Positional Rep Name beneath the view.
	Sets the display to show hidden lines.
	Sets the display to remove hidden lines.
	Sets the display to a shaded rendering.
	When enabled, uses the same display style as the base view.

Drawing Management

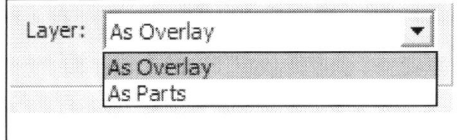	Sets the line style to be used for the view. May be set to As Overlay or As Parts. If set to As Overlay, uses the line style set to the Overlay Layer. If set to As Parts, uses the line style assigned to standard parts.

To create an Overlay View:

From the Browser	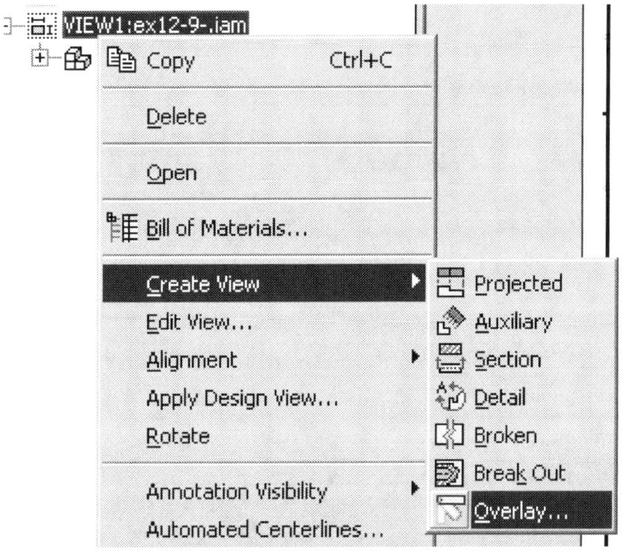	Highlight the source view. Right click and select **Create View→Overlay**.
From the Menu	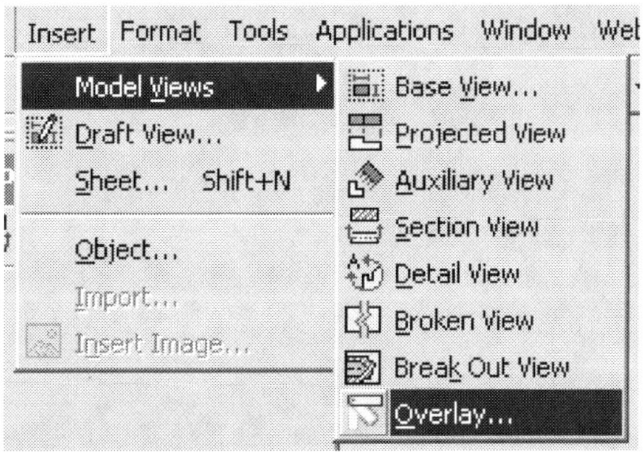	Go to **Insert→Model Views→Overlay**.
From the graphics window	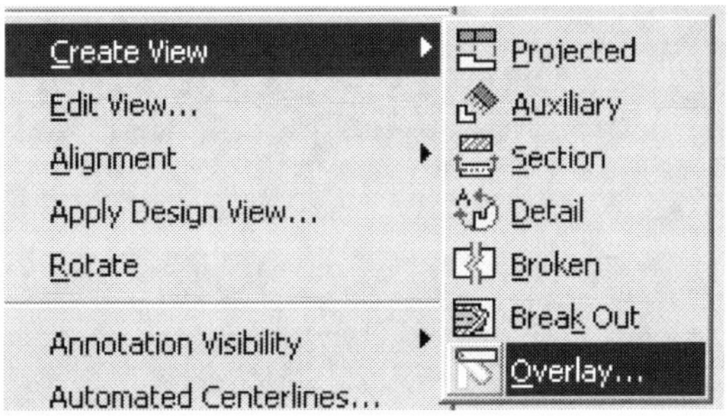	Highlight the source view. Right click and select **Create View→Overlay**.
Drawing Management toolbar	 Overlay...	Overlay

12-39

Exercise 12-9:
Creating an Overlay View

File: New
Estimated Time: 10 minutes

1. Start a new drawing.

2. Select the **Base View** tool.

3. Browse to the *Arbor Press* folder under *Program Files\Autodesk\Inventor 11\Samples\Models\Assemblies*.

4. Locate the ***Arbor Press.iam*** file.

   Press **Open**.

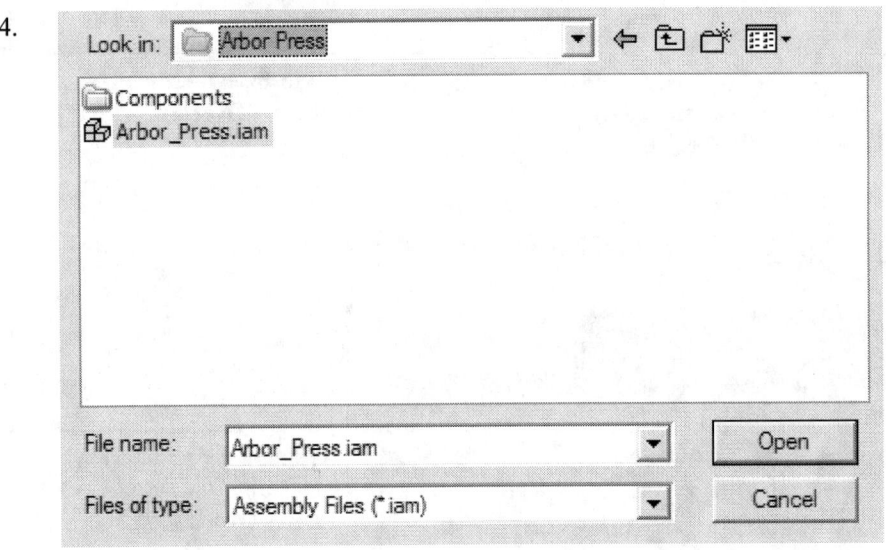

12-40

5.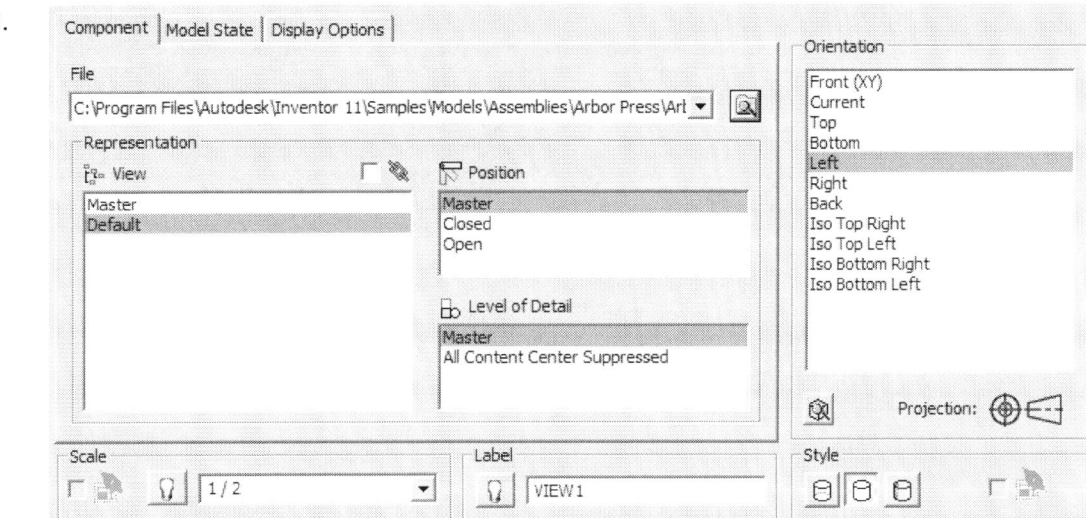

Set the **View** to **Default**.
Set the **Position** to **Master**.
Set the **Level of Detail** to **Master**.
Set the **Orientation** to **Left.**
Set the **Style** to **Hidden**.
Set the **Scale** to ½.
Place the view on the sheet.

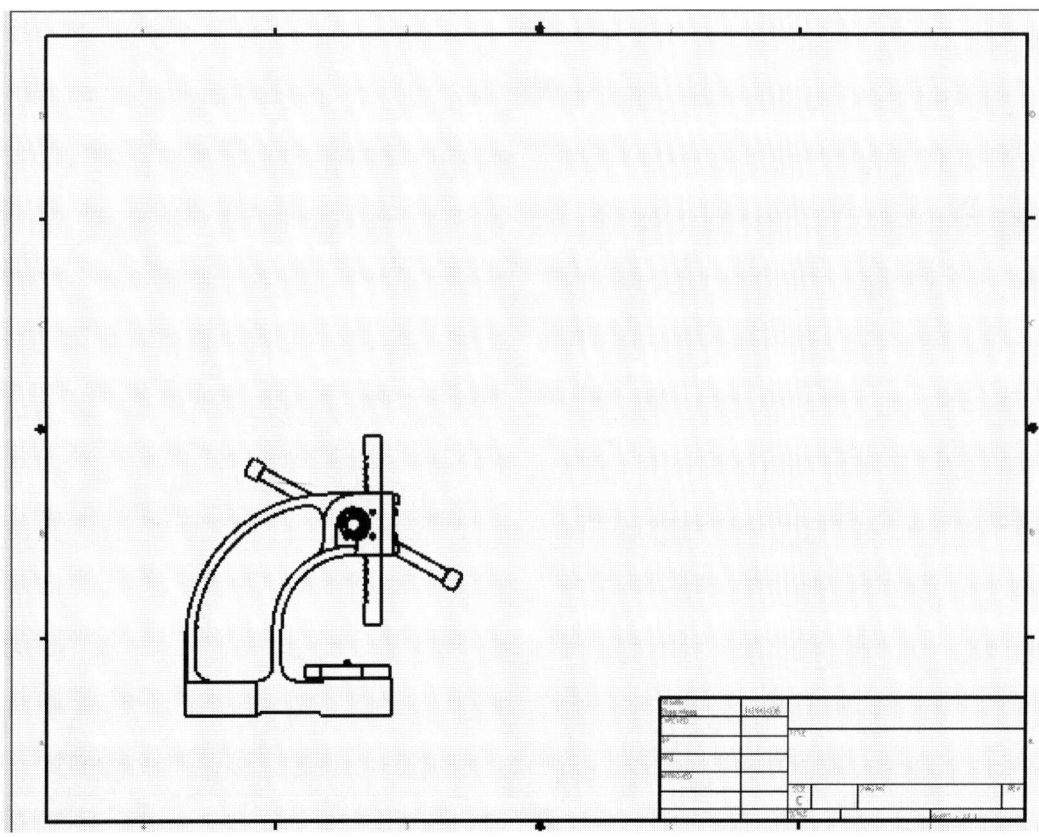

6. Select the **Overlay** tool.
Select the Base View.

7.

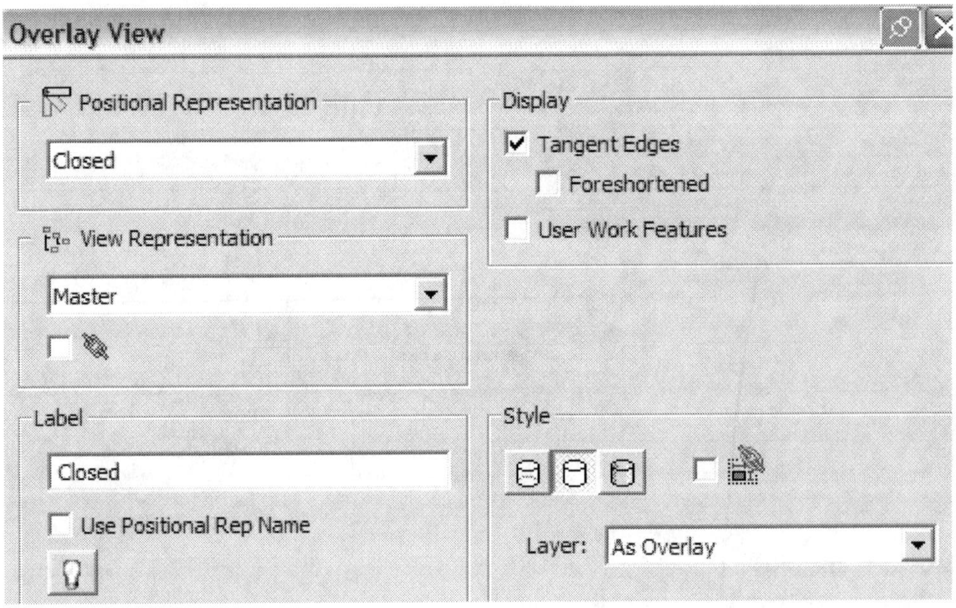

Select **Closed**.
Enable **Tangent Edges**.
Disable **Foreshortened**.
Set **View Representation** to **Master**.
Disable **Use Positional Rep Name**.
Enable **Hidden Style**.
Set **Layer** to **As Overlay**.
Press **OK**.

Drawing Management

8. 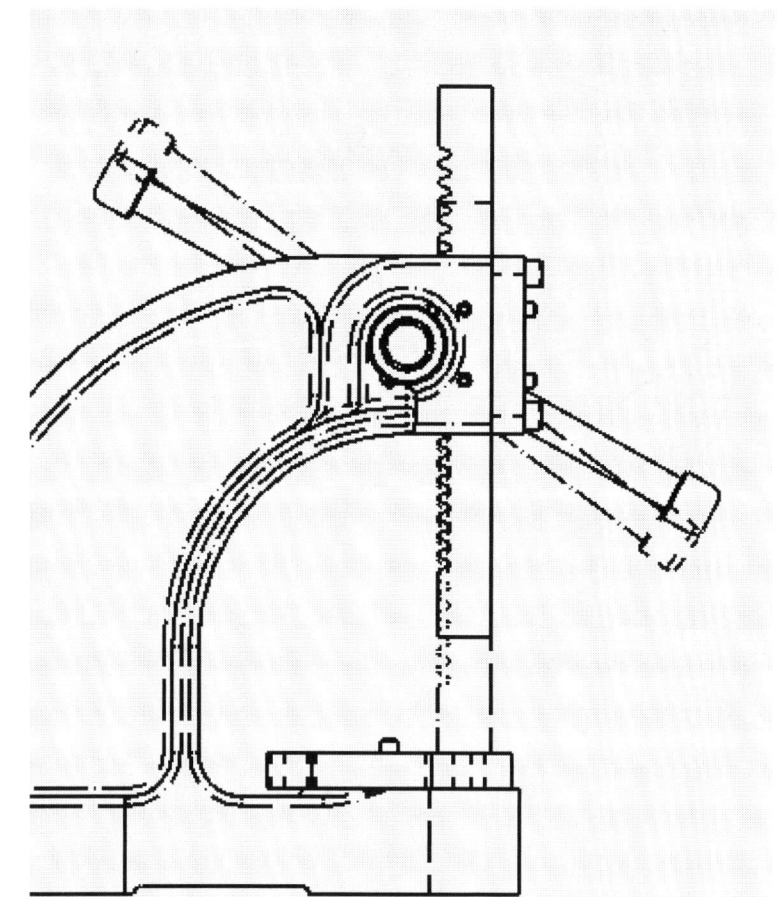 Zoom in to see the first overlay.

9. ![Overlay] Select the **Overlay** tool.
Select the Base View.

12-43

10.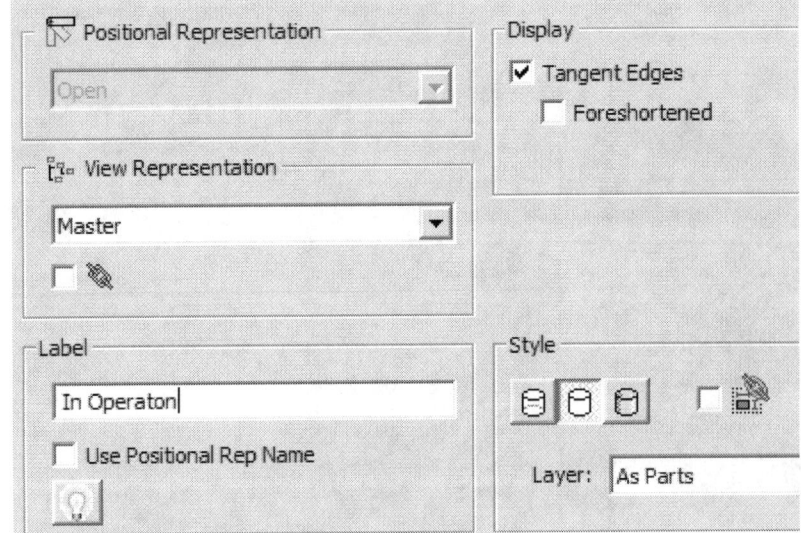

Select **Open**.
Enable **Tangent Edges**.
Disable **Foreshortened**.
Set **View Representation** to **Master**.
Disable **Use Positional Rep Name**.
Type **In Operation** in the **Label** name field.
Enable the Label **Visible**.
Enable **Hidden Style**.
Set **Layer** to **As Parts**.
Press **OK**.

11. 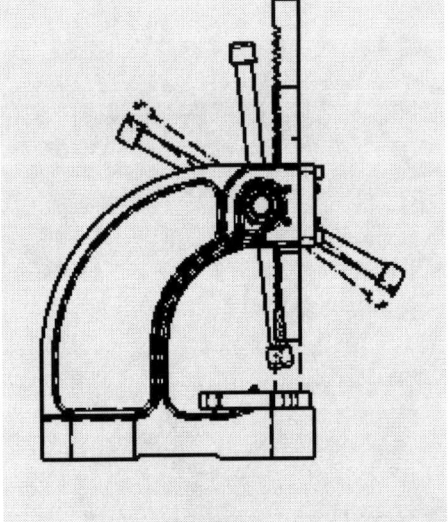   The next overlay is placed.

12. Save the file as *ex12-9.idw*.

 TIP: You will learn how to create the positional representation views in Lesson 17.

Drawing Management

New Sheet

Adds an additional sheet or page to the drawing layout.

Draft View

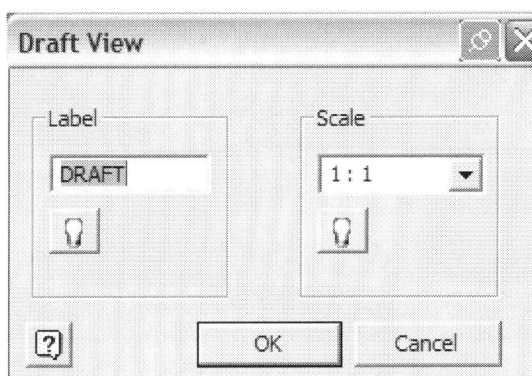

The Draft View is actually a layer that can be used to store redlines, notes, and additional geometry.
Users can use the Draft View to mark up drawings for engineering changes.
The Draft View automatically enables the Sketch Toolbar.

Edit Sheet

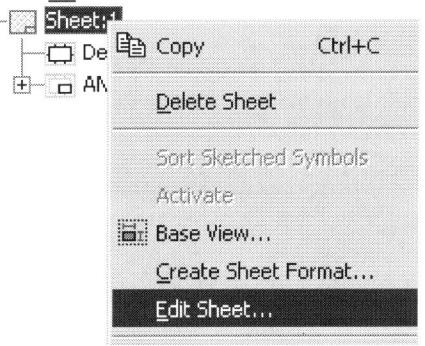

You can change the sheet size or settings by selecting the Sheet in the Browser, right click and select 'Edit Sheet'.

TIP: The name of the sheet can be changed by clicking in the Sheet Name edit box and typing in a new name.

12-45

Autodesk Inventor Fundamentals

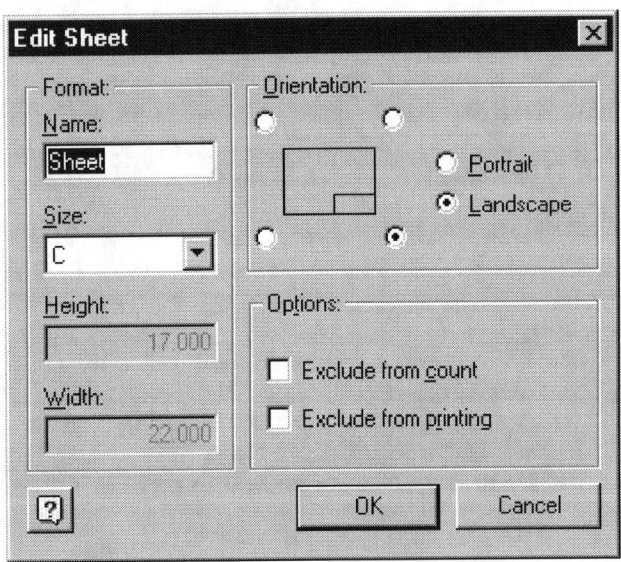

Size	Specifies a standard sheet size or format. Click the arrow and select the size or sheet format from the list. The standard sheet sizes are at the top of the list and current sheet formats are at the bottom of the list. Changing the sheet size changes the settings in the Height and Width boxes. Select Custom Size from the list to enter a different height and width.
Height	Sets the height of the sheet in drawing units. If you specify a standard size or sheet format in the Size box, this value is set automatically and the box is dimmed. To set a non-standard Height, select Custom Size and then enter a value in this box.
Width	Sets the width of the sheet in drawing units. If you specify a standard size or sheet format in the Size box, this value is set automatically and the box is dimmed. To set a non-standard Width, select Custom Size and then enter a value in this box.
Orientation	Sets the orientation of the page.
	Portrait	Sets the short edges of the paper at the top and bottom of the page.
	Landscape	Sets the long edges of the paper at the top and bottom of the page.
	Title block location	Edits the title block location. Click one of the 4 locators to set the new location.
Options specify whether the page is to be counted and printed with the rest of the drawing.	
Exclude from count	Specifies whether to exclude the selected sheet in the count of sheets in the drawing. Select the check box to exclude the sheet from the count; clear the check box to include the sheet in the count.
Exclude from printing	Specifies whether to exclude the selected sheet when printing the drawing. Select the check box to exclude the sheet from printing; clear the check box to print the sheet with the drawing.

Drawing Management

Create a Sheet Format

You can define sheet formats and add them to the Drawing Resources folder. Once you add a sheet format to a drawing, it can be used to add new sheets to that drawing.

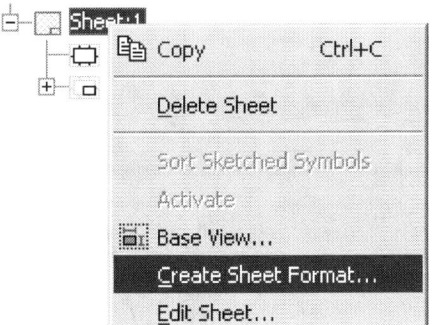

Right-click the sheet and select **Create Sheet Format** from the menu.

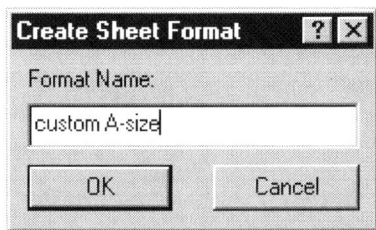

Enter the name for the new sheet format in the edit box.

NOTE: When you save the sheet format, it is added to the Drawing Resources folder in the Browser. To add a sheet using the new format, expand Drawing Resources and Format, and then double-click the desired sheet.

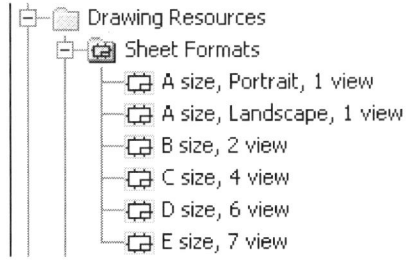

Drawing Resources stores all the sheet formats, title blocks, borders, and special symbols.

Define New Title Block

Inventor automatically places a Border and a title block when we start a new drawing.

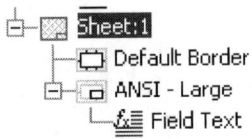

The ANSI – Large title block is the default title block placed when a new drawing files is created. You can change which title block is placed by default by modifying the *standard.idw* template.

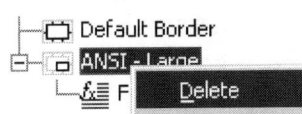

To remove the title block from the drawing, select in the Browser, right click and select '**Delete**'.

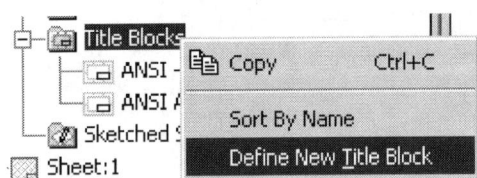

To create a new title block, highlight the Title Blocks folder in the Browser.

Right click and select **Define New Title Block**.

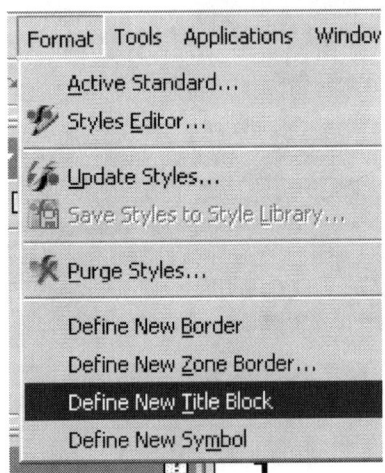

The Format menu is another method to create a new title block or border.

TIP. When defining the title block, it really doesn't matter where you position it on the sheet. The sketch will be inserted into the proper location.

Fill/Hatch Sketch Region

This tool allows the user to create special paint effects by adding color to any closed geometry, such as circles and rectangles. This tool is greyed out until a closed polygon is added to the Title Block.

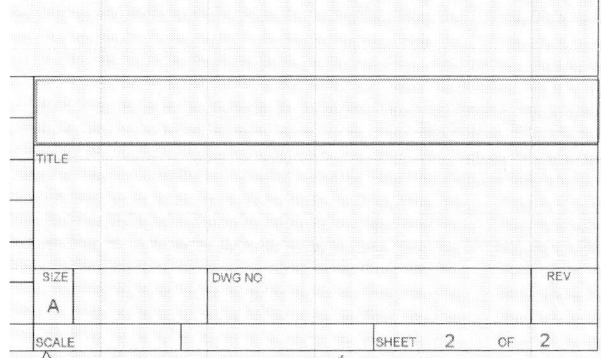

Draw a rectangle in the title block area.

Select the **Fill/Hatch Sketch Region** tool. Select the rectangle or the profile you wish to fill with color.

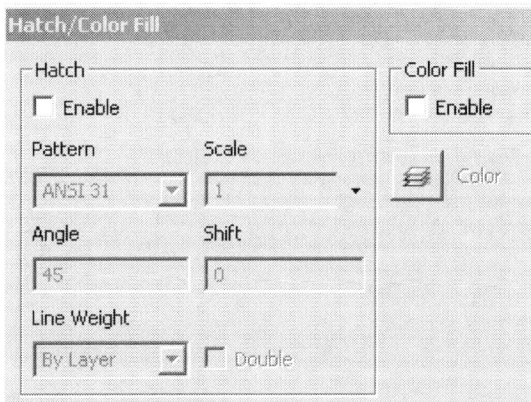

The Hatch/Color Fill dialog appears.
Pick the Enable box to add a hatch pattern.
To add color fill, place a check in the Enable box.
Press the square next to the word Color.

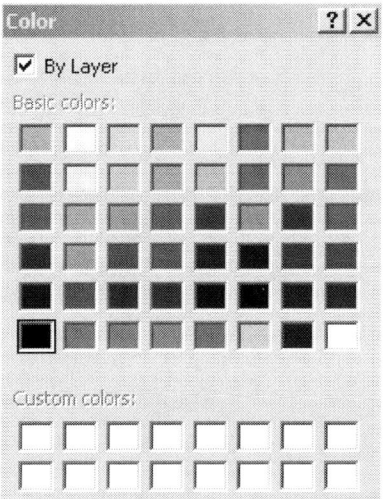

Select a color and press '**OK**'.

12-49

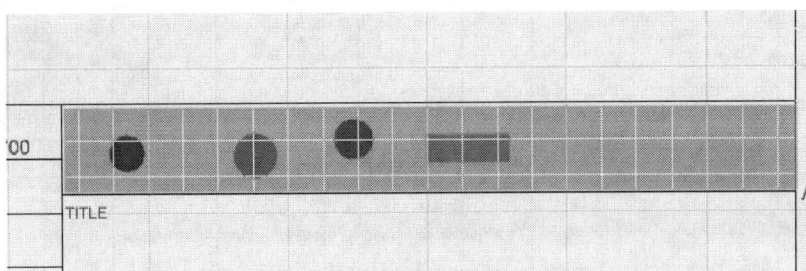

The user can use the sketch tools to create multi-colored graphics for the title block.
Simply create the desired shapes and then fill in the color.

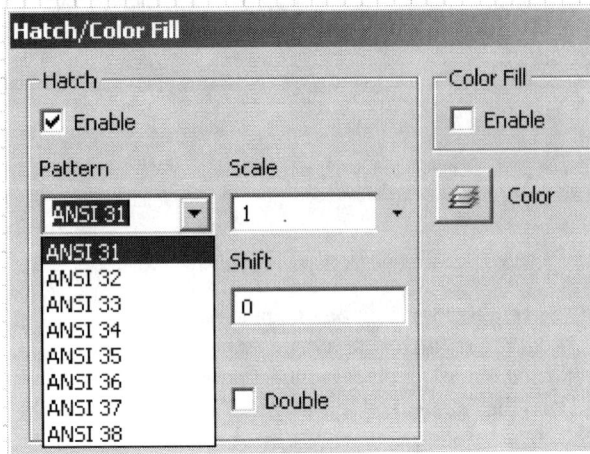

In order to create a Hatch pattern, enable the Hatch pattern by placing a check in the box beneath the Hatch.

The Pattern dropdown contains a list of available hatch patterns.

To change a hatch pattern that has been placed, simply pick on the hatch pattern.

 TIP: If you don't delete the existing title block before you start creating a new title block, the existing title block will remain in the drawing.

Drawing Management

Exercise 12-10:
Creating a Custom Title Block

File: New using Standard (inches)
Estimated Time: 60 minutes

This exercise will review the sketch tools used to create a custom title block.

1. Open a new Drawing file. Save as *ex12-10.idw*.

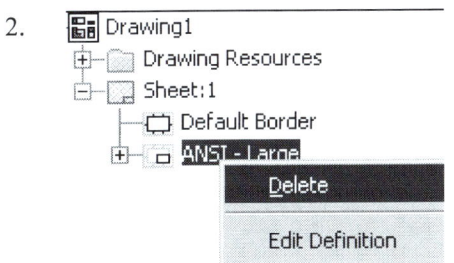

2. Highlight the ANSI-Large Title block in the Browser. Right click and select **Delete**.

3. Go to **Format→Define New Title Block**.

4. Use **Zoom Window** to zoom into the title block area.

12-51

5.

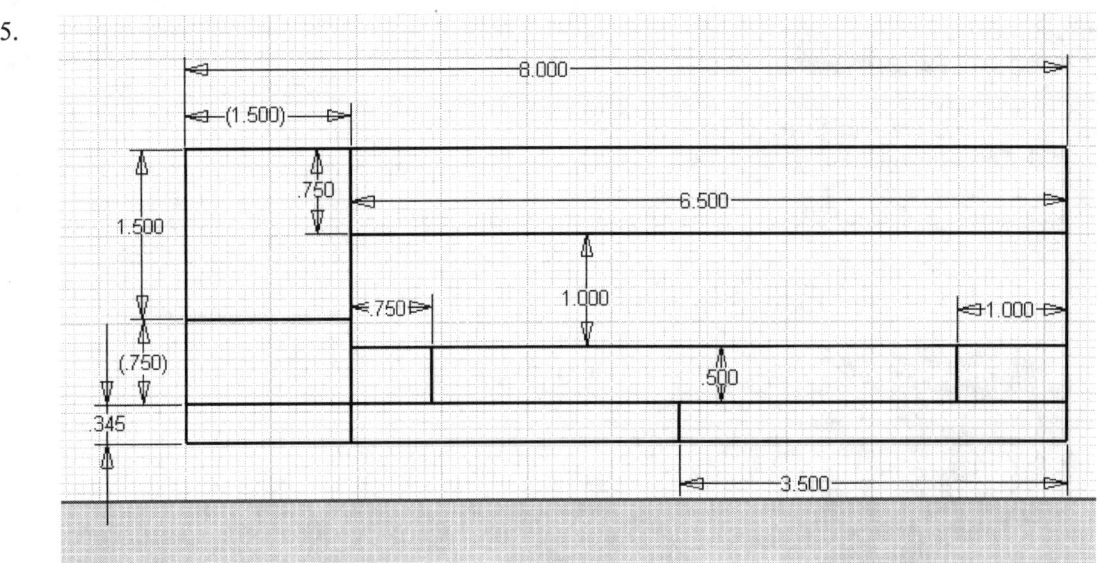

Draw the lines as shown using the **Line** tool.

6. 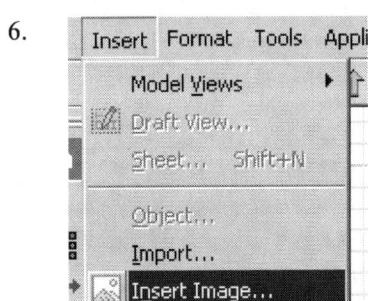 To insert a company logo, use **Insert→Insert Image**.

Next, indicate the area the picture is to be placed by picking two points to form a rectangle.

7. 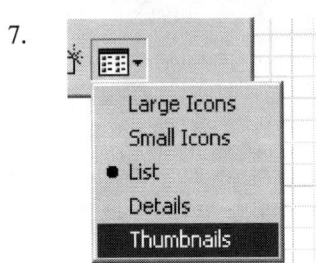 To see a preview of the image, set your view to **Thumbnails**.

8. 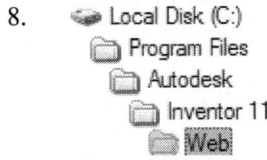 There are several bmp files in the Web folder under Inventor you can use if you don't have a bmp file available.

Drawing Management

9. 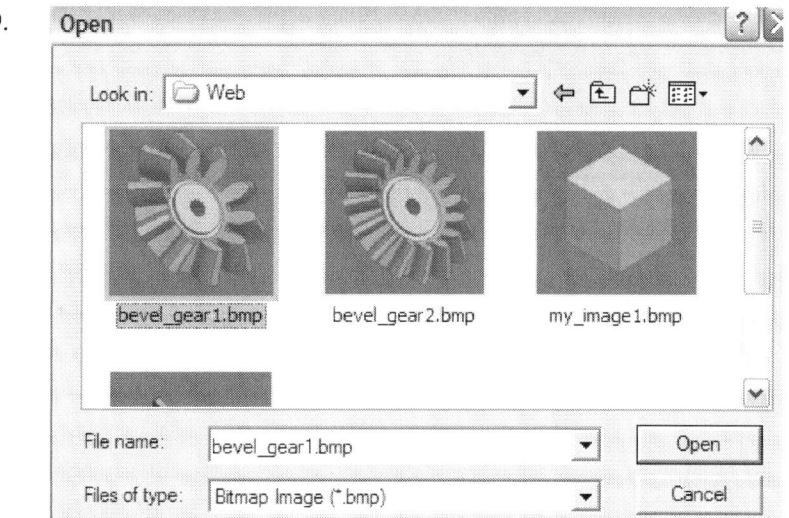 Browse to the Web folder. Select the *bevel_gear1.bmp*. Press **Open**.

10. 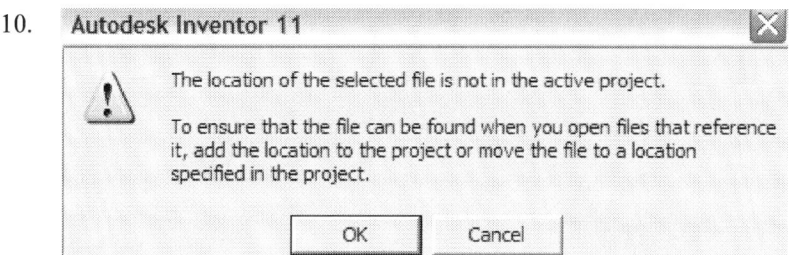 If you haven't saved your file (see Step 1), you will see this warning dialog.

    Press **OK**.

11. Right click and select **Done**.

12.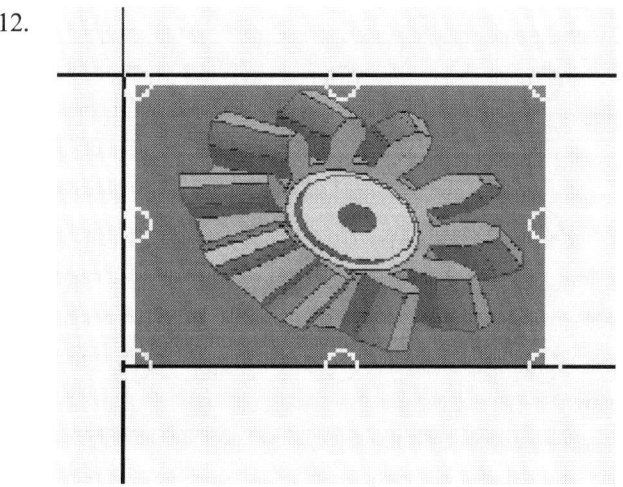

    You can use the corner grips on the bitmap to reposition and resize so it fits properly.

13. ![Text] Select the **Text** tool.

12-53

14. 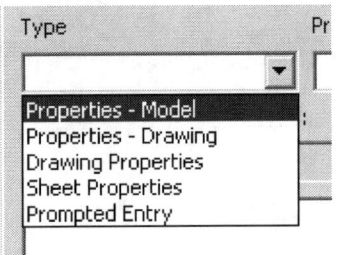 Draw a rectangle next to the logo to indicate the location for company information.

15. Select **Properties – Model** under **Type**.

16. Select **COMPANY** under **Property**.

    The Company name that is placed in the File Properties will automatically fill in the title block.

17. Set the text height to **.218**.

18. Press the **Add Text Parameter** button.

19. The property will appear in the text window.
    Press **OK**.

20. Note that you are still in Text mode.
    Draw a rectangle next to the Company field.

21.  Leave Type and Property blank.

    Enter in an address and website URL.

    This is a static value which can not be changed.

    Press **OK**.

22. Right click and select **Done**.

23. 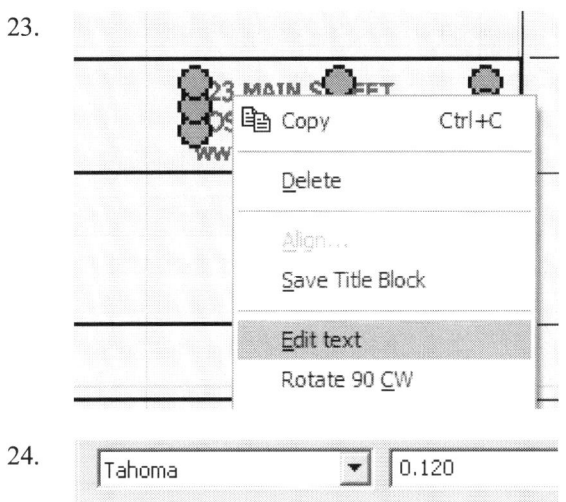 With **Text Box** enabled, right click and select **Edit Text**.

24. Set the text height to **0.120**.

TIP. When you use dimensions to set the size of elements in a title block or border, the dimensions are hidden when you finish editing.

25. Select the **Text** tool.

26. 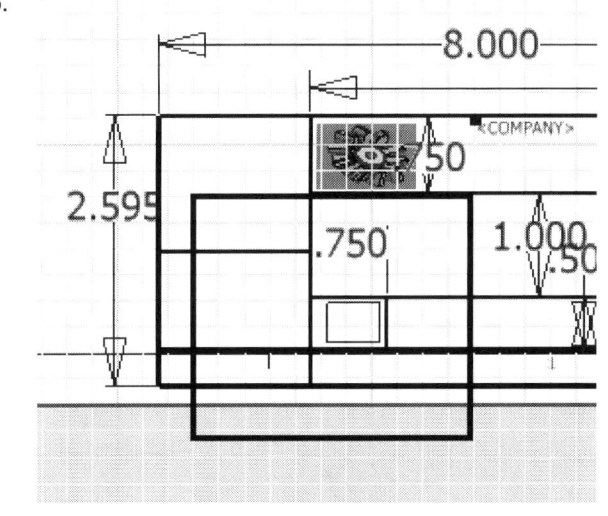 Draw a rectangle in the box for Sheet Size.

27. Set **Type** as **Sheet Properties**. Set **Property** as **Sheet Size**.

28.  Set font height to **0.120**. Press **OK**.

29. Press the **Add Text Parameter** button.

Autodesk Inventor Fundamentals

30. 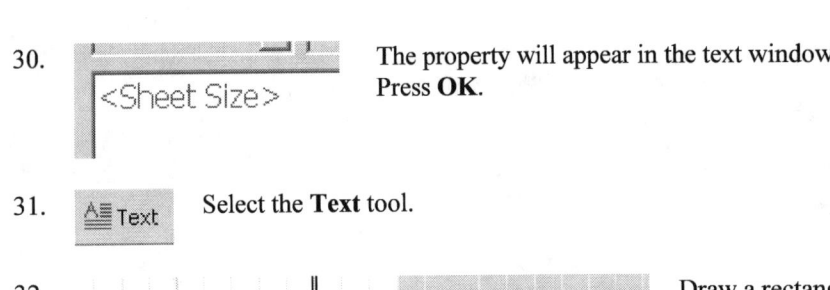 The property will appear in the text window.
    Press **OK**.

31. ![Text tool] Select the **Text** tool.

32. 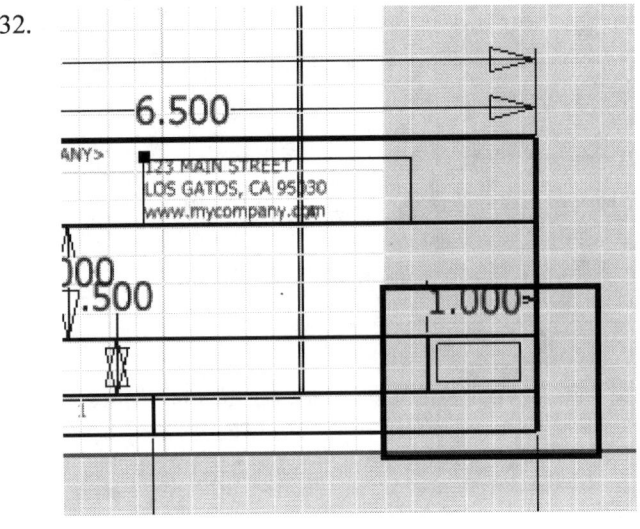 Draw a rectangle in the revision space.

 TIP: By using Sheet Properties instead of using a Prompted Entry, the value will automatically update when the user redefines the sheet properties in the Browser.

33. 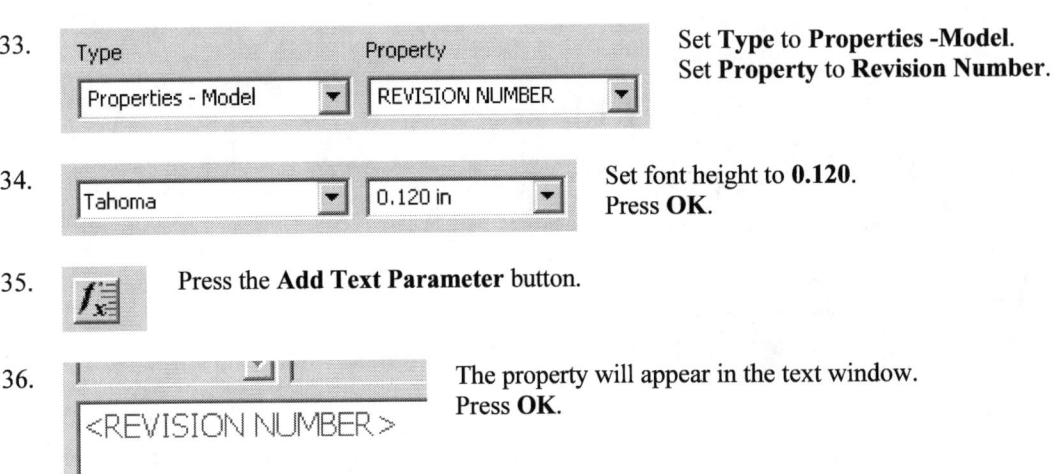 Set **Type** to **Properties -Model**.
    Set **Property** to **Revision Number**.

34. Set font height to **0.120**.
    Press **OK**.

35. Press the **Add Text Parameter** button.

36. The property will appear in the text window.
    Press **OK**.

12-56

37. 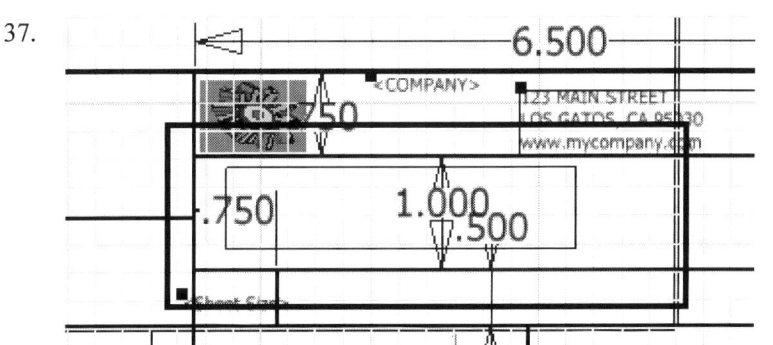 Draw a rectangle in the title space.

38. 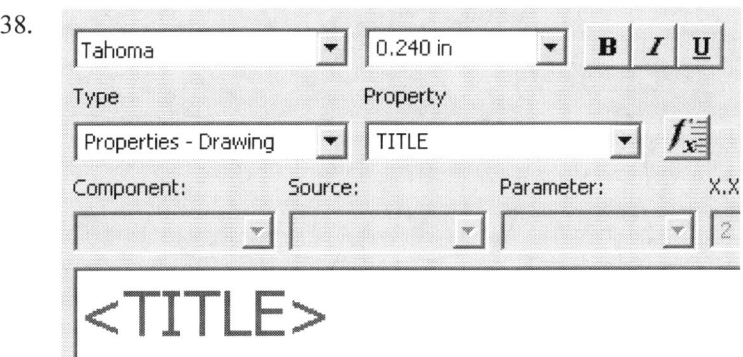 Set **Type** to **Properties - Drawing**.
Set **Property** to **Title**.
Set the text height to **0.240**.

39. Press the **Add Text Parameter** button.
The property will appear in the text window.
Press **OK**.

40. 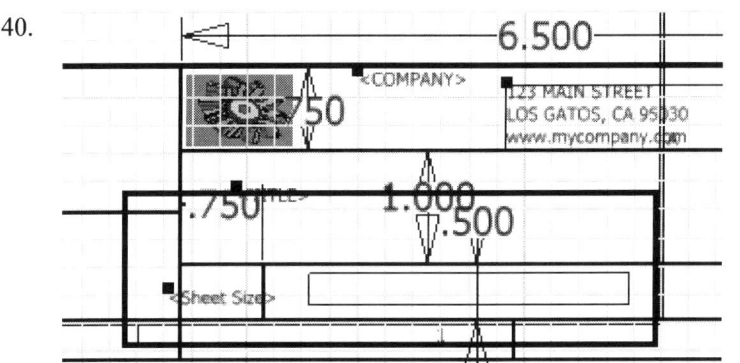 Draw a rectangle in the part number space.

41. 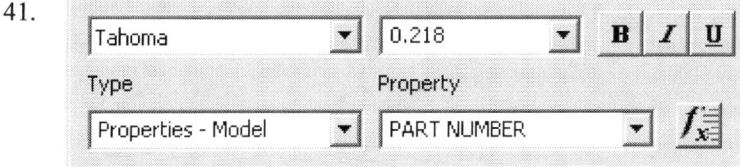 Set **Type** to **Model Properties**.
Set **Property** to **Part Number**.
Set text height to **0.218**.

42. Press the **Add Text Parameter** button.
The property will appear in the text window.
Press **OK**.

43.  Draw a rectangle in the sheet number space.

44. 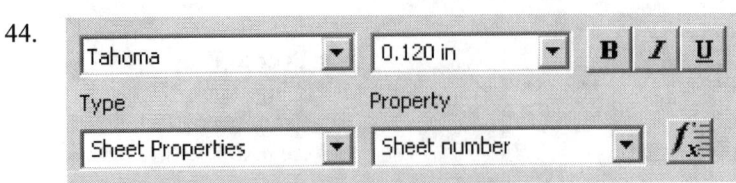 Set **Type** to **Sheet Properties**.
Set **Property** to **Sheet Number**.
Set **Height** to **0.120**.

45. Press the **Add Text Parameter** button.
The property will appear in the text window.
Press **OK**.

46. 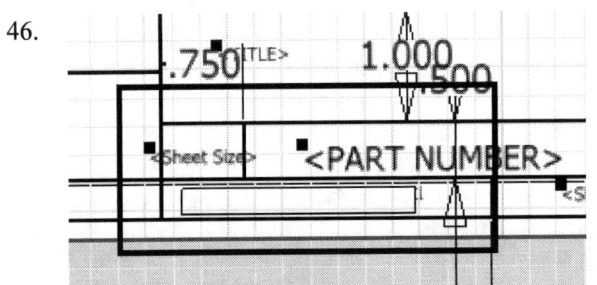 Draw a rectangle in the file name space.

47.  Set **Type** to **Properties - Drawing**.
Set **Property** to **FILENAME AND PATH**.
Set **Height** to **0.120**.

48. Press the **Add Text Parameter** button.
The property will appear in the text window.
Press **OK**.

49. 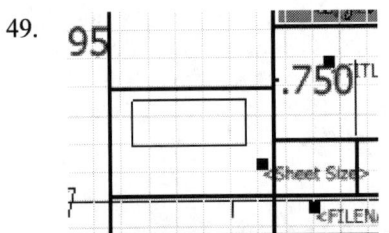 Draw a rectangle in the drafter name space.

50.  Set **Type** to **Properties - Drawing**.
Set **Property** to **AUTHOR**.
Set **Height** to **0.156**.

Drawing Management

51. Press the **Add Text Parameter** button.
    The property will appear in the text window.
    Press **OK**.

52. Draw a rectangle in the Date space.

53. To add a prompted entry (similar to an attribute):
    Set **Type** to **Prompted Entry**.
    Set **Prompt** to '**Enter Date**'.
    Set **Height** to **0.128**.

    Press **OK**.

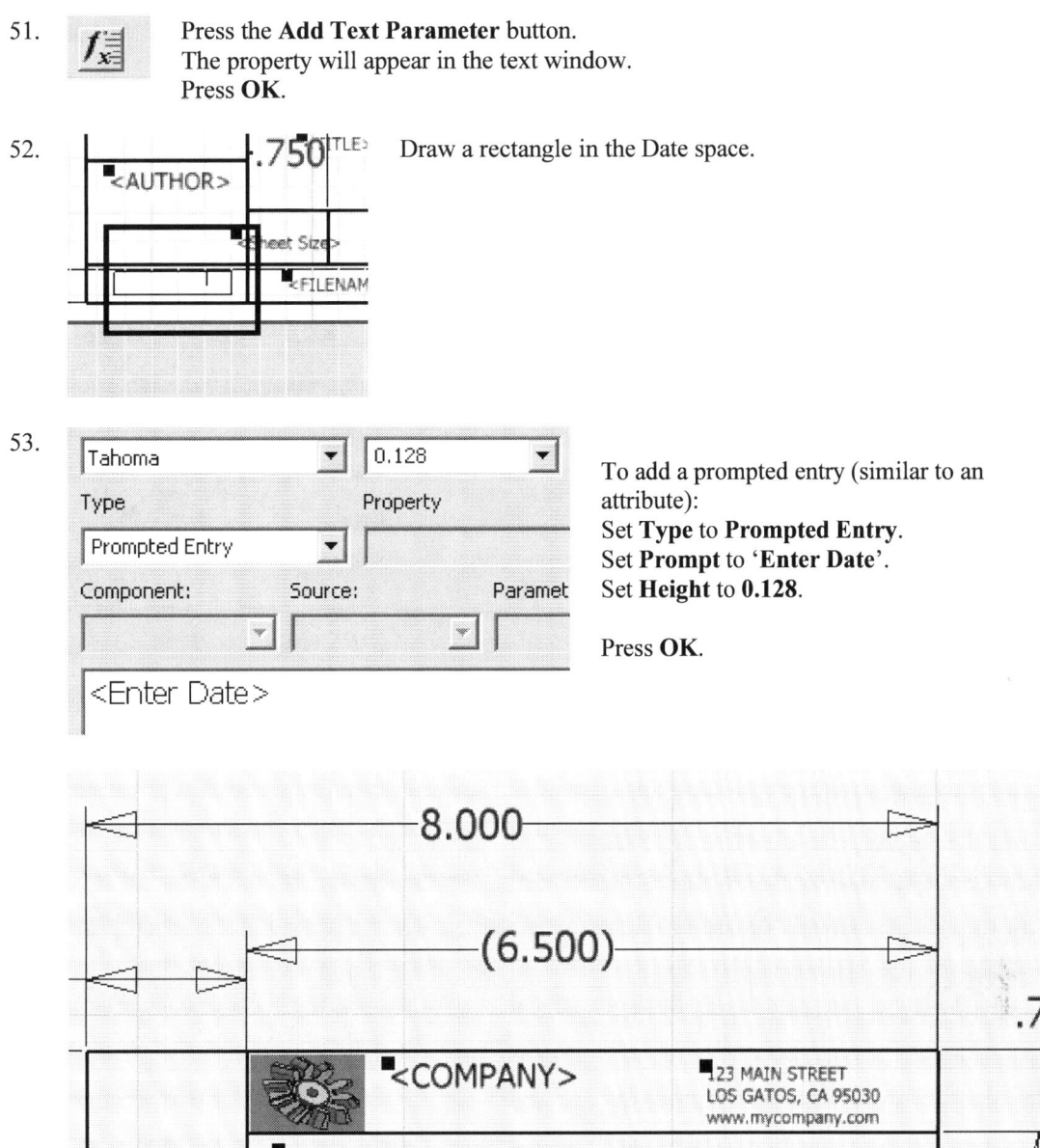

The title block automatically inserts values into the Model Properties fields using the values stored in the Properties dialog box for each model.

TIP: Inventor will only accept bitmaps for insertion, so your file must have a *.bmp extension.

12-59

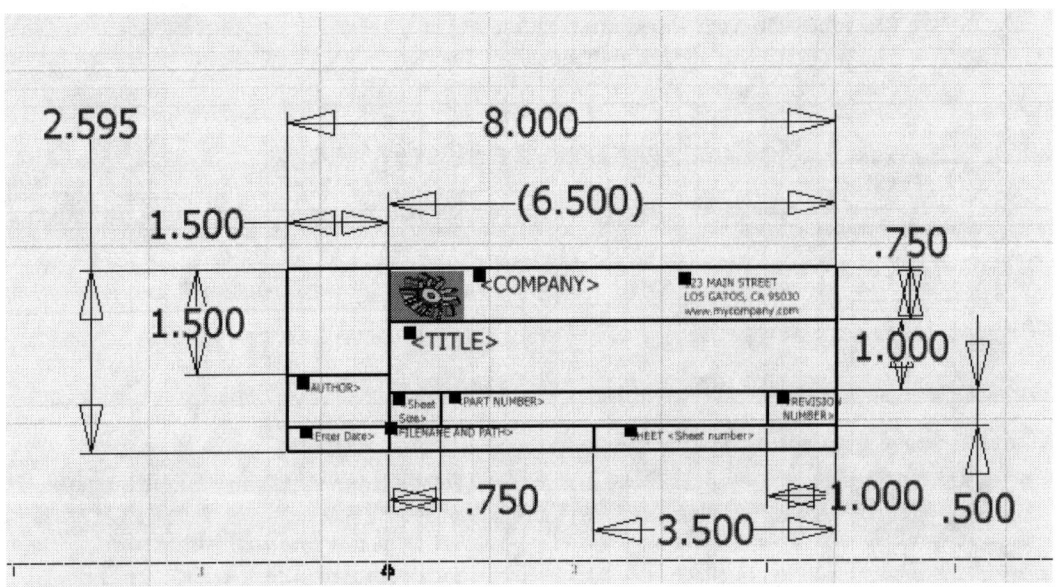

The completed title block.

54. Once the title block is complete, right click and select **Save Title Block**.

55. Assign a name to your title block. Do not use punctuation marks. Spaces are OK.

56. Your new title block name automatically appears in the Browser.

To insert, select and double click.

Use File → iProperties (to fill the blank)

12-60

57. 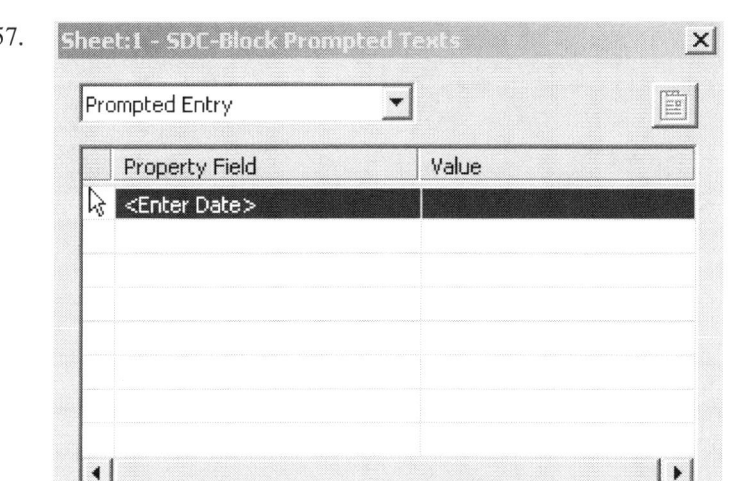 You will then be prompted for any prompted entry fields that were defined.

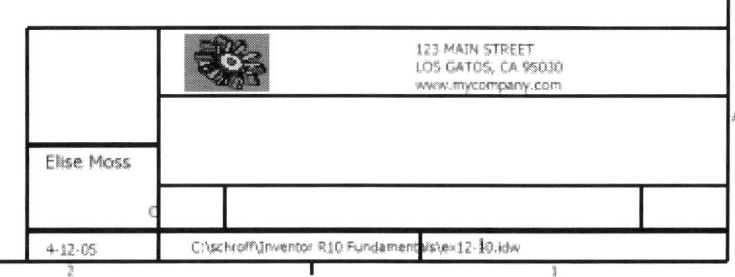 When you first place the title block, most of the fields will show blank. The only ones filled in are those linked to **Sheet Properties**.

Remember most of your fields are set by **Model Properties** and will vary depending on the model you use for your drawing.

Once you place a view in the drawing with the associated model properties defined, the title block should fill in correctly.

58. Save the file as *ex12-10.idw*.

If you would like to add fields to be used in your title block, you can add custom fields under the **Custom** tab in the **iProperties** dialog box.

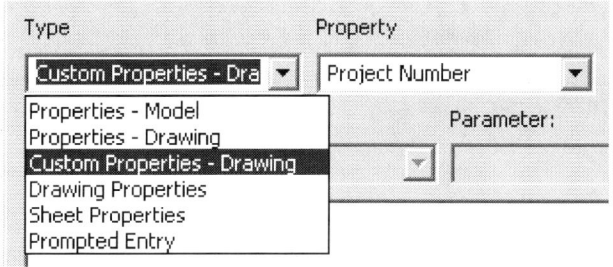 The **Custom** fields will then appear in the **Text** dialog box.

12-61

Autodesk Inventor Fundamentals

Exercise 12-11:
Adding Custom properties

File: Standard (inches).idw
Estimated Time: 30 minutes

This exercise will demonstrate how to add custom properties to a drawing template. If you add custom properties to the template, the properties will be available to you for any new drawing using that template.

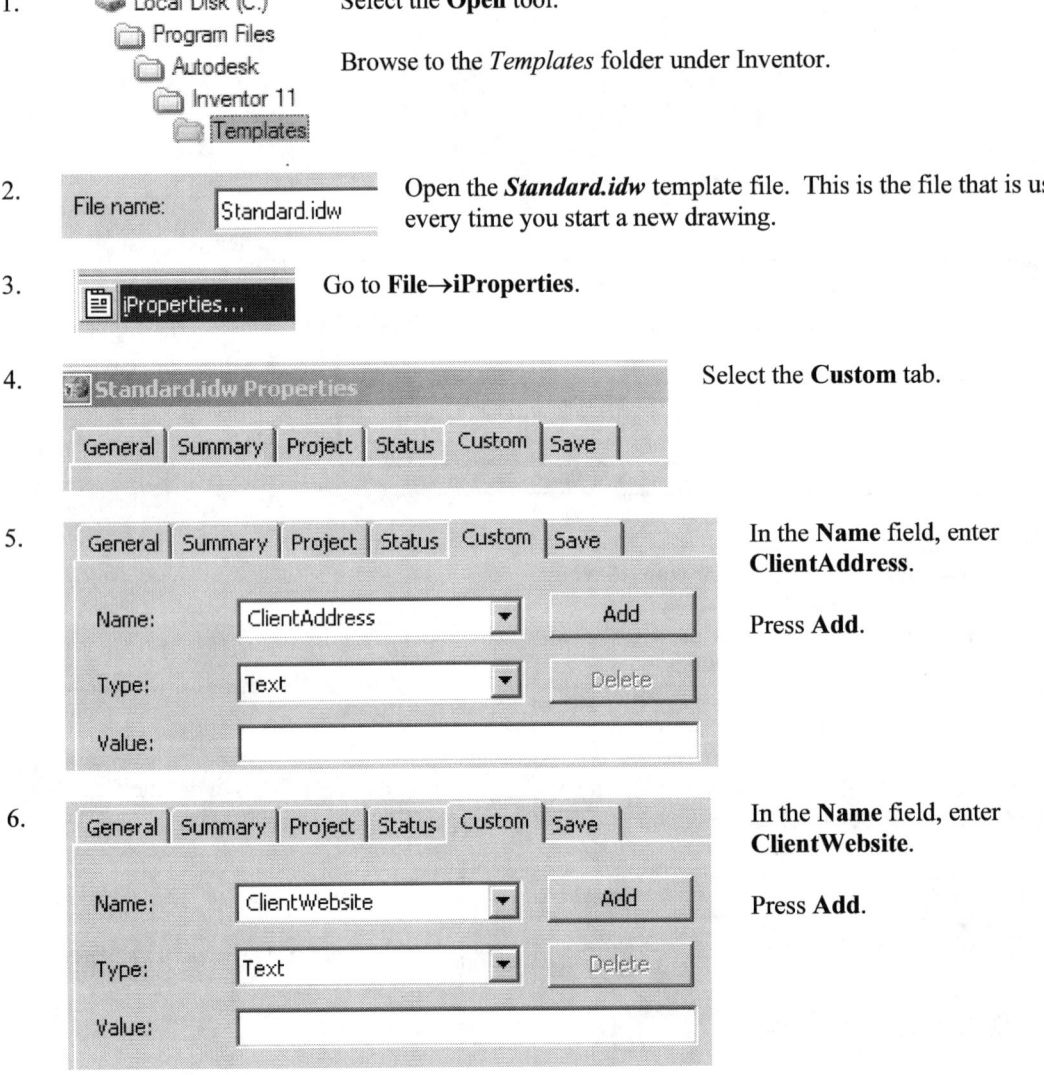

1. Select the **Open** tool.

   Browse to the *Templates* folder under Inventor.

2. Open the ***Standard.idw*** template file. This is the file that is used every time you start a new drawing.

3. Go to **File→iProperties**.

4. Select the **Custom** tab.

5. In the **Name** field, enter **ClientAddress**.

   Press **Add**.

6. In the **Name** field, enter **ClientWebsite**.

   Press **Add**.

12-62

Drawing Management

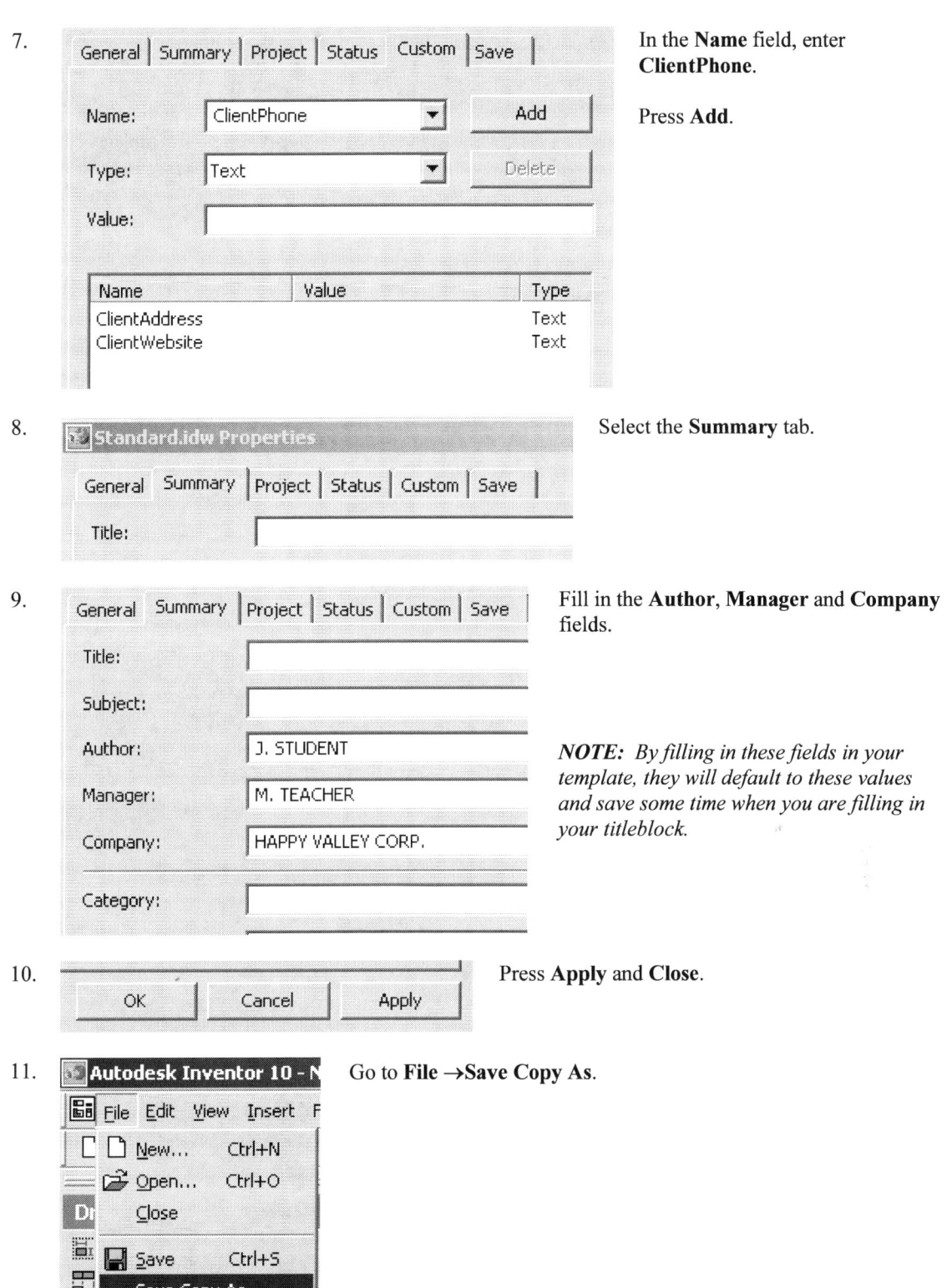

7. In the **Name** field, enter **ClientPhone**.

   Press **Add**.

8. Select the **Summary** tab.

9. Fill in the **Author**, **Manager** and **Company** fields.

   NOTE: By filling in these fields in your template, they will default to these values and save some time when you are filling in your titleblock.

10. Press **Apply** and **Close**.

11. Go to **File → Save Copy As**.

12-63

12. 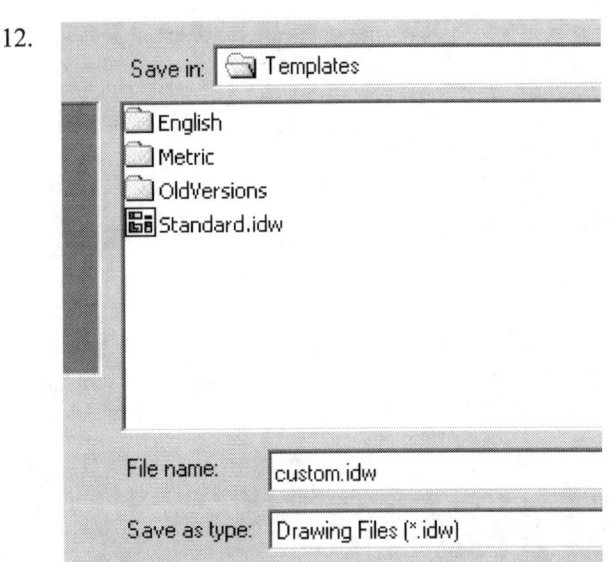 Name your file *custom.idw*.

    This file needs to be saved in the Templates folder or you can store your templates on the file server and re-direct your templates path in your options.

13. 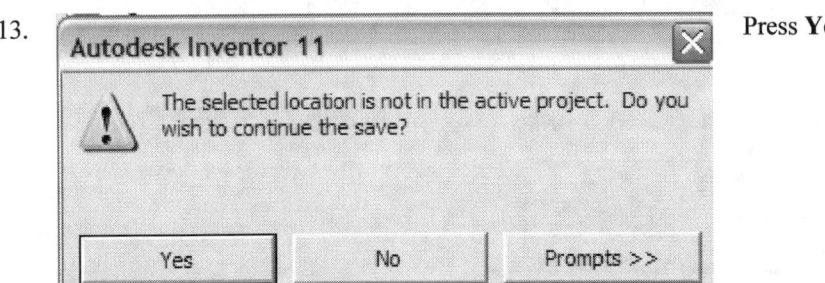 Press **Yes**.

14. Go to **Tools→Document Settings**.

15. Select the **Drawing** tab.

16.  Under **Properties in Drawing**, select **Browse**.

12-64

Drawing Management

17. Browse to the *Templates* folder where you stored your custom properties.

    Local Disk (C:) → Program Files → Autodesk → Inventor 10 → **Templates**

18. File name: custom.idw

    Select the *custom.idw* file.
    Press **Open**.

19. 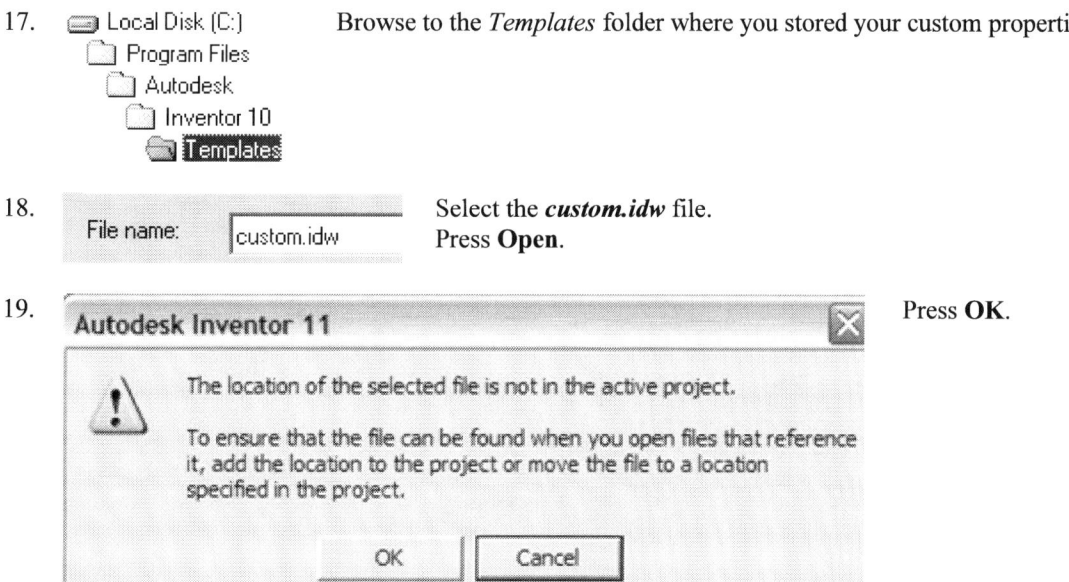 Press **OK**.

    Autodesk Inventor 11: The location of the selected file is not in the active project. To ensure that the file can be found when you open files that reference it, add the location to the project or move the file to a location specified in the project.

20. Press **Apply** and **OK**.

21. Go to **File→iProperties**.

22. Verify that the properties you created in the *custom.idw* are now available in this file.

    Standard.idw Properties — Summary tab:
    - Author: J. STUDENT
    - Manager: M. TEACHER
    - Company: HAPPY VALLEY CORP.

23. Close without saving.

12-65

Autodesk Inventor Fundamentals

Exercise 12-12:
Copy a Titleblock

File: Standard (inches).idw
Estimated Time: 10 minutes

Learn how to copy a titleblock from one drawing into another.

1. 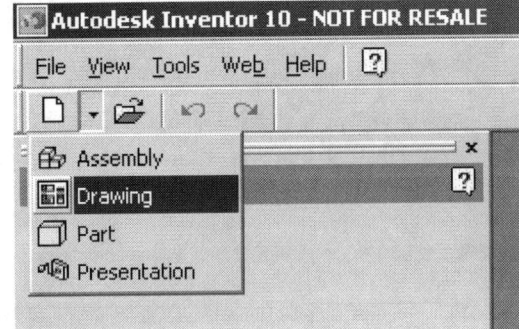 Start a new drawing.

2. Under the File menu, there is a list of recently opened files.

   Locate and open *ex12-10.idw*... the file where you created your custom title block.

3. 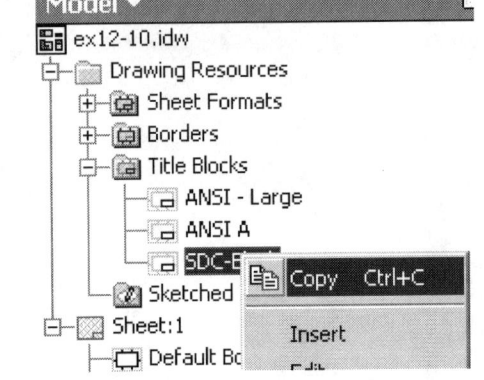 Expand the *Title Blocks* folder under Drawing Resources.
   Highlight the **SDC-Block**.
   Right click and select **Copy**.

4. 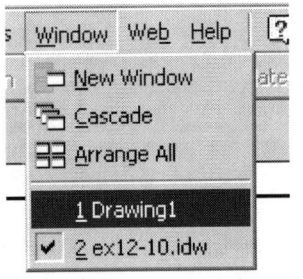 Go to **Window**.
   Select the other drawing you have open.

12-66

Drawing Management

5. 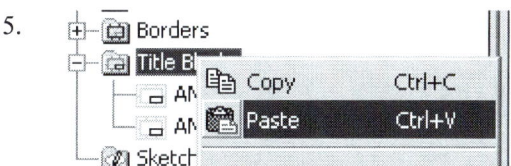 Expand the *Title Blocks* folder under Drawing Resources.
Highlight the Title Blocks category in the Browser.
Right click and select **Paste**.

6. 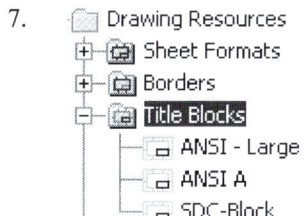 Press **OK**.

7. The block is now available to be inserted.

   NOTE: You must delete any existing title block in the sheet before you can insert a new title block.

8. Close both drawings without saving.

TIP: A single model dimension cannot be used in multiple views on the same sheet.

Sketch Overlay

You can add overlay sketches that are associated with the underlying sheet. A sketch can contain geometry, such as lines and arcs, or text. If a drawing view is selected when you activate the Sketch tools, the sketch is attached to the view. If you move the view, the sketch moves with it.

An overlay sketch cannot be saved as part of a drawing template, or copied between drawings. However, if you copy a view or sheet that has a sketch associated to it, the sketch is copied as well.

Use the tools on the Sketch toolbar to add sketched elements to a drawing. Sketches on a drawing sheet reside on overlays that are associated with the underlying sheet. If a drawing view is selected when you activate the sketch tool, the resulting sketch is associated with the selected view.

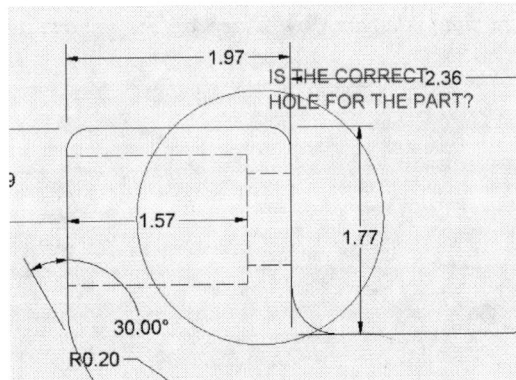

TIP: To edit a completed sketch, select a sketch element in the graphics window or Browser, right-click, and select edit to reactivate the sketch.

- Use the Zoom Window button on the Standard toolbar to zoom in on the area where you are working.
- Set the grid to the spacing needed to quickly line up the sketch elements.
- Check the Snap to Grid setting to more easily place sketch elements.
- To select a group of sketch elements, activate the Select tool, then click in the graphics window and drag a box around the elements.
- Use the dimension tools to set the size of sketched geometry or to add dimensions between the geometry in a sketch and elements in the underlying drawing view.

Sketch overlays are a useful tool for redlining drawings without affecting views or drawing data.

Drawing Management

Sketched Symbols

Sketched Symbols are one of the items listed under Drawing Resources

You can add a sketched symbol to the active sheet in the drawing. The symbol is associated to the sheet on which you place it. If you delete the sheet, the sketched symbol is deleted. If you copy the sheet, the sketched symbol is copied.

The Browser for each drawing or drawing template contains a Sketched Symbols folder in the Drawing Resources folder. You can create custom sketches and add them to Sketched Symbols to use in the drawing.

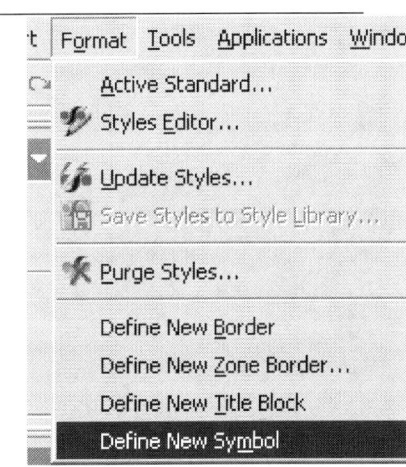
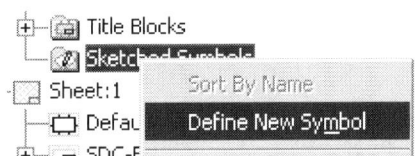

There are two methods to create a new symbol:

Under the Menu, go to **Format→Define New Symbol**.
In the Browser, under Drawing Resources, go to Sketched Symbols, right click and select 'Define New Symbol'.

TIP: You can use a property field in a sketched symbol to create a block with attributes.

Autodesk Inventor Fundamentals

Exercise 12-13
Define New Symbol

File Name: New (Standard using inches) idw
Estimated Time: 10 minutes

1. Start a New Drawing file.

2. 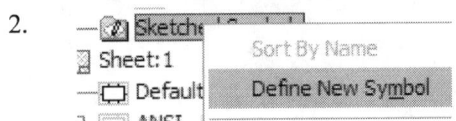 In the Browser, under Drawing Resources, go to **Sketched Symbols**, right click and select **Define New Symbol**.

3. 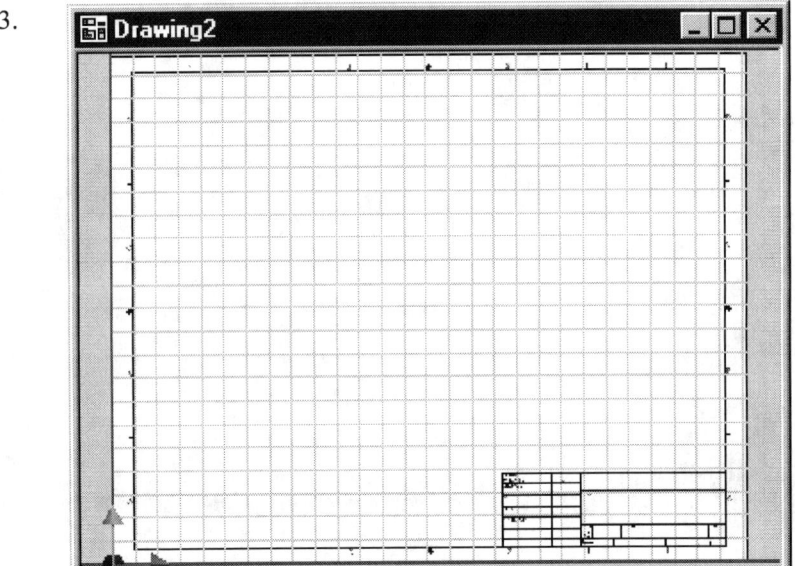 Your window will change to Sketch Mode.

4. 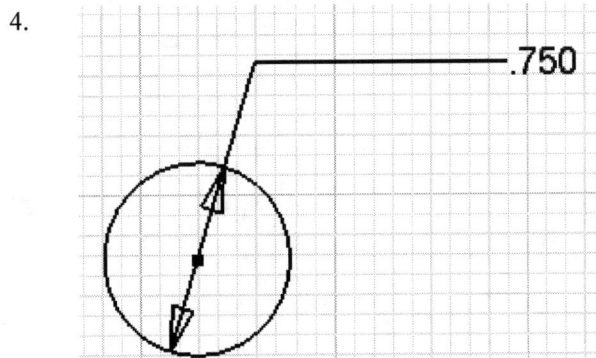 Draw a circle with 0.750 diameter.

12-70

Drawing Management

5. 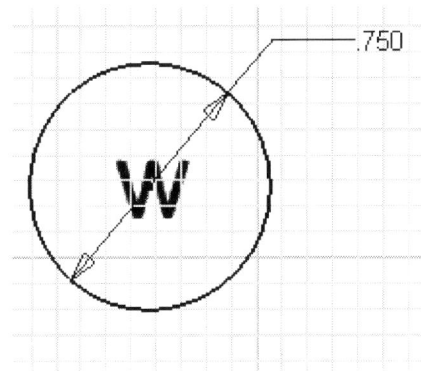 Use the **Text** tool.

   Place a **W** in the center of the circle.

6. 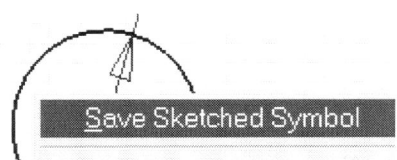 Right click anywhere in the graphics window.

   Select **Save Sketched Symbol**.

7. 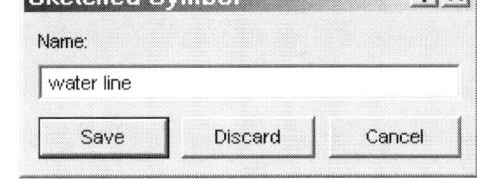 Enter the name for the new symbol in the dialog box.
   Type **water line**.
   Press **Save**.

8.  The sketched symbol appears under Sketched Symbols in your Browser.

   Save your file as *ex12-13.idw*.

TIP: You cannot edit the default border after it is placed. To change the border, delete it and insert a new border with the desired properties.

12-71

Autodesk Inventor Fundamentals

To add a sketched symbol to a drawing, double-click the symbol name in the Browser.

Sketched symbols are either associated with a sheet or with a view. If a sketched symbol is associated with a sheet, it is considered a symbol. If a sketched symbol is associated with a view, it is considered a callout.

You can add a sketched symbol to a drawing view as a callout, where a leader is automatically added. The symbol is associated to the view. If you delete the view, the sketched symbol is deleted. If you copy the view, the sketched symbol is copied.

Exercise 12-14
Inserting a Symbol

File: Ex12-13.idw
Estimated Time: 15 minutes

1. Open *ex12-13.idw*.

2. 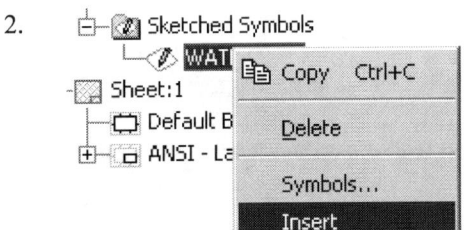   Highlight the sketched symbol called **water line**.

   Right click and select **Insert**.

3. 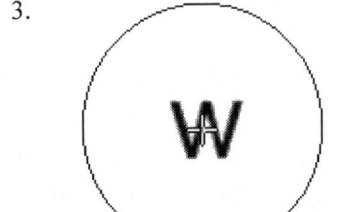   The symbol appears at the end of your cursor.

   Left pick to place.

   You can continue placing symbols as a multiple action.

4. 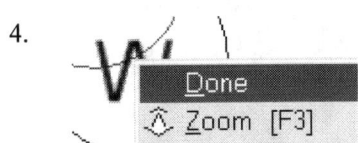   Right click and select **Done** when you are finished placing callouts.

5. 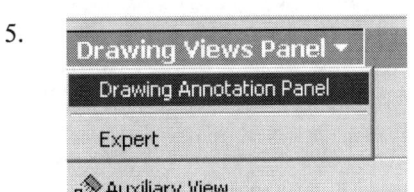   Left click the arrow on the **Drawing Views Panel**.
   Select **Drawing Annotation Panel**.

6. Select the **Symbols** tool.

12-72

7. 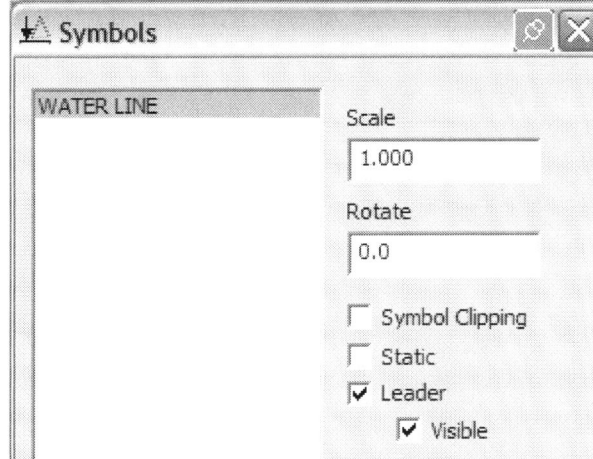 A dialog appears.

A list of all the sketch symbols available in the active drawing file appears on the left.

Symbol Clipping is used to trim existing dimensions, leaders, and extension lines that cross over the leader applied with the symbol.

Leader is enabled to automatically add a leader to the symbol.

Press **OK**.

8. 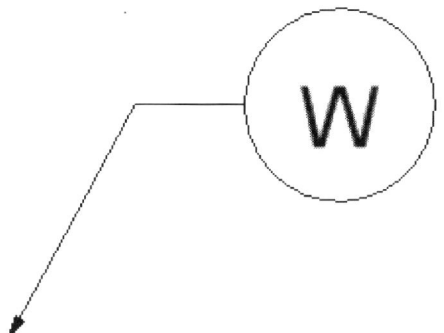 Select a point for the start of the leader (where the arrowhead is to be placed).
Select a second point for the start of the shoulder.
Select a third point for the end of the shoulder (and to place the symbol.)

Right click and select **Continue**.

(Select **Done** if you do not want to place a second call-out.)

9. The callout symbol is placed.

You can place another or right click and select **Done**.

10. 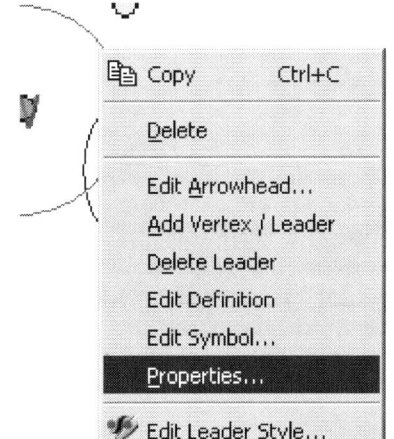 Select the callout.

Right click.
Notice that in the menu you can edit the **Arrowhead**, **Add Vertex/Leader/Delete Leader**.

11. Select **Delete Leader** and pick the leader.

12. 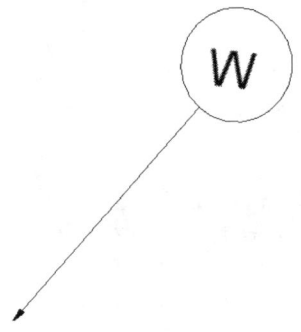 You can adjust your callout since you removed a leg in the leader.

13. Save the file as *ex12-14*.

Exercise 12-15
Creating a Symbol with Attributes

File: Ex12-14.idw
Estimated Time: 30 minutes

1. Open *ex12-11.idw*.

2. 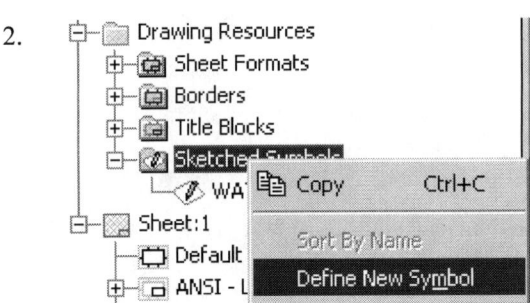 Highlight Sketched Symbols in the Browser. Right click and select **Define New Symbol**.

3. 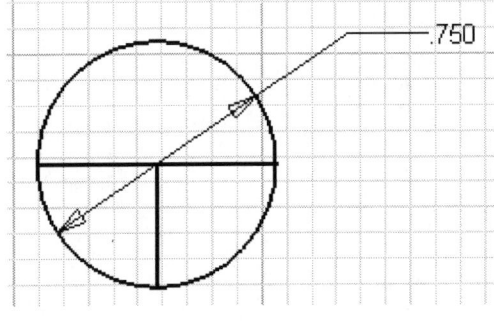 Draw a circle with a 0.750 diameter.
Draw a horizontal line across the diameter.
Draw a vertical line from the center point down to the lower quadrant.

4. Use the **Text** tool to create three **Prompted Entries**.

12-74

5. 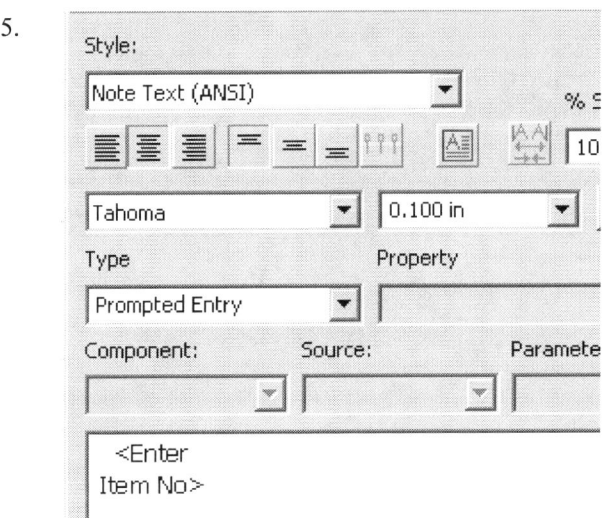 Set each field so it will be centered in its area.

6. 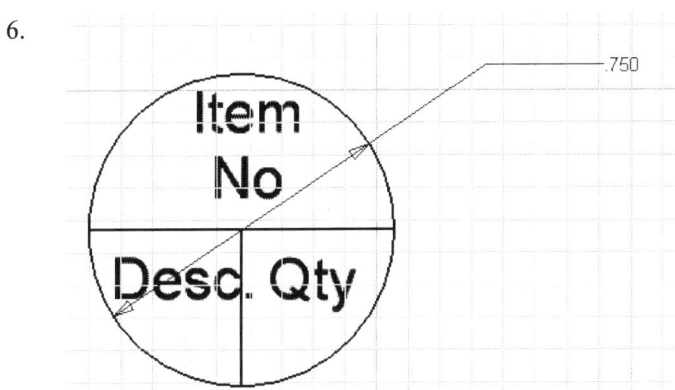 Create a prompted entry for **Item No**, **Qty**, and **Description**.

7.  Right click and select **Save Sketched Symbol**.

8. Edit the **Name** to 'item-balloon'.

9. Save as *ex12-15.idw*.

Autodesk Inventor Fundamentals

Exercise 12-16
Editing Symbols

File: Ex12-15.idw
Estimated Time: 15 minutes

1. Open the *ex12-12.idw* file.

2. 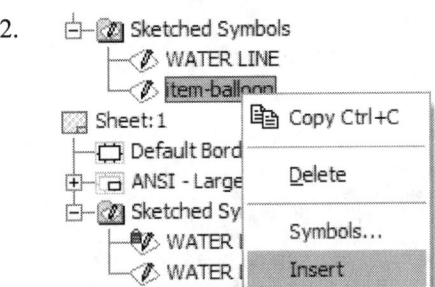 Highlight the item-balloon in the Browser.
   Right click and select **Insert**.

3.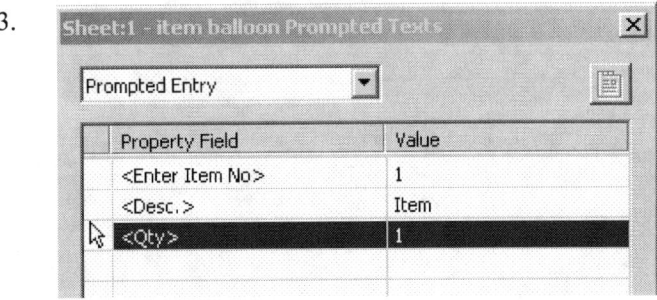

   A field entry dialog will pop up for each prompted entry.
   Type in **1** for the **Item No** field.
   Type **1** for the **Qty** field.
   Type **ITEM** for the **Description** field.

   Press **OK**.

4. 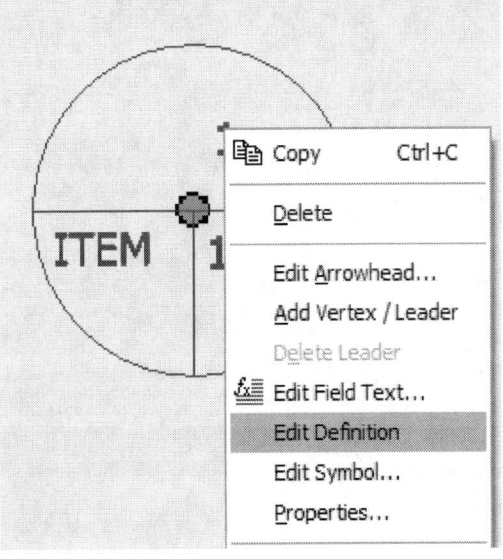 Select the symbol.

   Right click.

   To modify the symbol, you can select **Edit Definition**.

   To modify the attributes, select **Edit Field Text**.

   Select **Edit Field Text**.

12-76

Drawing Management

5. 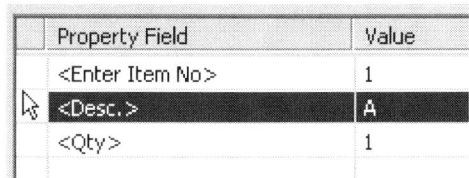 A dialog box will appear where it is easy to change the values for each field.

    Change the value of the **Description** to **A**.

    Press **OK**.

6. 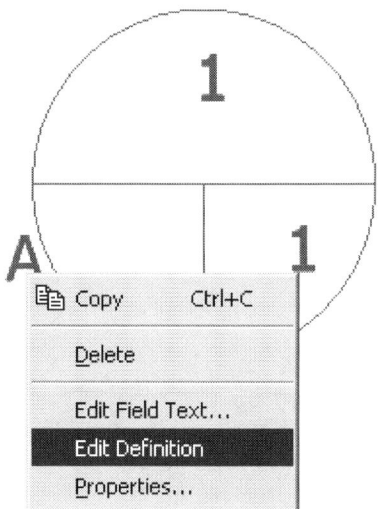 The field is not justified properly.

    Select the symbol.

    Right click and select **Edit Definition**.

7.  Highlight the field that is not justified properly.

    Right click and select **Edit Text**.

8. Set the **Justification** to **Top Center**.
    Press **OK**.

9. Shift the text so it will be located properly.

10. 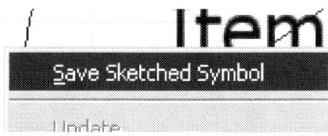 Right click and select **Save Sketched Symbol**.

11. 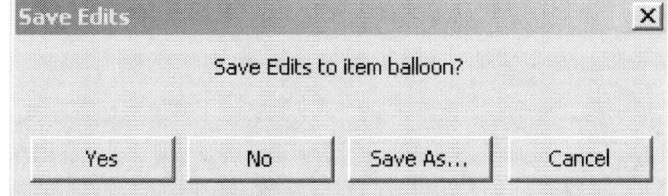 Press **Yes**.

12-77

12. 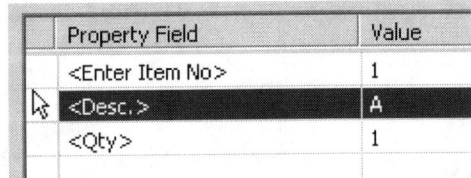 Enter **A** for the **Description** value.

    Press **OK**.

13. 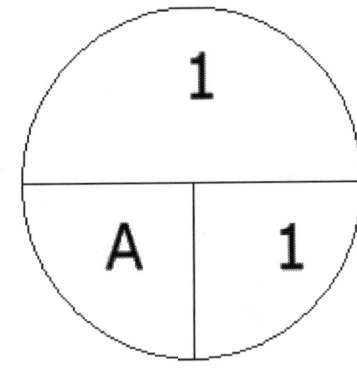 The symbol updates and is corrected.

    Save the file as *ex12-16.idw*.

Drawing Borders

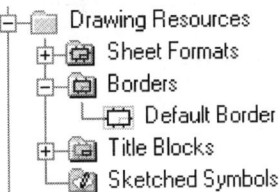

Drawing Borders are located under the Drawing Resources folder.
Your first drawing sheet automatically has a border applied.

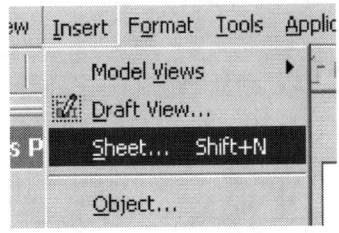 However, if you use Insert →Sheet to add a new sheet to your drawing file, no border or title block is automatically placed. This is because many companies use a different format for sheets other than Sheet 1.

12-78

Insert Drawing Border

To add the Default Border, highlight it in the Browser, right click and select 'Insert Drawing Border'.

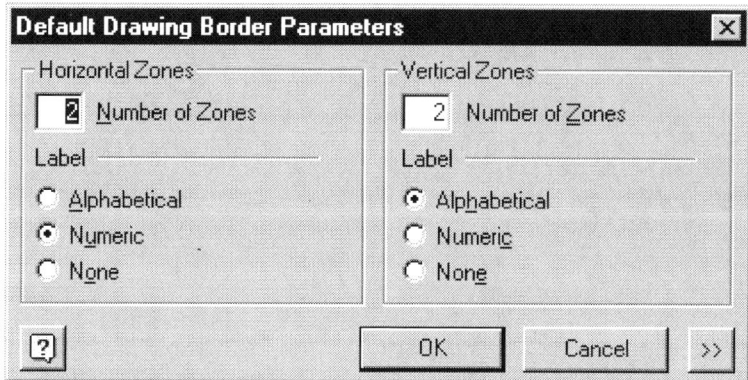

The Default Drawing Border Parameters dialog box appears.

Zones set the number and style for the zones that the border defines.	
Horizontal Zones	Sets the number and style for the horizontal zones. Number of Zones sets the number of horizontal zones. Enter the number in the box. Zone Labeling Sets the label style for horizontal zones. Click to select an option.
Vertical Zones	Sets the number and style for the vertical zones. Number of Zones sets the number of vertical zones. Enter the number in the box. Zone Labeling sets the label style for vertical zones. Click to select an option.

Autodesk Inventor Fundamentals

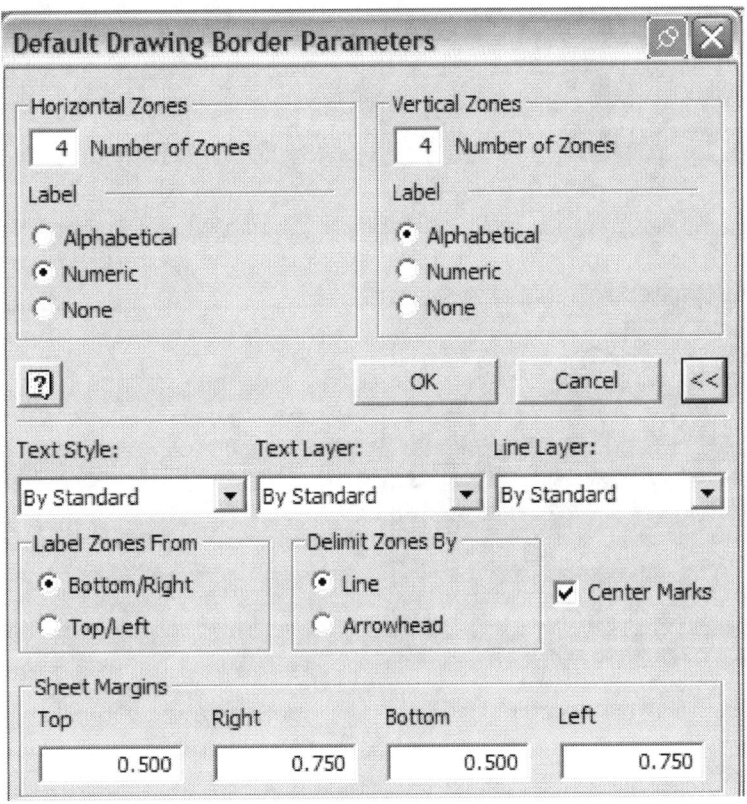

Selecting the More button ![>>] expands the dialog box so that you can set the text size and style, line width, zone properties, and page margins. Click to expand or collapse.

Text Style	Sets the typeface for the zone labels. Click the arrow and select the font from the list.
Text Layer	Sets the layer used by text objects. This is helpful if you import your drawings into AutoCAD.
Line Layer	Sets the layer used by the border lines.
Center Marks	Specifies whether to a incorporate center marks into the border. Select the check box to add center marks; clear the check box to omit them.
Delimit Zones By	Specifies the mark used to show zone boundaries. Click to select an option. Line sets lines to indicate the boundaries. Arrowhead sets arrows to indicate the boundaries.
Label Zones From	Sets the starting point for zone labels. Click to select an option. Bottom/Right starts the labeling at the bottom, right corner of the page. Horizontal labels proceed from right to left and vertical labels proceed from bottom to top. Top/Left starts the labeling at the top, left corner of the page. Horizontal labels proceed from left to right and vertical labels proceed from top to bottom.
Sheet Margins	Sets the space between the edge of the page and the border line on each side of the sheet. Enter the size for each margin.

Drawing Management

Define New Border

 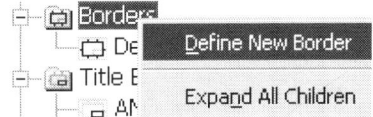

You can also create your own border formats.

Go to **Format→Define New Border**.
In the Browser, right click on the Borders folder and select 'Define New Border'.

Drawing Management Toolbar

Button	Tool	Function	Special Instructions
	Base View	Creates a view of a 3D model	The user must select the 3D model file to be used
	Projected Views	Creates an orthographic view	Requires a base view to have been defined
	Auxiliary View	Creates an auxiliary view	Select an edge to project a view
	Section Views	Creates a section view	User must define a section line before the view can be created
	Detail View	Adds a detail view	
	Break-out View	Creates a break-out view	User must create a sketch to indicate the exposed area
	Overlay View	Creates a view of a positional representation	The assembly file must have the positional representations already defined.
	New Sheet	Adds a new layout sheet	
	Draft View	Adds a sketch overlay to a drawing	Used to mark up or redline a drawing

Review Questions

1. Deleting a base view automatically deletes all dependent views.

    A. True
    B. False

2. A single model dimension cannot be used in multiple views on the same sheet.

    A. True
    B. False

3. Sketch overlays are used to clip or edit drawing views.

    A. True
    B. False

4. Drawing views cannot be copied from one sheet to another.

    A. True
    B. False

5. You can only create a view using the default orientations, i.e. Front, Top, Right.

    A. True
    B. False

6. 'Show Contents' is used to:

    A. List the views in a drawing
    B. List the features in the part
    C. List the format of a title block
    D. List the format of a sheet

7. The 'Fill Sketch Region' tool is used to:

    A. Add color to a profile in a title block
    B. Add color to a sketch overlay
    C. Add color to a view
    D. Add hatching to a section view

8. If you change the scale of the base view from 1:2 to 1:1, the scale of the isometric view:

    A. will change to 1:1
    B. will change to 3:4
    C. will remain 1:2
    D. will change to 4:3

9. The Section View tool creates all the section view types listed EXCEPT:

    A. Full
    B. Half
    C. Offset
    D. Revolved

ANSWERS: 1) B; 2) A; 3) B; 4) B; 5) B; 6) B; 7) A; 8) C; 9) D

Lesson 13
Drawing Annotation Toolbar

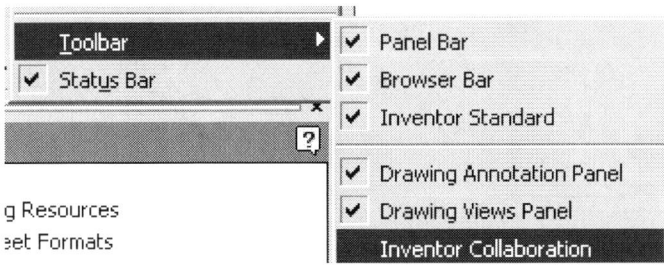

The Drawing Annotation toolbar can be accessed through the **View→Toolbar** menu.

Right clicking on the Standard toolbar also allows you to enable available toolbars.

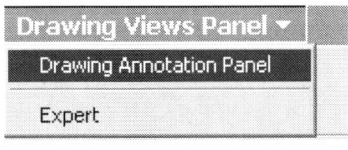

Left clicking on the arrow at the top of the Panel Bar can also activate the Drawing Annotation toolbar. Using this method will retire the Drawing Management Panel bar.

Autodesk Inventor Fundamentals

The Drawing Annotation toolbar contains tools that add annotations to a drawing. It is divided into four sections: Dimension, Symbol, Text, and Parts List tools.

- Dimension tools add general drawing dimensions and ordinate dimensions to a selected element.
- The center mark tools add center marks, center lines, center line bisectors, and centered patterns to a drawing.
- The symbols tools add a variety of GDT and other symbols to a drawing.
- The text tools add notes to a drawing.
- Parts list and balloon tools add those annotations to a drawing.
- The Retrieve Dimensions tool allows you to get model dimensions for a selected view.

General Dimension

Use the General Dimension button on the Drawing Annotation toolbar to add drawing dimensions to a view. Drawing dimensions do not affect the size of the part, but they provide documentation in the drawing.

As you develop a drawing, you can use two types of dimensions to document your model. Model dimensions are the dimensions that control the feature size in the part. They were applied during the sketching or creation of the feature. In a drawing view, you can display model dimensions that are planar to the view.

Drawing dimensions are dimensions that you add to further document the model. Drawing dimensions do not change or control features or part size. You can add drawing dimensions as annotations to drawing views or geometry in drawing sketches.

If you change the size of a part that has multiple occurrences in an assembly or is used in multiple assemblies, all occurrences of the part change.

Dimension styles are controlled in the Style Library. No two styles of the same style can have the same name in a drawing file. There can only be one dimension style named "Default (ANSI)".

When Inventor compares two styles of the same name, it checks all the properties to see if they are equal.

Dimension styles have three subtypes:

> Primary text style: used to format general dimensions
> Tolerance text style: used to format tolerance units
> Leader Style: used on hole notes and leaders

Drawing Annotation

Exercise 13-1
Using the Style Manager Library

File: None
Estimated Time: 60 minutes

1. Close Inventor.

2.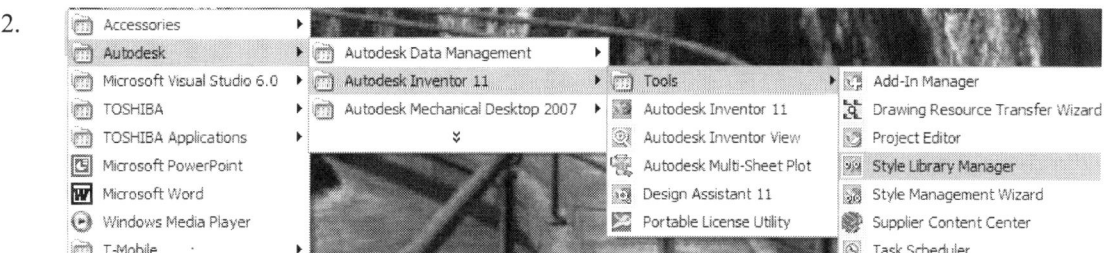

   Go to **Programs→Autodesk→Autodesk Inventor 11→Tools→Style Library Manager**.

3. Select **Create New Style Library** (located in the upper right corner of the dialog).

4.    Select **Create Empty Style Library** from the drop-down list.

5. 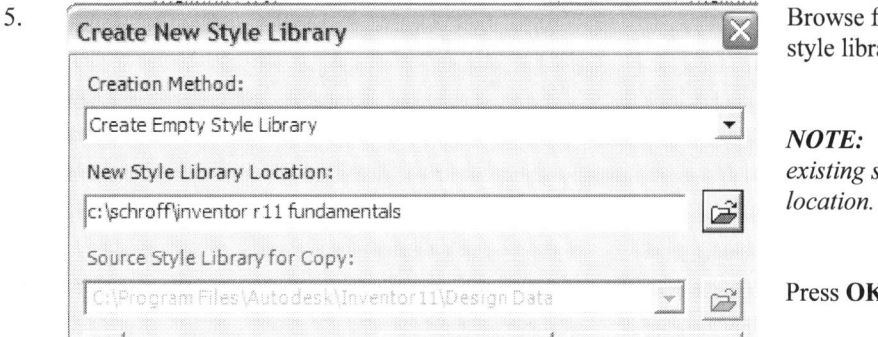   Browse for a folder to save the style library file.

   NOTE: You cannot have an existing style library in this location.

   Press **OK**.

6. Press **Exit** to close the Style Library Manager.

7.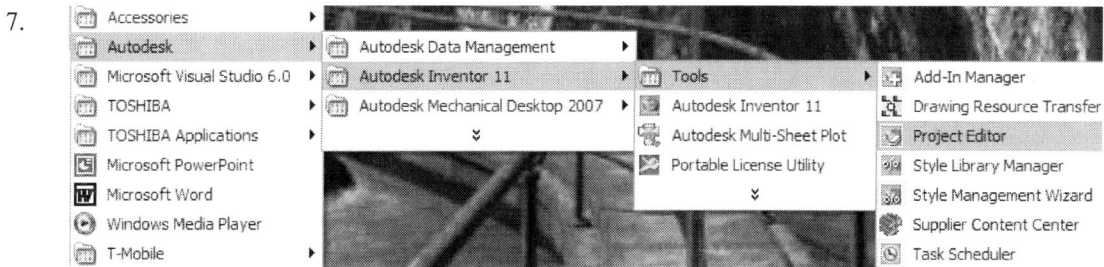

   Go to **Programs→Autodesk→Autodesk Inventor 11→Tools→Project Editor**.

13-3

8. 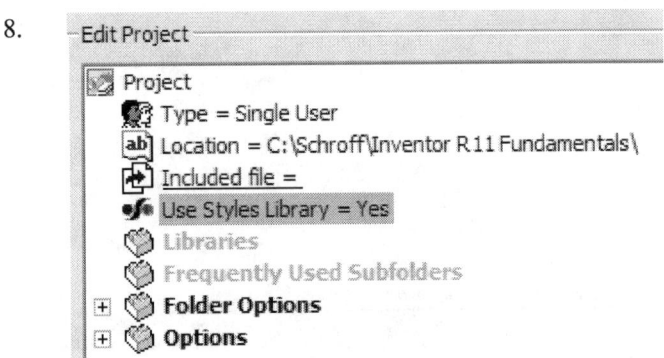 Set **Use Styles Library** to **Yes**.
   This allows you to modify the styles library.

TIP: When you set a Style Library to Yes, you have Read/Write functionality. When you set a Style Library to Read Only, styles are available to other users who are working on the project, but they can't modify the style. If you set a Style Library to No, then the document will not be accessing the style library; instead you will set up styles inside each document.

9. Set the folder options for styles to the path where you saved the new style.

10. Save the changes to the project and close the dialog.

11. Launch Inventor.

12. Close the Project Editor.

13. Close the dialog.

14. 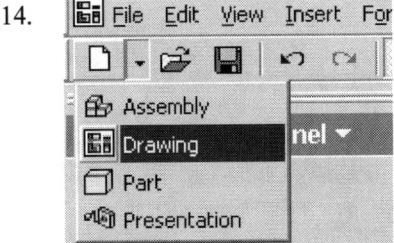 Create an empty idw file by holding down Ctl+Shift and selecting the New drawing icon under File.

    A drawing sheet will open with no border or titleblock.

Drawing Annotation

15. Go to **Tools→Application Options**.

16. Verify that the Design Data (and Default Style Library is set to the Styles path where you saved your styles.

    If it is not, reset it to the correct folder.

TIP: To ensure that you have no problems exporting, importing, copying, and deleting styles make sure that your project has the Style Library pointed to the path where you store your styles. The default path is *\Program Files\Autodesk\Inventor 10\Design Data*.

17. Press **Apply**.

18. Press **OK**.

    Close Inventor without saving the file and re-start it.

19. Create an empty idw file by holding down Ctl+Shift and selecting the New drawing icon under File.

    A drawing sheet will open with no border or titleblock.

20. Go to **Format→Styles Editor**.

13-5

Autodesk Inventor Fundamentals

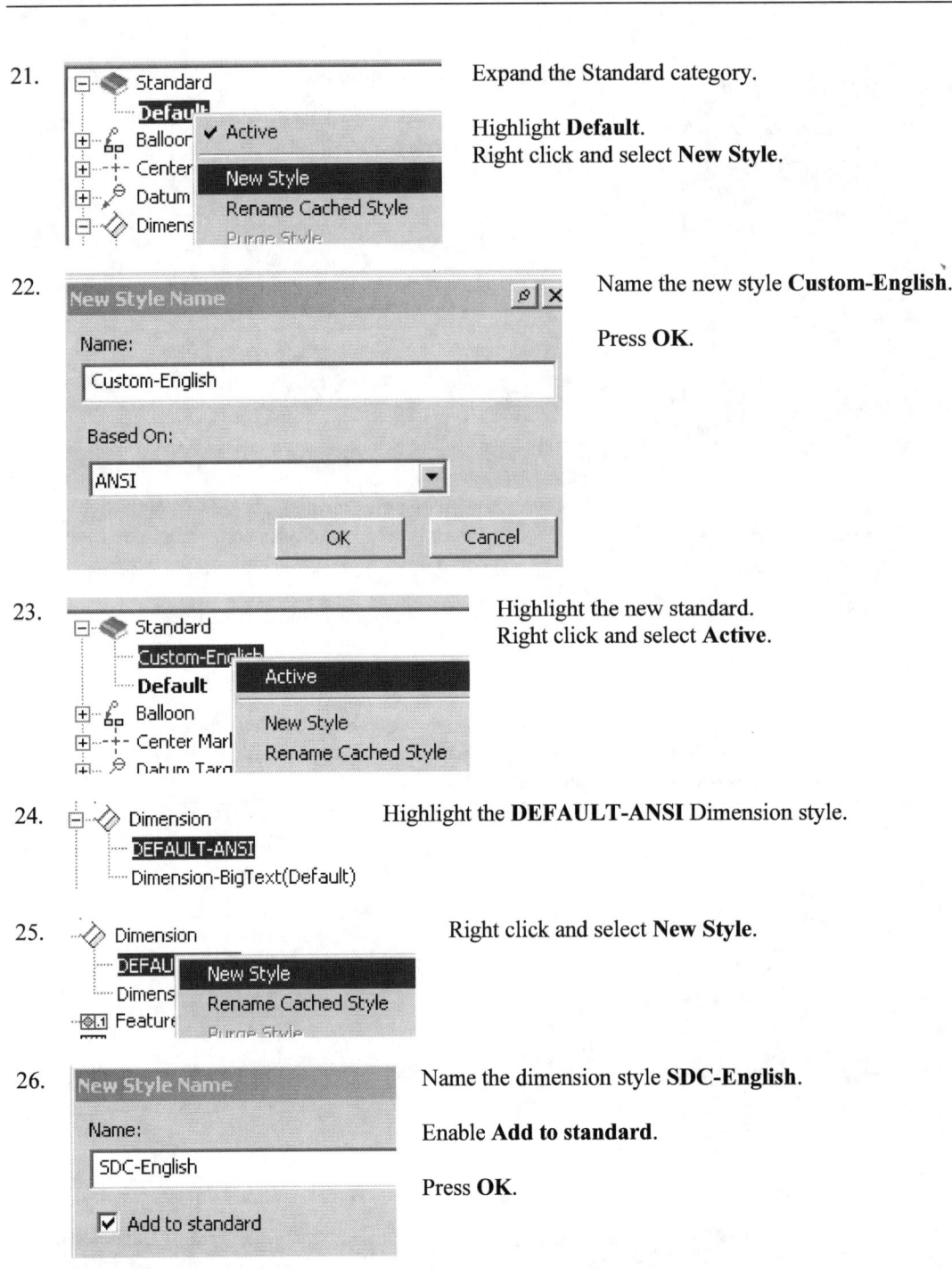

21. Expand the Standard category.

    Highlight **Default**.
    Right click and select **New Style**.

22. Name the new style **Custom-English**.

    Press **OK**.

23. Highlight the new standard.
    Right click and select **Active**.

24. Highlight the **DEFAULT-ANSI** Dimension style.

25. Right click and select **New Style**.

26. Name the dimension style **SDC-English**.

    Enable **Add to standard**.

    Press **OK**.

27. Set the precision for linear dimensions to three places.

Drawing Annotation

28. 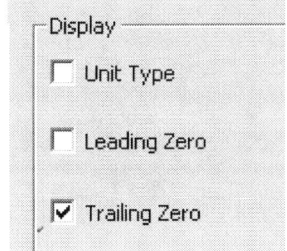 Disable the **Leading Zero** for linear dimensions to conform with ANSI standards.

29. Select the **Display** tab.

 TIP: The Thread.xls and Clearance.xls files should be copied into the Styles folder that is assigned to the project or you will not be able to use data from those files.

30.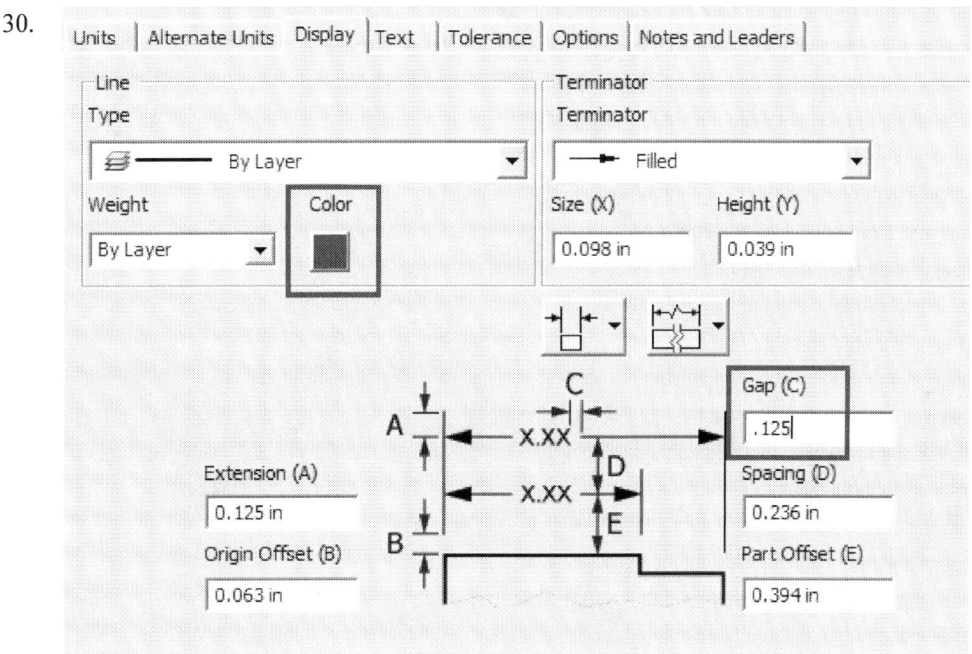

    Set the **Color** to **Red**.
    Set the **Gap** to **.125**.

    Select the **Text** tab.

31. 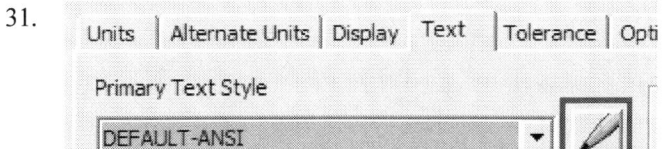 Select the **Edit button** next to **Primary Text Style**.

13-7

Autodesk Inventor Fundamentals

32. If you are prompted to save the style, press **Yes**.

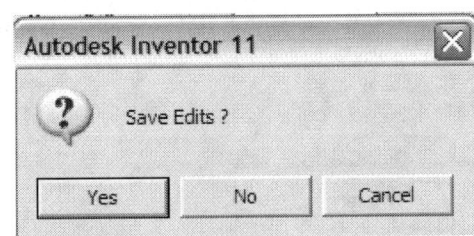

33.  Set the **Color** to **Red**.

34. 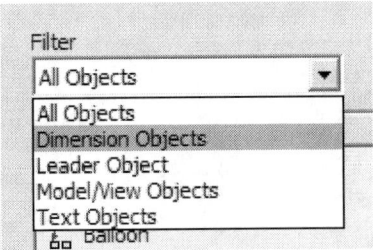 Press **Save**.

35. Locate the **Object Defaults** category. Highlight **Object Defaults (Default)**.

Select **Dimension Objects** under the **Filter** drop-down.

36. 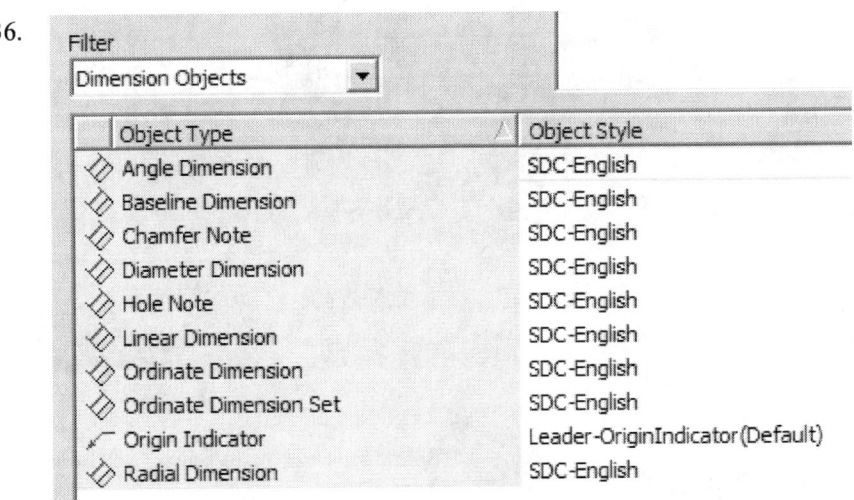 Select **SDC-English** as the default dimension style using the drop-down for all the dimension objects.

37. Press **Save**.

13-8

Drawing Annotation

38. Highlight **Custom-English** under the Standard category.

39. Select **All Styles** in the drop-down on the far right of the dialog box.

40. Select the **Available Styles** tab.

41. Highlight **Dimension** under the **Style Type**.

   All the dimension styles are listed and enabled.

42. Highlight the Default style under the Standard category.

43. Highlight **Dimension** under the **Style Type**.

   All the dimension styles are listed, but they are not all enabled.

44. Set the filter to only show the **Active Standard**.

45. Only the **Custom-English** style is now listed under the Standard category.

46. Select the **Available Styles** tab in the main dialog.
   Highlight **Dimension**.
   Disable all the dimension styles *except* for **SDC-English**.

13-9

Autodesk Inventor Fundamentals

47. 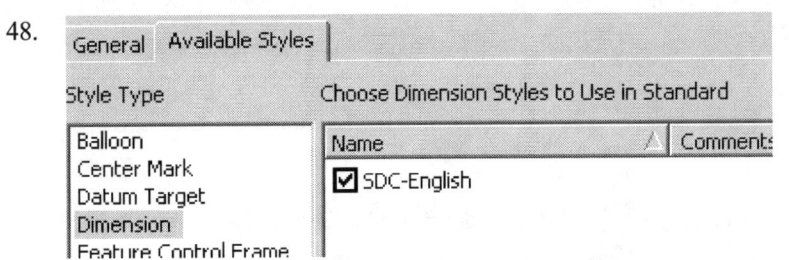 Select **All Styles** from the drop-down list.
This will refresh the list.
If prompted to save, press **Yes**.
Select **Active Standard** from the drop-down list.

48. 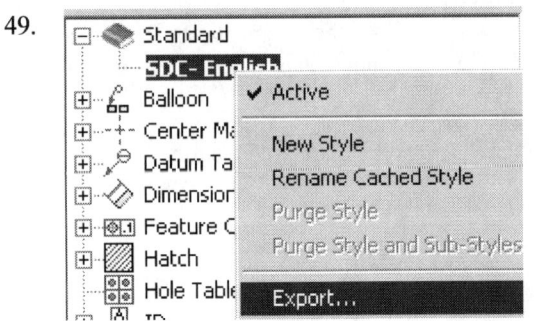 Select the **Available Styles** tab.

Highlight **Dimension**.

Only the dimension style you created is now listed.

49. 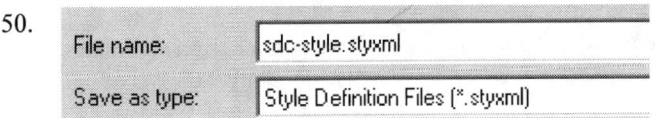 Highlight **SDC-English** under Standard.

Right click and select **Export**.

50. 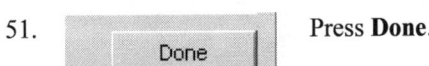 Browse to the folder where you are storing your lesson files.
Name your file *sdc-style.styxml*.
Press **Save**.

51. 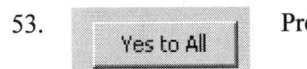 Press **Done**.

52. Go to **Format→Save Styles to Style Library**.

53. Press **Yes to All**.

54. Press **OK**.

13-10

55. ![Overwrite Style Library Information dialog] Press **Yes**.

56. Close the file without saving.

TIP: The reason it requires a change to the project before you can save styles is so that a CAD Manager in a department can control who has control over company styles and standards.

Autodesk Inventor Fundamentals

Exercise 13-2
Applying a General Dimension

File: Ex12-2.idw
Estimated Time: 20 minutes

You can add a General Dimension to a drawing view using the following methods:

From the Drawing Annotations Panel	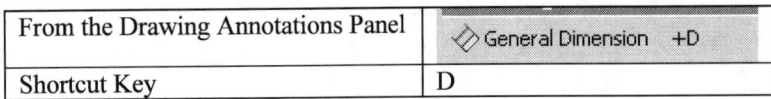
Shortcut Key	D

1. Locate *ex12-2.idw* and press **Open**.

2. Go to **Format→Styles Editor**.

3. Select the **Import** button.

4. Locate the *sdc-style.styxml* file saved in the previous exercise.
   Press **Open**.

5. Set the **Custom-English** standard as the active style.

   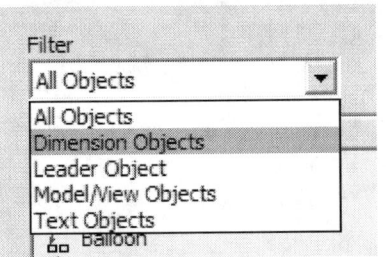

   Set the pane to filter for **Dimension Objects**.

6. Locate the **Object Defaults** category.
   Highlight **Object Defaults (Default)**.

13-12

Drawing Annotation

7. Note that **SDC-English** is already set as the default dimension style.

Object Type	Object Style
Angle Dimension	SDC-English
Baseline Dimension	SDC-English
Chamfer Note	SDC-English
Diameter Dimension	SDC-English
Hole Note	SDC-English
Linear Dimension	SDC-English
Ordinate Dimension	SDC-English
Ordinate Dimension Set	SDC-English
Origin Indicator	Leader-OriginIndicator(Default)
Radial Dimension	SDC-English

Press **Done**.

8. Left click the arrow on the **Drawing Views Panel**. Activate the **Drawing Annotation Panel**.

9. Use **Zoom Window** to zoom into the top view.

10. Select the **General Dimension** tool.

11. Select **SDC-English** from the style drop-down to use that style to apply dimensions.

13-13

12.

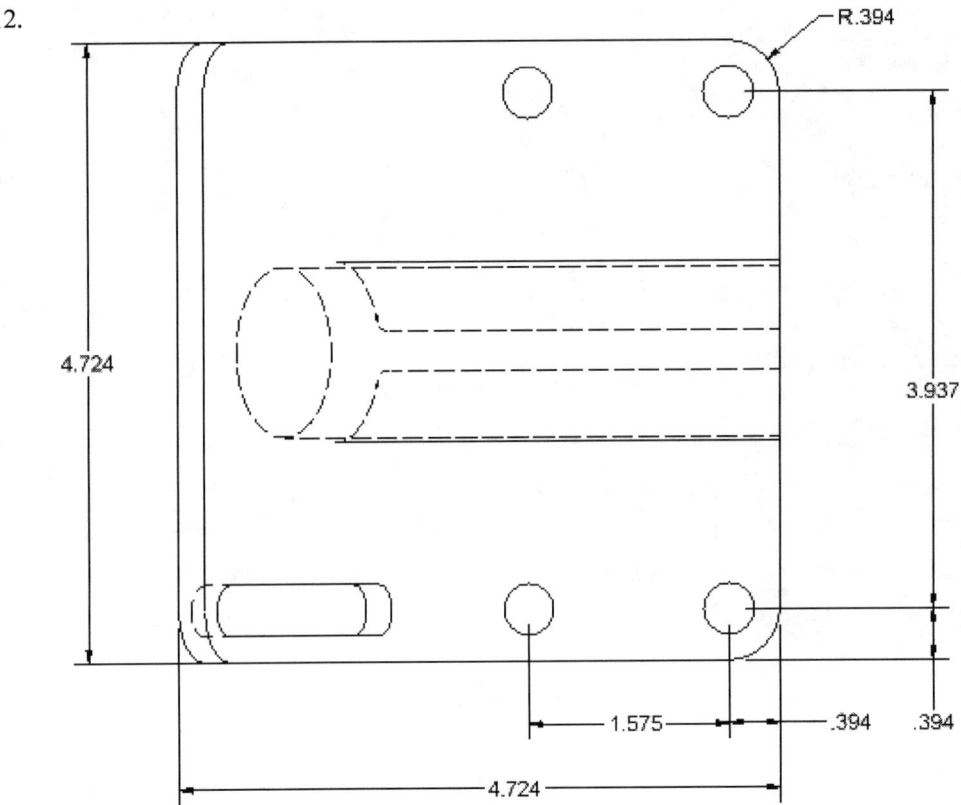

Using the **General Dimension** tool, apply the dimensions shown. The dimensions should appear with the style changes you applied.

13. Save the file as *ex13-2.idw*.

Drawing Annotation

Baseline Dimension

Exercise 13-3
Applying a Baseline Dimension

File: Ex13-2.idw
Estimated Time: 15 minutes

1. Locate *ex13-2.idw* and press **Open**.

2. Zoom into the right side view.

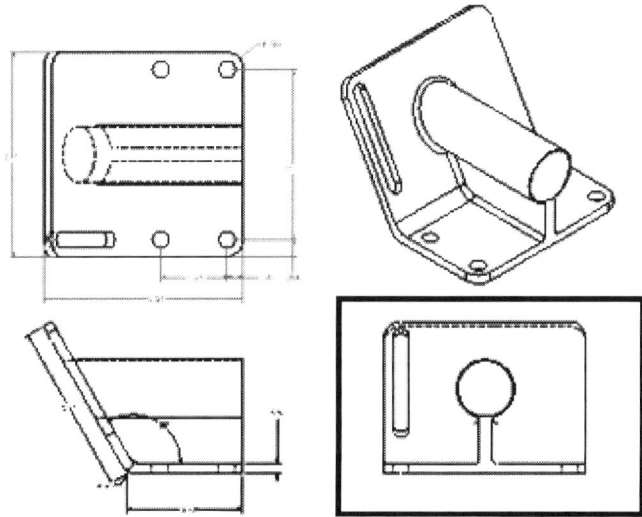

3.    Select the **Baseline Dimension Set** tool.

   Note that the active standard is **SDC-English**.

TIP: You can also type 'A' as a shortcut to activate the Baseline Dimension Set command.

13-15

4. 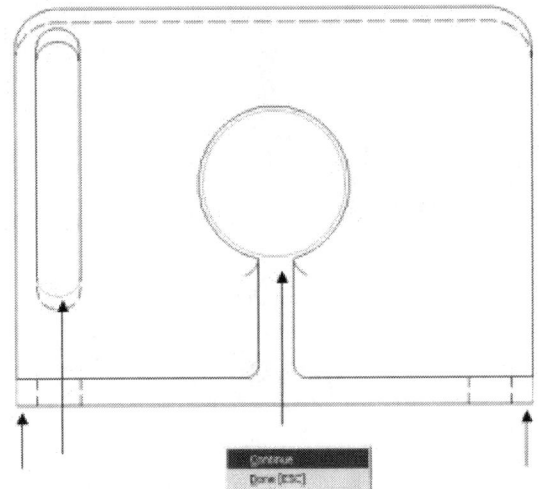 You'll be prompted to select an edge to be included in the baseline dimension.

Select the edges and arcs indicated by the arrows.

Right click and select **Continue**.

5.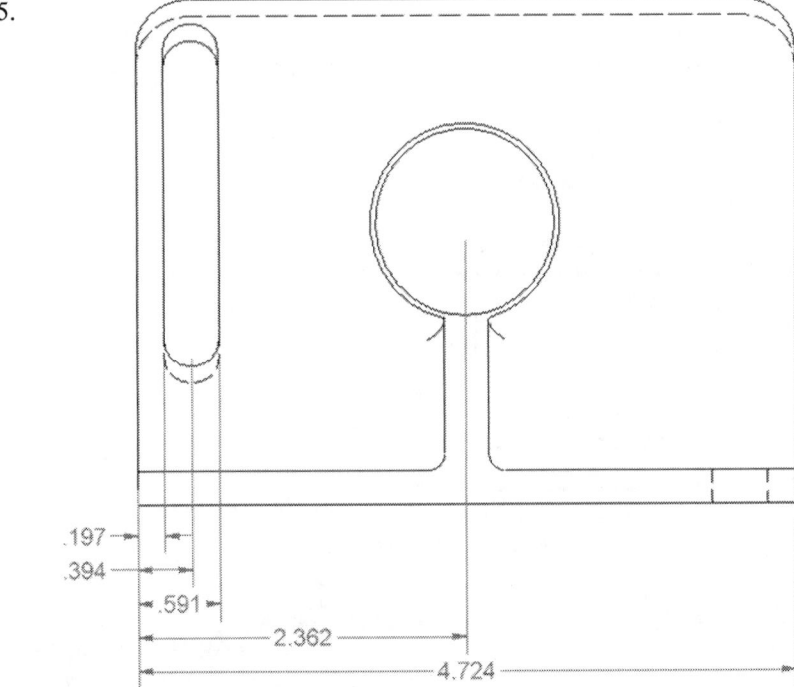

Pick a point below the view.
This will place the baseline dimensions.

6. 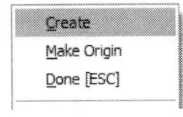 Right click and select **Create**.

Drawing Annotation

7. 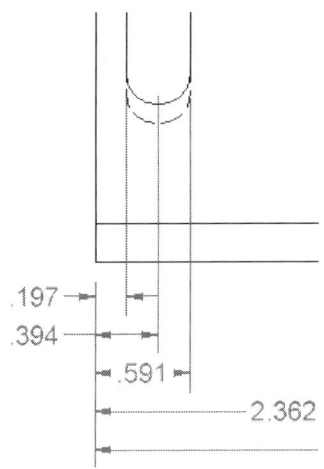 Normally when dimensioning a slot, we would only have the dimension to the center of the slot. This means that we should delete the .591 dimension and the .197 dimension.

8. Select the .591 dimension and right click.

9. 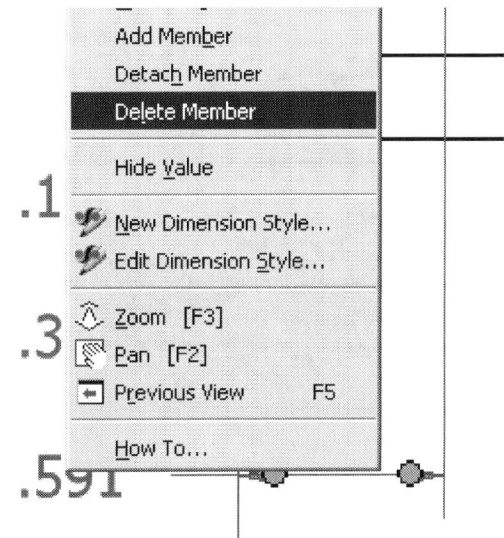 If you select '**Delete**' you will delete the entire set of dimensions.

Instead, select the '**Delete Member**' option.

10. 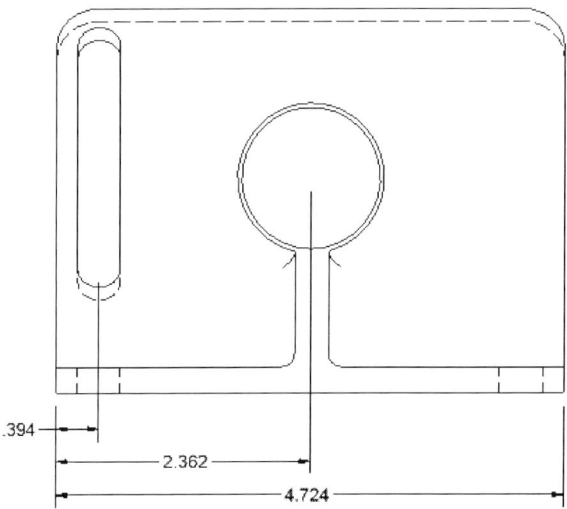 Repeat to delete the .197 dimension.

You can reposition the dimensions so they are spaced properly by simply picking to activate the grips and moving them into the correct location.

11. Save as *ex13-3.idw*.

Ordinate Dimension Set

Use the Ordinate Dimension Set button on the Drawing Annotation toolbar to add ordinate dimensions. This tool links all the dimensions placed as a group. The first point selected is automatically assumed to be the origin. You must place all the dimensions along one axis, right click and select 'Create'. Then repeat to place dimensions along the other axis.

The dimensions automatically align and adjust to avoid interference as you place them.

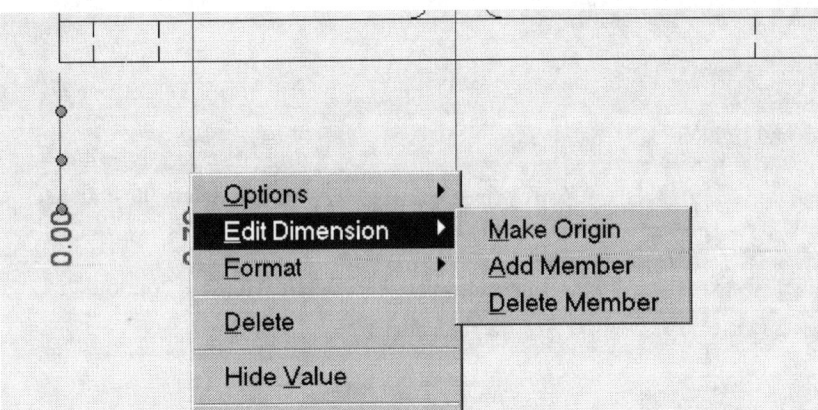

To modify ordinate dimensions, select the dimension set in the graphics window and right-click to display the menu. You have to option to reset the Origin, Add or Delete a dimension.

TIP:
The first point picked when placing an ordinate dimension is automatically assigned the 0 value. Place all the dimensions desired for that axis (whether horizontal or vertical), then right click to select Create. Then place the dimensions for the other axis.

Pressing **O** on the keyboard will also initiate the Ordinate dimension command.

Ordinate Dimension

The Ordinate Dimension Set tool places multiple ordinate dimensions along a single axis. The Ordinate Dimension tool allows the user to place one or more dimensions on one or more axis.

The other difference between the tools is that ordinate dimensions placed with the Ordinate Dimension Set are considered a group so if you delete one dimension in the set, you delete all the dimensions along the same axis in that set. Dimensions placed with the Ordinate Dimension tool are considered independent of all the other ordinate dimensions along the same axis.

The Ordinate Dimension tool also requires that the user identify the origin of the part before any ordinate dimensions are placed.

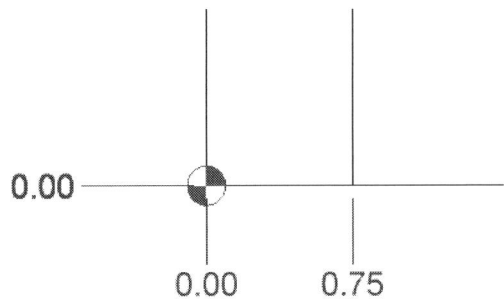

The tool places a marker at the origin point.

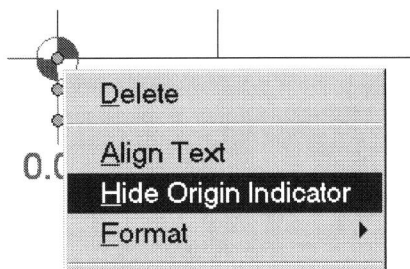

To hide the marker, select it, right click and select **Hide Origin Indicator**.

 TIP: Hole notes can be added only to hole features created using the Hole feature tool in parts.

Exercise 13-4
Applying Ordinate Dimensions

File: Ex13-2.idw
Estimated Time: 15 minutes

1. Open the *ex13-2.idw* file.

2. Zoom into the right side view.

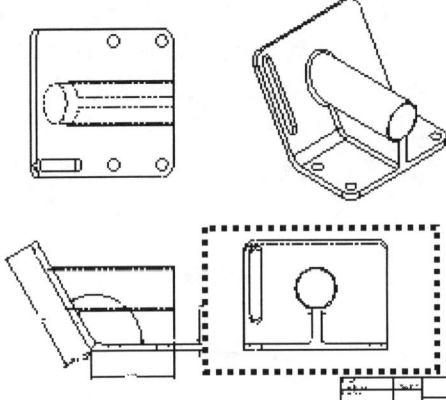

3. 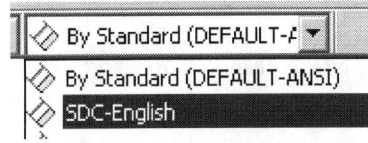 Select the **Ordinate Dimension Set** tool.

4. Select **SDC-English** from the style drop-down list.

5.
   Pick the lower left corner of the bracket.
   Pick to select the center of the arc.
   Pick to select the center of the cylinder.
   Pick the far right corner of the bracket.
   Right click and select **Make Origin**.
   Select the lower left corner of the bracket.
   Right click and select **Create**.

   Your ordinate dimensions are placed.

Drawing Annotation

When you place ordinate dimensions using the ordinate dimension set command, the dimensions are linked as a group. To add or delete members of the set, you need to right click and use the right click shortcut menu.

TIP: Once you select the first point to place the first ordinate dimension, all ordinate dimensions placed following will be placed automatically based on the selection.

6.  Select the **Ordinate Dimension** tool.

7. Select **SDC-English** from the style drop-down list.

8. You'll be prompted to select a view to dimension. Select the right side view.

9.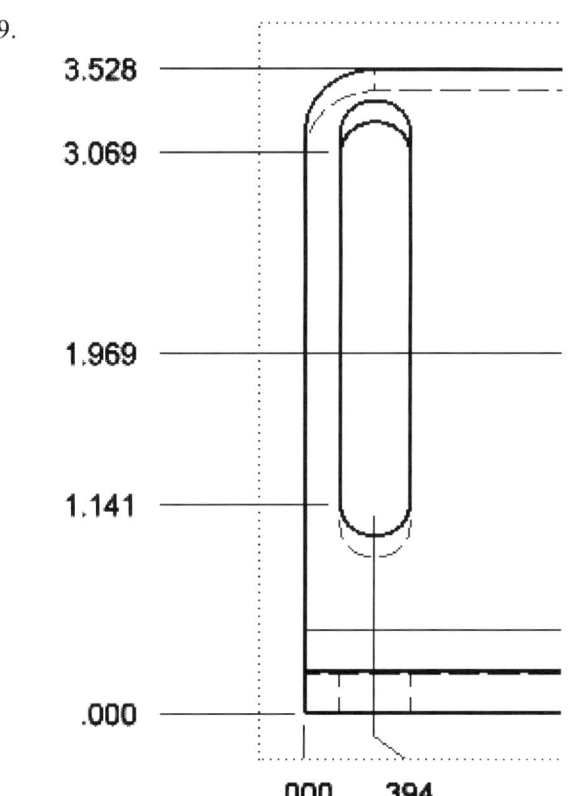

    Select the left lower corner of the bracket.

    You are still in Ordinate Dimension mode.

    NOTE: Verify that you are in the correct dimension style.

    If you look in the panel, the Ordinate Dimension button is still depressed.

    Select the center of the lower arc to start the ordinate dimension.

    Select a point to the left of the view.

13-21

10.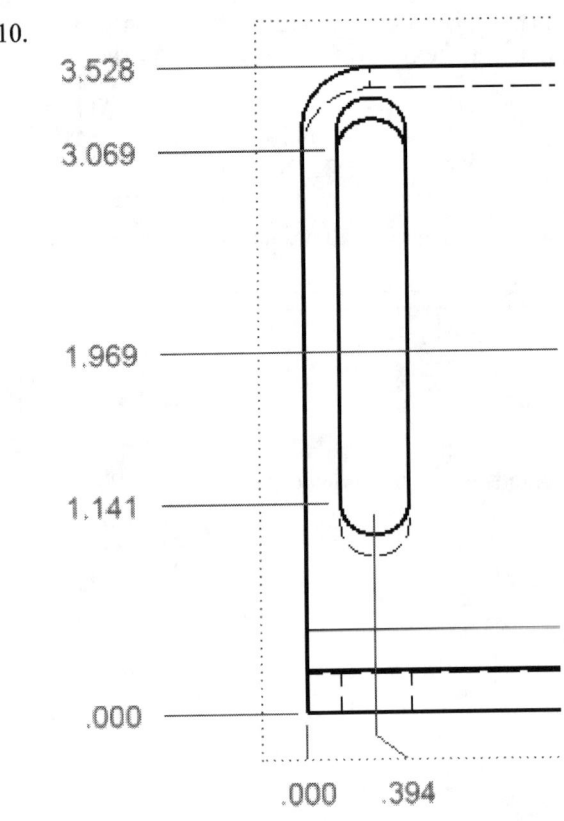

Select the top arc to start a second ordinate dimension.

Select a point to the left of the view.

Select the top of the bracket.

Select a point to the left of the view.

Right click and select **Make Origin**.

Select the lower left corner of the bracket.

Right click and select **Create**.

11. Save as *ex13-4.idw*.

Drawing Annotation

Hole/Thread Notes

Use the Hole Notes button on the Drawing Annotation toolbar to add a hole or thread note with a leader line. The text, style, color, and line weight for the notes are determined by the active drafting standard.

> **TIP:** You can automatically calculate the number of holes in a view using the <QTY> parameter. Simply type <QTY> below the hole note.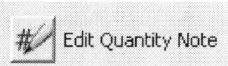

Exercise 13-5
Adding a Hole Note

File: Ex13-4.idw
Estimated Time: 15 minutes

1. Open the *ex13-4.idw* file.

2. Zoom into the top view.

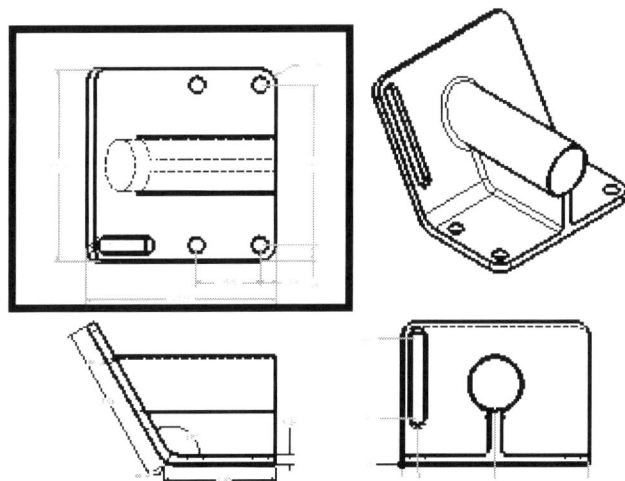

3. 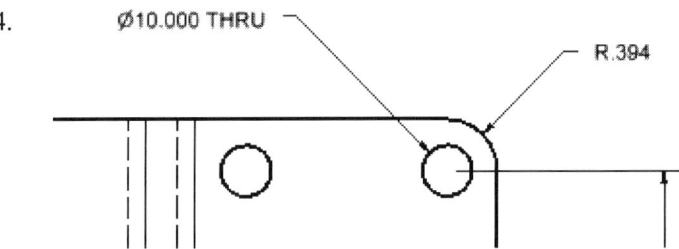 Select the **Hole/Thread Notes** tool.

4. Select the top right circle in the view.

   Place the dimension as shown.

Autodesk Inventor Fundamentals

5. 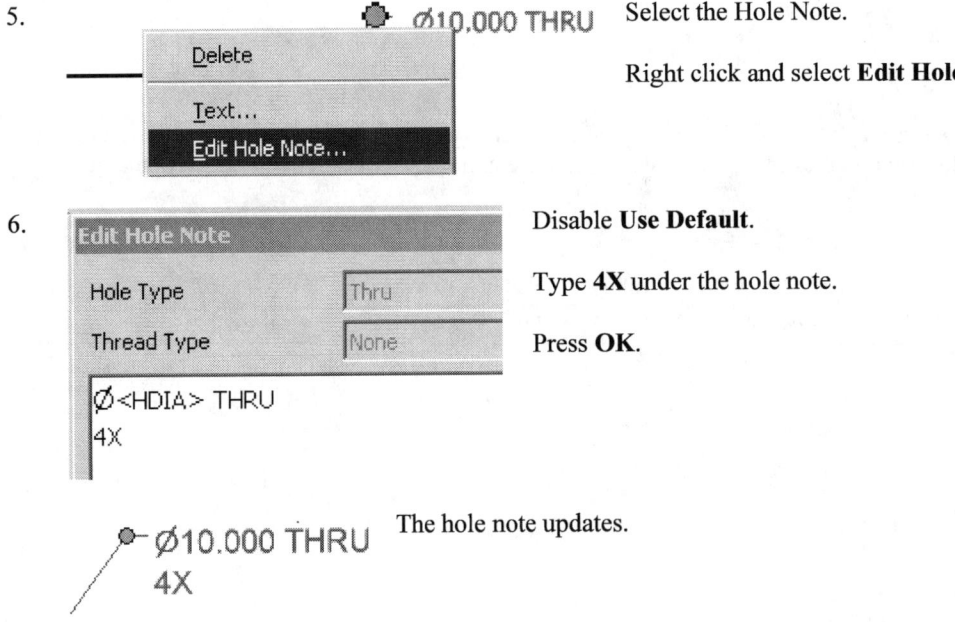 Select the Hole Note.

Right click and select **Edit Hole Note**.

6. Disable **Use Default**.

Type **4X** under the hole note.

Press **OK**.

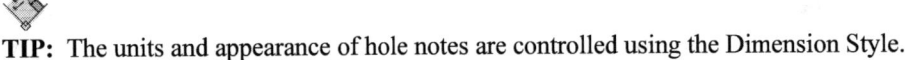 The hole note updates.

7. Save the file as *ex13-5.idw*.

TIP: The units and appearance of hole notes are controlled using the Dimension Style.

13-24

Drawing Annotation

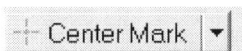

Center Mark

There are four types of center markings that you can add to a drawing view; center marks, center line bisectors, center lines, and centered patterns. To choose a center marking option, click the arrow next to the active center marking button, and select from the pop-up menu.

TIP: Pressing 'C' on the keyboard will also activate the Center Mark command.

This is a pull-down with other options available.

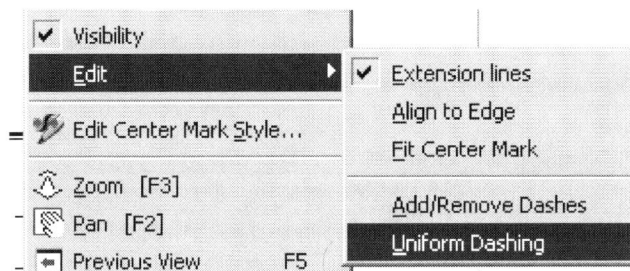

You can modify a center mark by selecting and then right clicking.

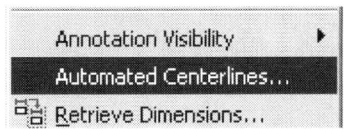

To automatically add centerlines to any arc or circle in a view, select the view, right click and select '**Automated Centerlines**'.

13-25

Autodesk Inventor Fundamentals

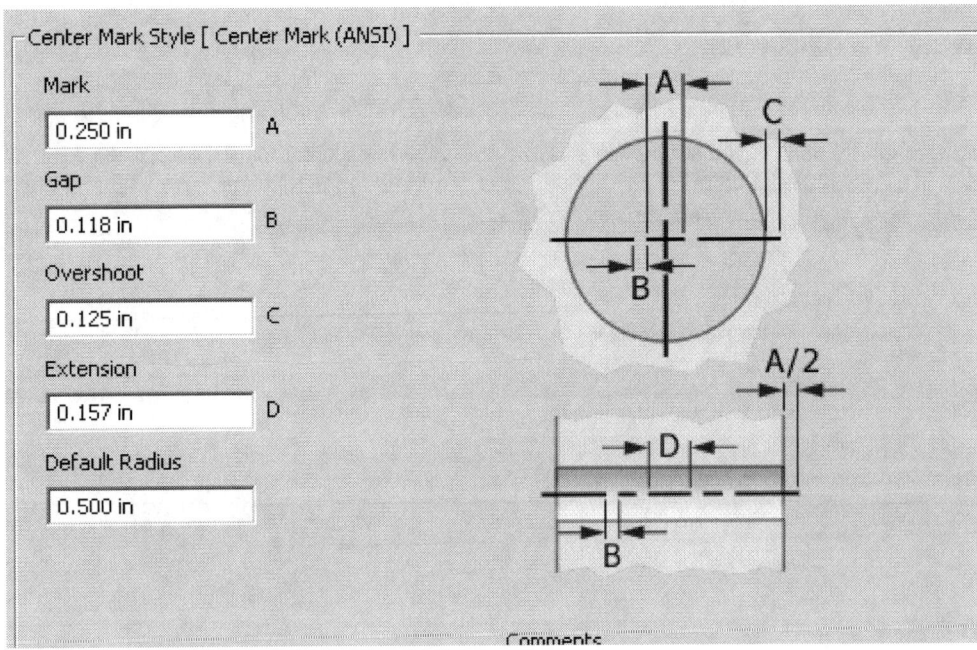

The Center Mark Standards are accessed in the Styles Format Dialog under the Center Mark category.

Mark	Sets the size of the center indicator mark.
Gap	Sets the gap distance between the center indicator and the extension line.
Overshoot	Sets the distance that center mark extension lines extend beyond the edges of the features that they define.
Extension	Sets the minimum length of center mark extension lines.
Default Radius	Sets the size of center marks for suppressed pattern features.

You can lengthen or shorten an extension line for any center line or center mark by dragging its endpoint. In addition, you can edit center marks in the several ways.

Exercise 13-6
Adding Center Marks

File: Ex13-5.idw
Estimated Time: 15 minutes

1. Open the *ex13-5.idw* file or continue working in the open drawing.

2. Select the top view.

   Right click and select **Automated Centerlines**.

3. A dialog appears.

   You can apply center marks to holes, fillets, cylinders, revolves, patterns, bends, punches or sketches. You can set the projection for a top or projected view.

   Under **Apply To**:
   enable **Holes**.

   Under **Projection**:
   enable **Top View**.

   Press **OK**.

13-27

4.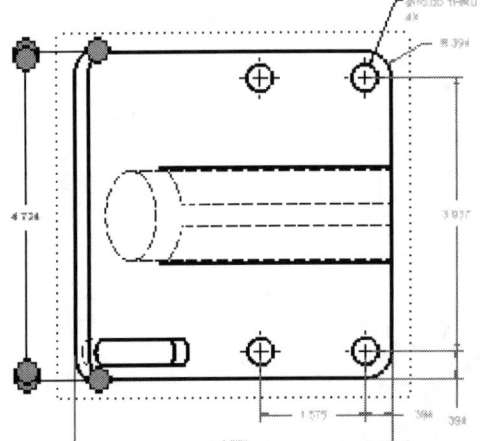

Center marks are added to the four holes in the view.

5.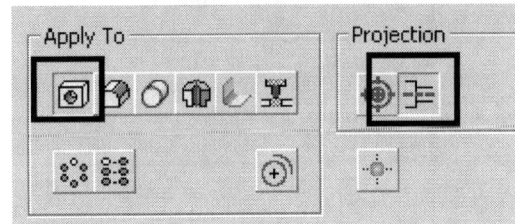

Select the Front view.

Right click and select **Automated Cenerlines**.

Enable **Holes**.
Enable **Projected**.
Press **OK**.

6.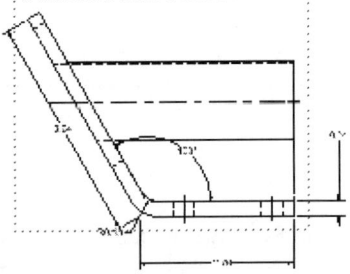

Center lines are added to the holes.

7. Save the file as *ex13-6.idw*.

Drawing Annotation

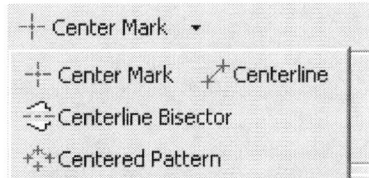

Center Line

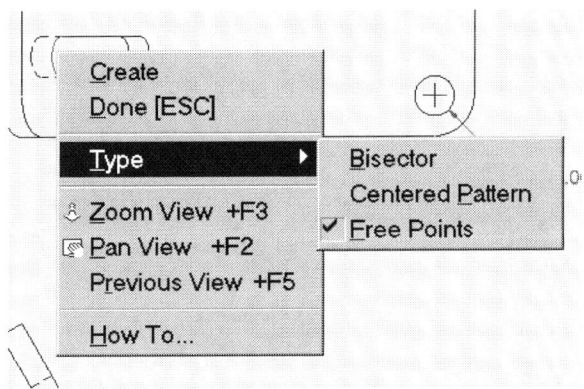

> **TIP:** If the features form a circular pattern, the center mark for the pattern is automatically placed when you have selected all of the members.

Exercise 13-7
Adding Center Lines

File: Ex13-6.idw
Estimated Time: 15 minutes

1. Open the *ex13-6.idw* file or continue working in the open drawing.

2. Zoom into the right side view.

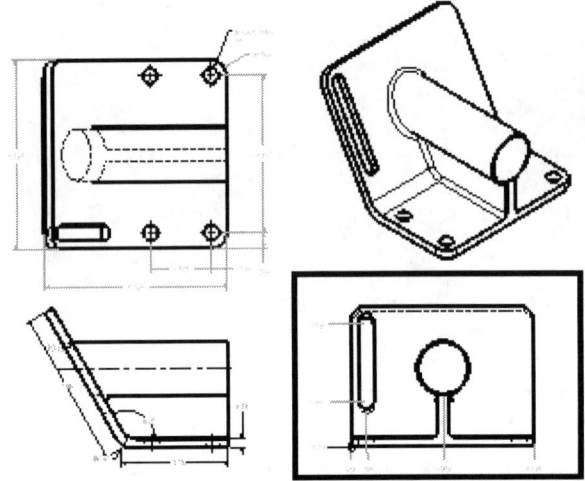

3. Select the **Centerline** tool.
   (This is part of the Center Mark flyout.)

4. Select the center point of the top arc.

   Drag the line down vertically and select the center point of the bottom arc.

5.  Right click and select **Create**.
   Right click and select **Done**.

13-30

Drawing Annotation

6. 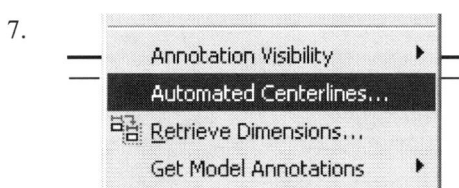   A center line appears in the slot.

7. Select the Right Side view.
   Right click and select **Automated Centerlines**.

8. 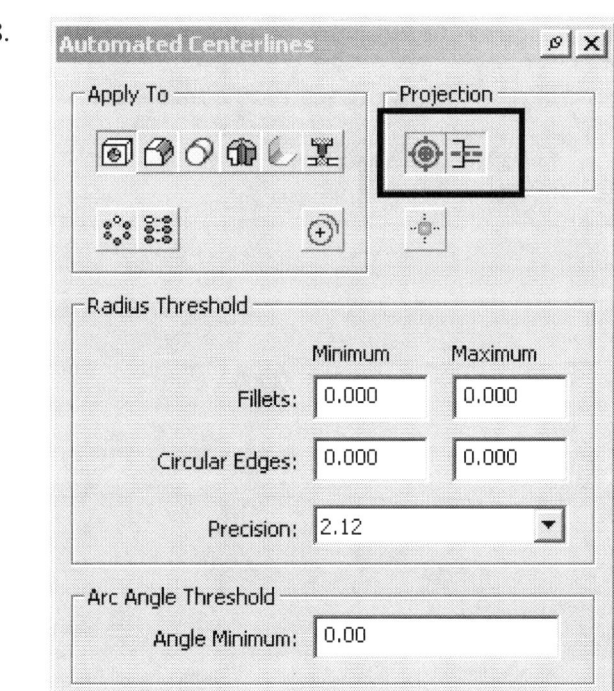   In this view, we want to place center marks at the cylinder and for the holes indicated by the arrows.

   Enable '**Hole**'.
   For Projection, enable both top and side projections.

   Press '**OK**'.

13-31

9. 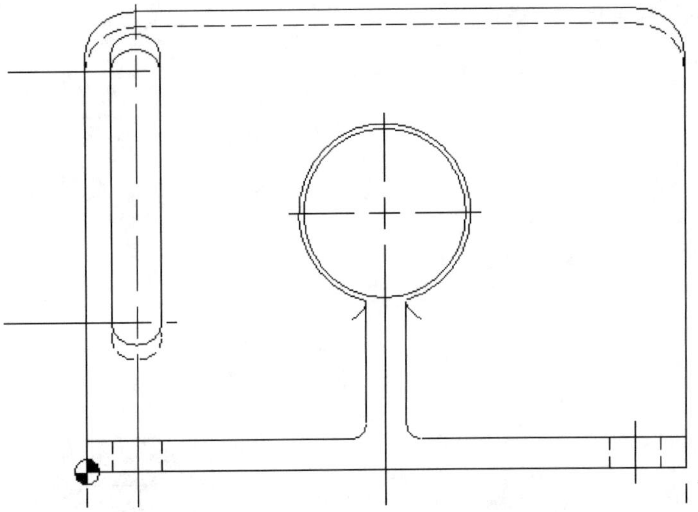 Center lines are added to the holes.

10. 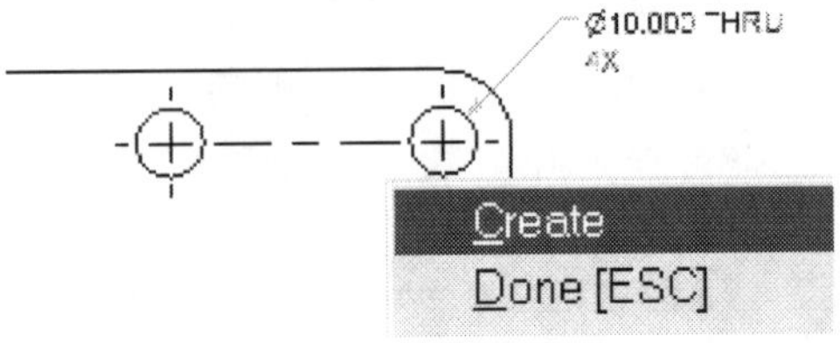 Zoom into the Top View.

   Place a centerline between the two top holes.

   Right click and select **Create**.

11. 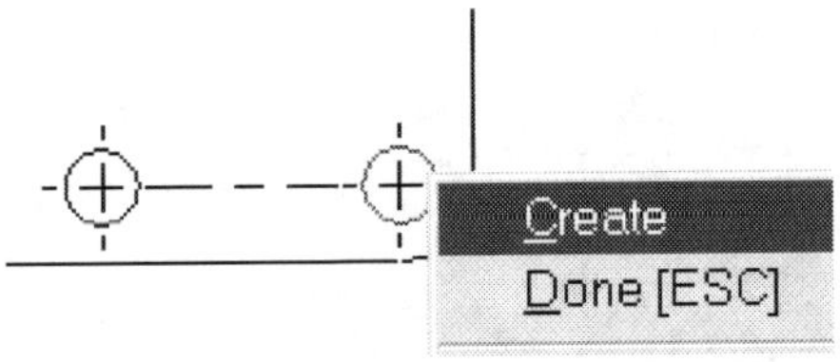 Place a centerline between the two bottom holes.

   Right click and select **Create**.
   Right click and select **Done**.

12. Save as *ex13-7.idw*.

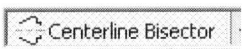

Center Line Bisector

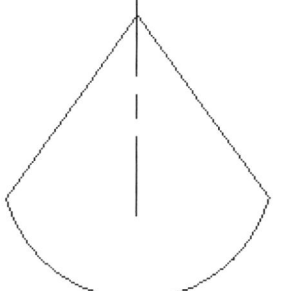
The Centerline Bisector tool is used to add a centerline to an angle.

To add the center line:
Select the view.
Select the Center Line Bisector tool.
Select the two lines/edges that form the angle you want to bisect.

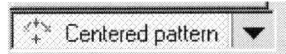

Centered Pattern

TIP: The drawing annotation tools can be selected before or after you select the view. You can select the tool and then the view or vice versa.

TIP: No centerline is added between the final feature in the pattern and the beginning feature. You can manually extend the line if needed.

Exercise 13-8
Adding a Centered Pattern

File: Views-6.ipt (This file is available in the Tutorial Files folder under Inventor or as a free download from the publisher's website.)
Estimated Time: 15 minutes

1. 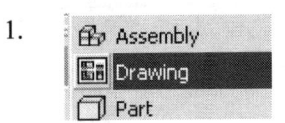 Start a New drawing.

2. Select the **Base View** tool.

3. Locate the *views-6.ipt* file.

   It is located under the Tutorial Files folder or can be downloaded from the publisher's website.

4. Select the **Top** view and place it in your sheet.

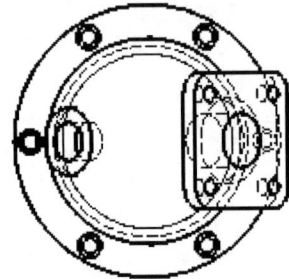

5. Switch to the **Drawing Annotation Panel**.

6. 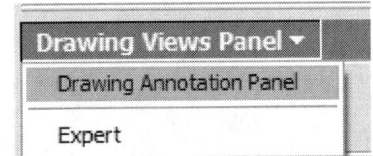 Select the **Centered Pattern** tool.

13-34

Drawing Annotation

7. 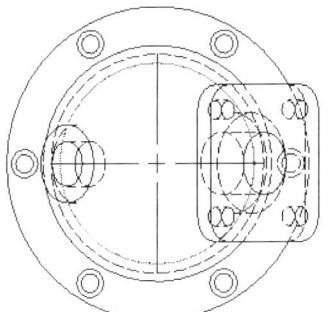 Select the center hole to indicate the center point for your centered pattern.

8. 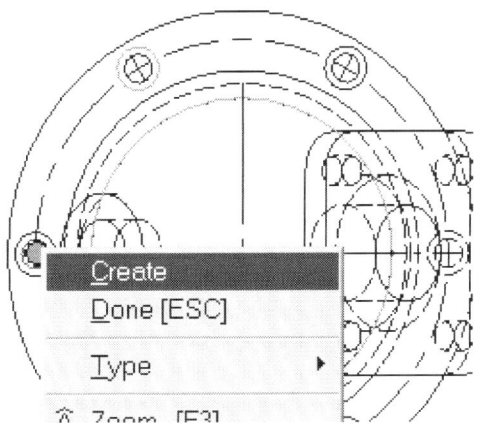 Inventor will add a center mark to the pattern center.

   Next select the center for each hole in the pattern.

   Move around the circle clockwise selecting each circle's center point in order.

   Be sure you select the circle you started with to close the pattern or you will have a gap.

   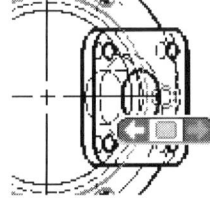 You can use the **Select Other** tool to tab through the circles until the correct circle is highlighted.

9.  Right click and select **Create**.
   Then right click and select **Done**. click once more on the dwg To save the ⊄

10. 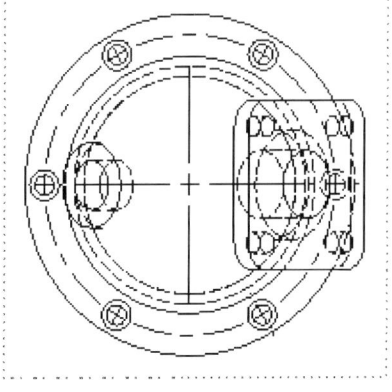 Save as *ex13-8.idw*.

13-35

Surface Texture Symbol (use the icon under (Dwg Annotation Panel)
RMB → continue

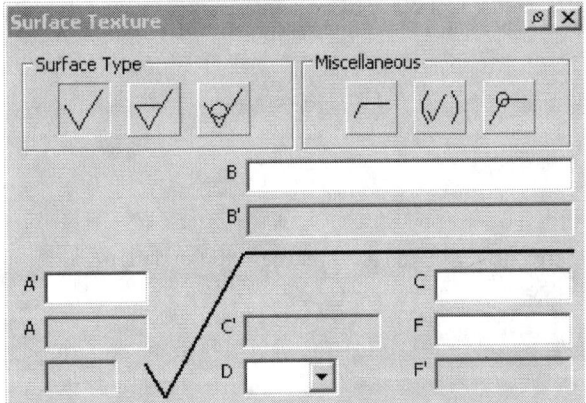

Specifies the content of a surface texture symbol. The options in the dialog box are determined by the active drafting standard.

The symbols can be created as stand-alone objects, be placed directly on the geometry in a drawing view, or reside on top of a leader line which points to the geometry. Surface Texture Symbols may also be used in notes and in the tolerance block.	
Surface Type	Click a button to select the desired surface type. Basic Material removal required Material removal prohibited
Miscellaneous	Specifies the general attributes of the symbol. Click the button to add or remove each attribute. Force tail adds a tail to the symbol. Majority indicates that this symbol specifies the standard surface characteristics for the drawing. All-around adds the all-around indicator to the symbol.
Surface characteristics	Defines the values for the surface characteristics. Enter the appropriate values in the boxes. **A** specifies the roughness value, roughness value Ra minimum, minimum roughness value, or grade number. **A'** specifies the roughness value, roughness value Ra maximum, maximum roughness value, or grade number. Available only when A has an entered value. **B** specifies the production method, treatment, or coating. If the active drafting standard is based on ANSI, this box can be used to enter a note callout. **B'** specifies an additional detail for the production method if the drafting standard is based on ISO or DIN. This option is available only when B has an entered value. **C**, for ANSI, specifies the roughness cutoff or sampling length for roughness average; for ISO or DIN, specifies the waviness height or sampling length; for JIS, specifies the cutoff value and evaluation length. **C'**, for ANSI, specifies the roughness cutoff or sampling length for additional roughness value; for ISO or DIN, specifies the sampling length for additional roughness value; for JIS, specifies the reference length and evaluation length. **D** specifies the direction of lay. Click the arrow and choose the symbol from the list. This option is not available when the removal prohibited option is selected. **E** is not available when the Machining Removal Prohibited is selected. Specifies the machining allowance. **F** specifies the roughness value other than Ra or the parameter value other than Ra. For ANSI, can also specify the waviness height. **F'** specifies the surface waviness for JIS. This option is not used for ANSI, ISO, or DIN standards.

Drawing Annotation

The user picks points to place the symbol. The first point selected indicates the location of the arrowhead. The user can then select any number of points to control the placement of the leader and symbol. When finished, right click and select 'Continue'. Selecting 'Done' is equivalent to canceling out the command. Selecting 'Back up' allows the user to back up one selected point and select a new point (similar to the Undo command in AutoCAD).

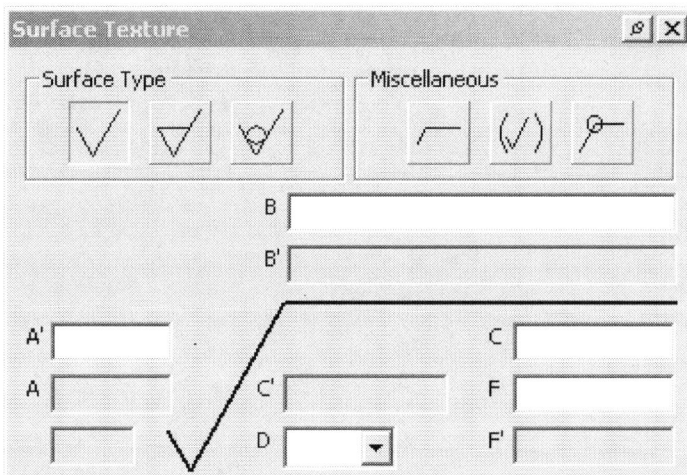

The Surface Texture Symbols are controlled in the Styles Editor dialog box under the Surface Texture category.

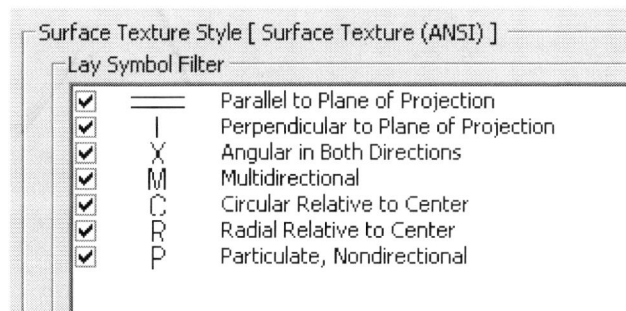

TIP: To place a symbol without a leader, double-click to set the symbol location and open the dialog box.

The options in the dialog box are determined by the active drafting standard.

If you place the first point on a highlighted edge or point, the leader line is attached to the geometry.

If you turn off a symbol in the Standards dialog that is already used in a drawing, the existing symbols will continue to display, but the symbol will not be available when new symbols are added.

13-37

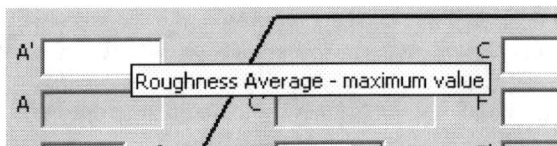

Moving your mouse over the dialog box will bring up a help tip to assist in filling out the dialog box.

To edit an existing symbol:

Select the placed symbol.
Right click and select **Edit**.

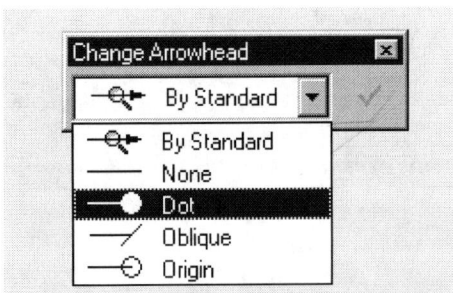

Selecting the Edit Arrowhead option will bring up a dialog where the user can select the current standard or a different arrowhead.

Selecting the 'Add Segment or Vertex option' activates the Grip mode and the user can use the mouse to stretch or modify the leader for the symbol.

Drawing Annotation

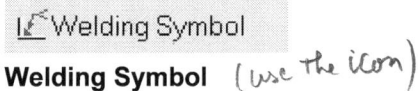
Welding Symbol (use the icon)

Specifies the content of a welding symbol. The dialog box opens when you place a welding symbol. The options available are determined by the active drafting standard.

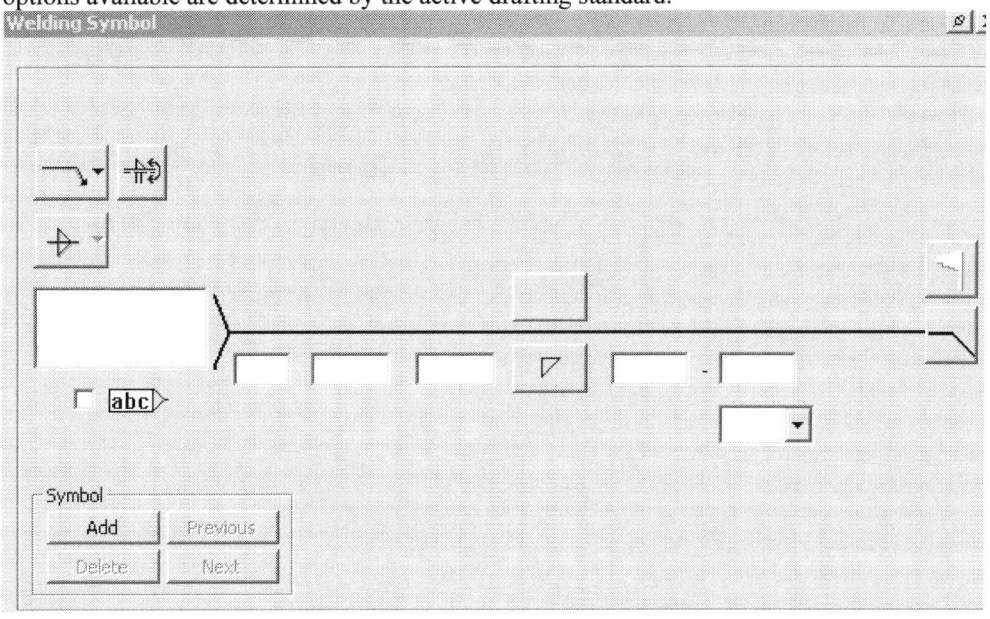

Orientation	Sets the orientation for the components of the welding symbol. Click the button to set the desired orientation for each component. ♦ Swap arrow side reverses the arrow side and other side for the selected reference line. ♦ Identification line Available for ISO and DIN only. Sets the location of the identification line to the arrow side or the other side for the selected reference line. ♦ Left/right orientation sets the location of the reference line relative to the leader line.
Stagger	Staggers the welding symbols for fillets. Available only if fillet welding symbols are set on both sides of the reference line.
Symbol	♦ Selects, adds, or deletes reference lines on the welding symbol. ♦ Add adds a reference line to the symbol. The selected reference line is moved toward the arrowhead and the new line is added further from the arrowhead than the selected reference line. You can set the orientation and other attributes for the new reference line. ♦ Delete removes the selected reference line from the symbol. ♦ Previous selects the next reference line away from the arrowhead. ♦ Next selects the next reference line toward the arrowhead.
Note	Adds text to the selected reference line. Enter the text in the box.
Arrow side	Opens the Arrow Side options so that you can set the symbol and values for the selected reference line.
Other side	Opens the Other Side options so that you can set the symbol and values for the selected reference line.
Flag	Specifies whether to add a flag indicating a field or site weld to the selected reference line. Click the button to turn the flag off or on.
All-around symbol	Specifies whether to use an all-around symbol on the selected reference line. Click the button to turn the symbol off or on.
Secondary fillet type	Specifies the type of weld for secondary fillets. Click the button to display and choose from the palette of available weld types. This button is available only when the active drafting standard is based on ANSI.

Angle	Specifies the angle between weldments.
Brazing	Specifies whether the weld is brazed. Select or clear the check box to add or remove the brazing symbol.
Clearance	Specifies the clearance for the braze.
Contour	Specifies the contour finish for the weld. Click the arrow and choose the contour from the list.
Depth	Specifies the depth of the weld.
Diameter	Specifies the diameter of the weld.
Gap	Specifies the space between weldments.
Height	Specifies the height of the weld.
Length	Specifies the length of the weld.
Method	Specifies the finish method for the weld. Click the arrow and choose the method from the list.
Middle	Specifies the type of inspection to perform on the weld.
Number	Specifies the number of welds.
Pitch	Specifies the distance between welds.
Root	Specifies the root thickness of the weld.
Root gap	Specifies the gap for the weld.
Size	Specifies the size of the weld.
Small leg	Specifies the thickness of the weld.
Spacing	Specifies the space between welds.
Thickness	Specifies the thickness of the weld.

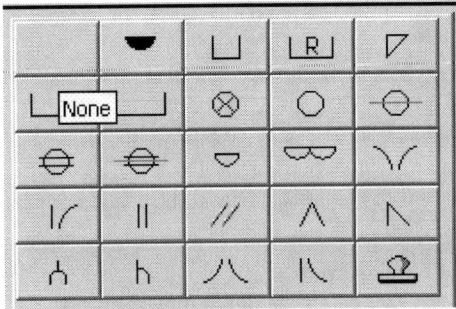

Weld type	Specifies the type of weld. Click the button to display and choose from the palette of available weld types. The weld types displayed on the palette are determined by the active drafting standard.

TIP: Running your mouse over the buttons without clicking will reveal a help indicating what each symbol represents.

To add or remove symbols from the palette, customize the drafting standard.

To associate the welding symbol with geometry, move the cursor over the desired edge and click when the edge highlights. Symbols that are associated with geometry will move if you move the drawing view.

Move the cursor and click to add a vertex to the leader line.

The line type, line weight, color, and gap for the symbol are determined by the active drafting standard.

Drawing Annotation

To place the symbol:

Continue placing welding symbols. When you finish placing symbols, right-click and select 'Done' from the menu to end the operation.

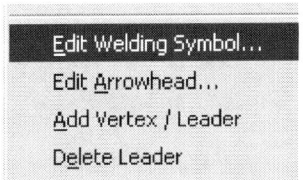

To edit an existing symbol:

Select the placed symbol.
Right click and select **Edit**.

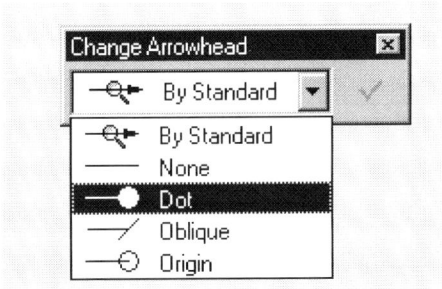

Selecting the Edit Arrowhead option will bring up a dialog where the user can select the current standard or a different arrowhead.

Selecting the 'Add Segment or Vertex option' activates the Grip mode and the user can use the mouse to stretch or modify the leader for the symbol.

13-41

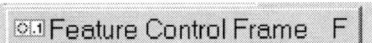

Feature Control Frame

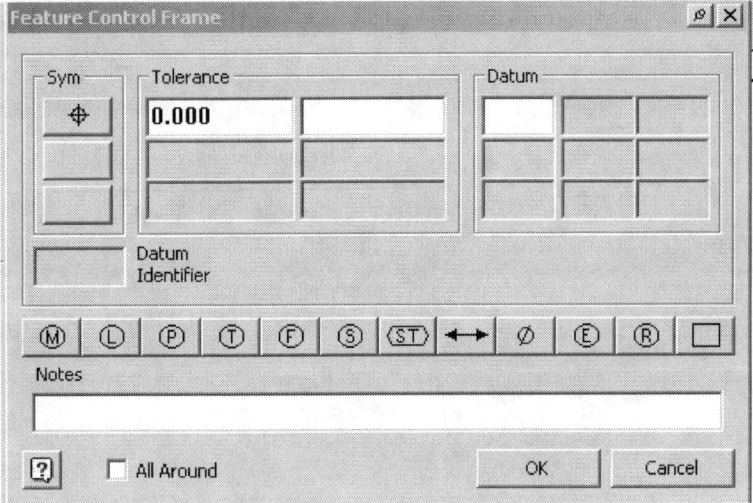

Specifies the content of a feature control frame. The options in the dialog box are determined by the active drafting standard.

Sym	Specifies the tolerance symbol. Click the button to open the symbol palette and select the desired symbol. The symbols on the palette are determined by the active drafting standard.
Tolerance	Sets the tolerance values. Tolerance 1 specifies the tolerance. Tolerance 2 specifies the lower tolerance (ANSI standard only).
Datum	Specifies the datums that affect the tolerance.
Insert modifier buttons	Adds modifier characters to the text at the insertion point. Click a button to add the desired character. The available characters are determined by the active drafting standard.
All Around	Adds an all-around character next to the feature control frame. Select the check box to add the character; clear the check box to remove it.

Continue placing feature control frames. When you finish placing symbols, right-click and choose 'Done' from the menu to end the operation.

TIP: The text boxes in the second row are available only if you select a tolerance-type symbol for the row. You can specify in the active drafting standard whether to merge corresponding fields when the information is the same.

You can create a feature control frame with a leader line or as a stand-alone symbol. The color and line weight of the symbol are determined by the active drafting standard.

To place a symbol without a leader, double-click to set the symbol location and open the dialog box.

The Style for the Feature Control Frame can be set using **Format→Styles Editor**.

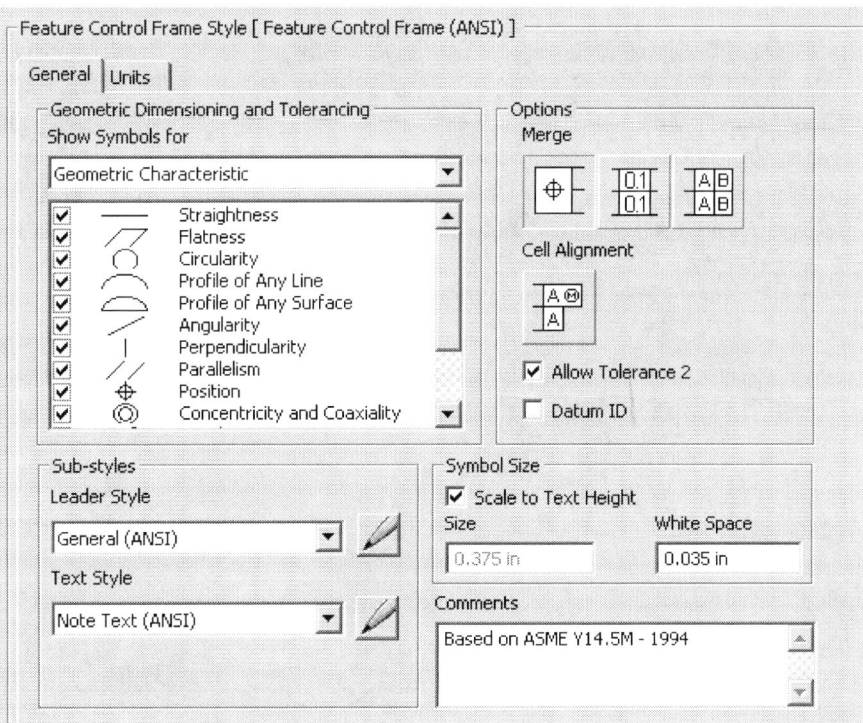

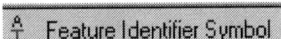
Feature Identifier Symbol

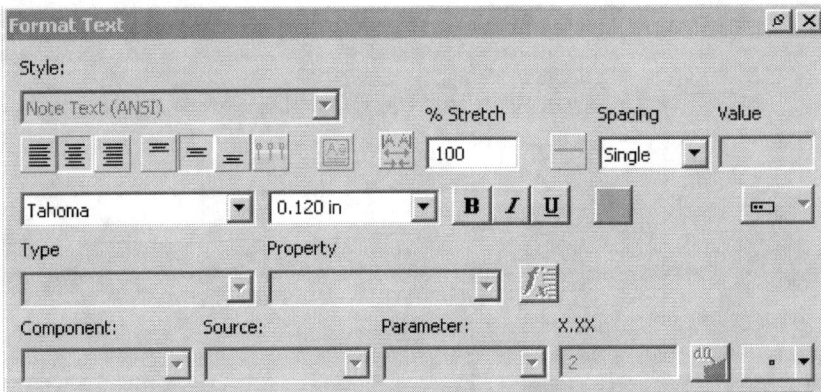

Style	Specifies the text style to apply to the text. Click the arrow and select from the list of available text styles. Specifies the paragraph properties for selected text.
Justification	Positions the text within the text box. 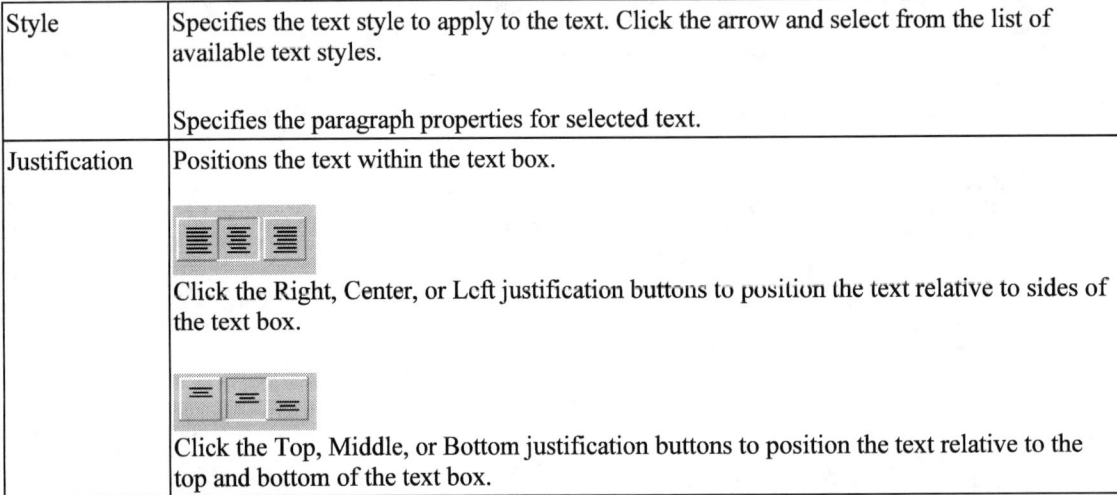 Click the Right, Center, or Left justification buttons to position the text relative to sides of the text box. Click the Top, Middle, or Bottom justification buttons to position the text relative to the top and bottom of the text box.

TIP: To associate the symbol with geometry, move the cursor over the desired edge and click when the edge highlights. Symbols that are associated with geometry will move if you move the drawing view.

Drawing Annotation

Color	Specifies the text color. Click the color button, and then select a color from the Color dialog box.
Rotation	Rotates the text. Select a button to specify the angle of rotation for the text. This option is not available for leader text, dimension text, and datum identifiers.
%Stretch	Specifies the text width. Enter 100 to display the text as designed, enter 50 to decrease the width of the text by 50%.
Line Spacing	Specifies the spacing between the bottom of one line of text and the bottom of the next line of text.
Value	Specifies the value for line spacing, when you set line spacing to Exactly or Multiple.

Selects a named parameter and inserts its value into the text at the insertion point. Parameter options are available only when adding or editing text in general drawing notes and dimension text.

Component	Specifies the model file that contains the parameter. If the drawing contains views of more than one model, click the arrow and select the file from the list. ***NOTE***: *If the drawing contains derived parts, the donor parts are also included in this list.*
Source	Selects the type of parameter to show in the Parameter list. Click the arrow and select from the list. Model Parameters lists the named parameters automatically added to the model when you add dimensions or features. User Parameters lists the user parameters added to the model.
Parameter	Specifies the parameter to insert into the text. Click the arrow and select from the list. The parameters in the list change, depending on the Source you selected.
X.XX	Sets the precision of the value display. Enter the number of decimal places to display.
Add Parameter button	Adds the selected parameter from the selected component to the text.
Font	Specifies the font. Click the arrow and select from the list of available fonts.
Font Size	Sets the height of the text in sheet units (inches or millimeters). Enter the size or click the arrow and select a size from the list.
Style	Sets the style. Click the Bold, Italic, or Underline buttons to apply the style to the text.
↑ ↓	Zooms in and out on the text so you can read the value easier.

Datum Identifier Symbol

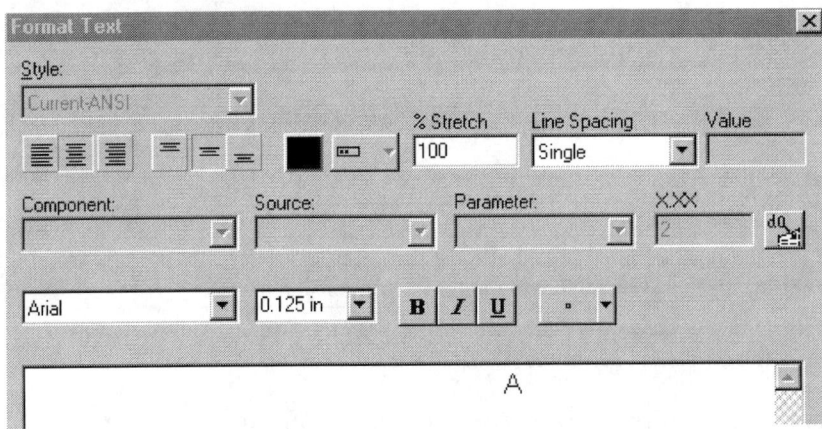

Use the Datum Identifier button on the Drawing Annotation toolbar to create one or more datum identifier symbols. You can create a datum identifier with a leader line or as a stand-alone symbol. The color and line weight of the symbol are determined by the active drafting standard.

Datum Target

Click the arrow next to the Datum Target Symbol button and choose the datum target type from the flyout menu.

Dimension	Specifies size of the target area. Enter the number.
Datum	Specifies the label for the target.

Target types are Leader, Line, Rectangle, Circle, and Point.

TIP: To stack datum targets, right-click an existing datum target and select Attach Balloon from the menu.

The color, target size, and line weight of the symbol are determined by the active drafting standard.

Drawing Annotation

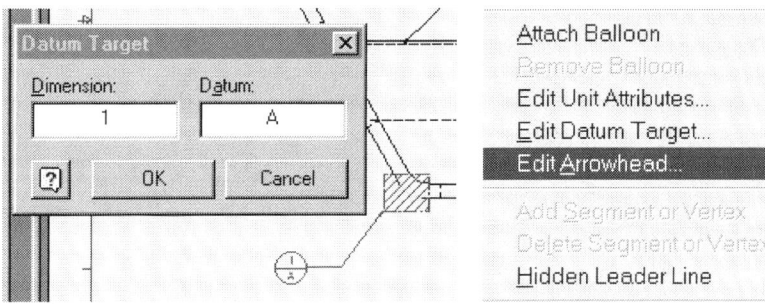

To edit an existing symbol:

Select the symbol.
Right click and select **Edit**.

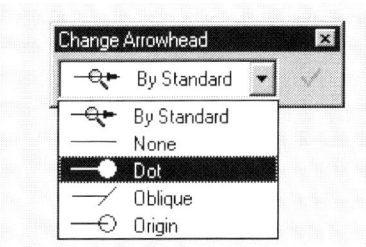

Selecting the Edit Arrowhead option will bring up a dialog where the user can select the current standard or a different arrowhead.

Selecting the 'Add Segment or Vertex option' activates the Grip mode and the user can use the mouse to stretch or modify the leader for the symbol.

To stack datum targets, select Attach Balloon from the menu.

Selecting 'Hidden Leader Line' changes the linetype of the leader to Hidden.

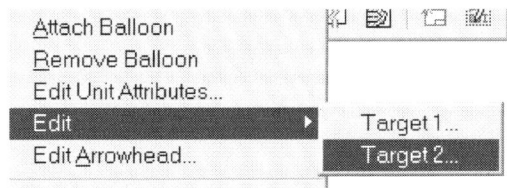

If you have stacked datum targets and need to make modifications, the right click menu adapts to give you the option of editing different targets.

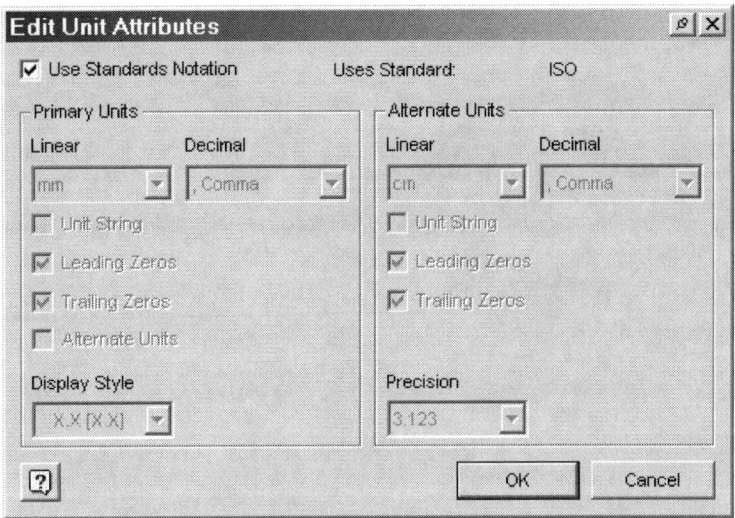

Selecting the Edit Unit Attributes option brings up this dialog.

Sets the format of the numerals displayed. Disable Use Standards Notation in order to use Alternate Units.

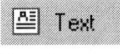
Text

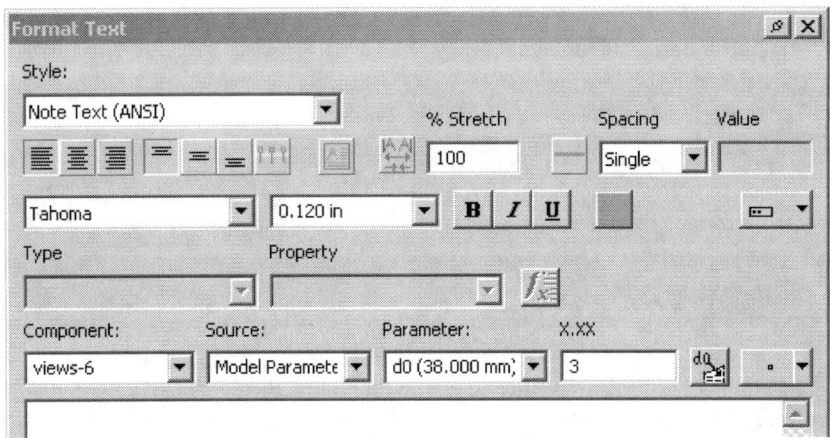

Changes the style and format for:

- notes in drawings
- text in drawing sketches
- text in title blocks, borders, datum identifiers, and sketched symbols
- text added to dimensions, view labels, and hole notes

Drawing Annotation

NOTE: The default text format is controlled by the active drafting standard. To change the default text for a drawing or template, modify the drafting standard.

Use the Text button on the Drawing Annotation toolbar to add general notes to a drawing. General notes are not attached to a view, symbol, or other object in the drawing.

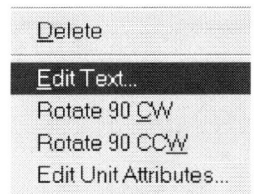

To modify a general note, select it in the graphics window, right-click, and select the operation from the menu.

- To copy the note, select copy. You can paste the text to another place on the drawing sheet, to a different sheet in the same drawing, or to a sheet in another drawing.
- To remove the note, select Delete.
- To edit or format the existing text, select Edit Text to open the Format Text dialog box.
- To rotate the note, select the desired rotate direction.

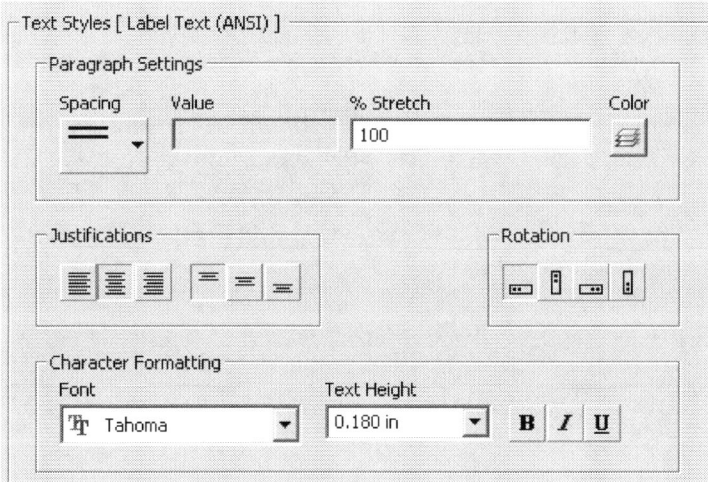

 TIP: The default text format is controlled by the active drafting standard. To change the default text for a drawing or template, modify the drafting standard. Go to Format→Styles Editor and select the Text category.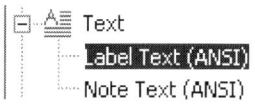

13-49

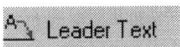
Leader Text

Use the Leader Text button on the Drawing Annotation toolbar to add notes with leader lines to a drawing.

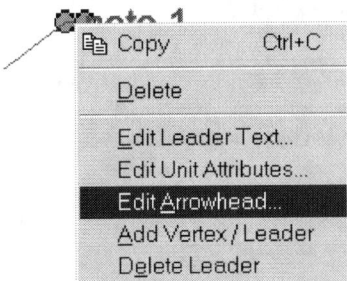

To modify a note, select it in the graphics window, right-click, and select the operation from the menu.

- To remove the note, select Delete.
- To edit or format the existing text, select Edit Text to open the Format Text dialog box.
- To change the arrowhead or other terminator, select Edit Arrowhead and select the new terminator from the dialog box.
- To add a point to the leader line, select Add Segment or Vertex, then click in the graphics window to place the point. You can drag the added point to redefine the leader line.

TIP: If you attach a note leader line to a view or to geometry within a view, the note is moved or deleted when the view is moved or deleted.

The active drafting standard controls the default text format. To change the default text for a drawing or template, modify the drafting standard. Go to Format→Standards and select the Common tab.

Balloon/ Auto Balloon

Use the desired Balloon button (Balloon/Auto Balloon) on the Drawing Annotation toolbar to add reference balloons to a drawing.

The default style for balloons is defined by the active drafting standard. After placing a balloon, you can change the arrowhead style, attach or remove additional balloons, or add segments to the leader line.

To remove a balloon from a stacked set, right-click the balloon and select 'Remove Balloon' from the menu. If there are several attached balloons, the last balloon to be attached is removed first.

A balloon can contain the item number of a first-level component in the assembly, or of an individual part, depending on the setting in the active drafting standard. You can change the level of the item number in a balloon.

To change the Level of a balloon reference, right-click the balloon and select Part to show the item number for the part to which the balloon is attached, or First-Level Component to show the item number for the subassembly that contains the part.

TIP: A first-level component can be either a subassembly or a part.
If you add balloons to a drawing before creating a parts list, the balloons will show the item numbers of first-level components. You can select a balloon and change it to show the item number of the part to which it is attached.
Typing 'B' will initiate the Balloon command.

TIP: To change the default heading location and title text for parts lists, use Format>Standards>Parts List to change the attributes of the active drafting standard.

Autodesk Inventor Fundamentals

Exercise 13-9
Adding Balloons

File: Ex8-2.iam
Estimated Time: 15 minutes

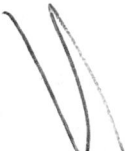

1. Start a New drawing.

2. Select the **Base View** tool.

3. Locate and select the *NewLinkRod.iam* file located in the Tutorial files under Inventor or from the publisher's website.

4.  Place a view as shown.

   Hint: Use the Front view.

5. Activate the Drawing Annotation Panel.

6. Select the **Auto Balloon** tool.

   This is located under the Balloon flyout.

7.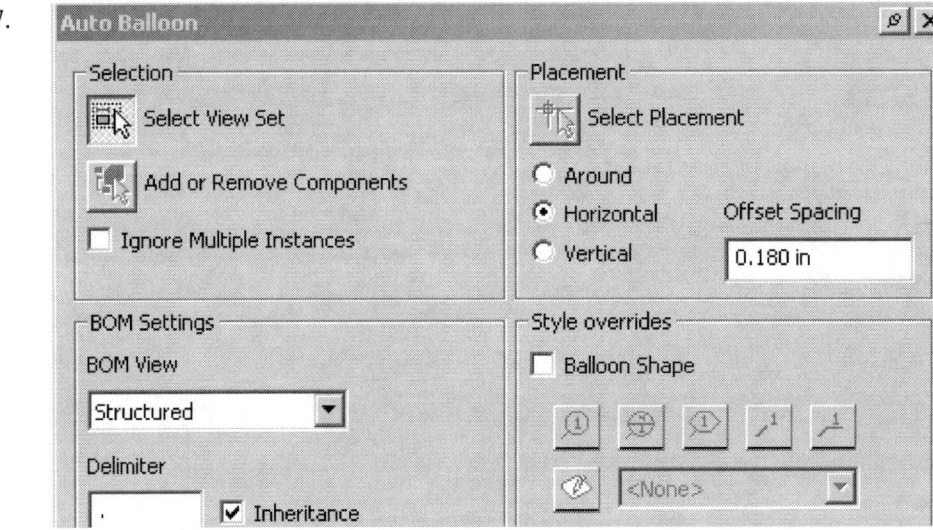

The first enabled button prompts you to select a view.
Select the base view you just placed.

13-52

Drawing Annotation

8. Window around the view.
Set the Placement to Horizontal.

9. Disable **Ignore Multiple Instances**.

10. Enable **Horizontal**.

11. Press the **Select Placement** button to place the balloons.

12. Pick a point above the view to place the balloons.

Press **Apply** and close the dialog

13. Balloons are added to each part.

14. To move the balloons, select to activate the grips and move to the desired location.

15. Select an Item 1 Balloon.
Right click and select **Edit Balloon**.

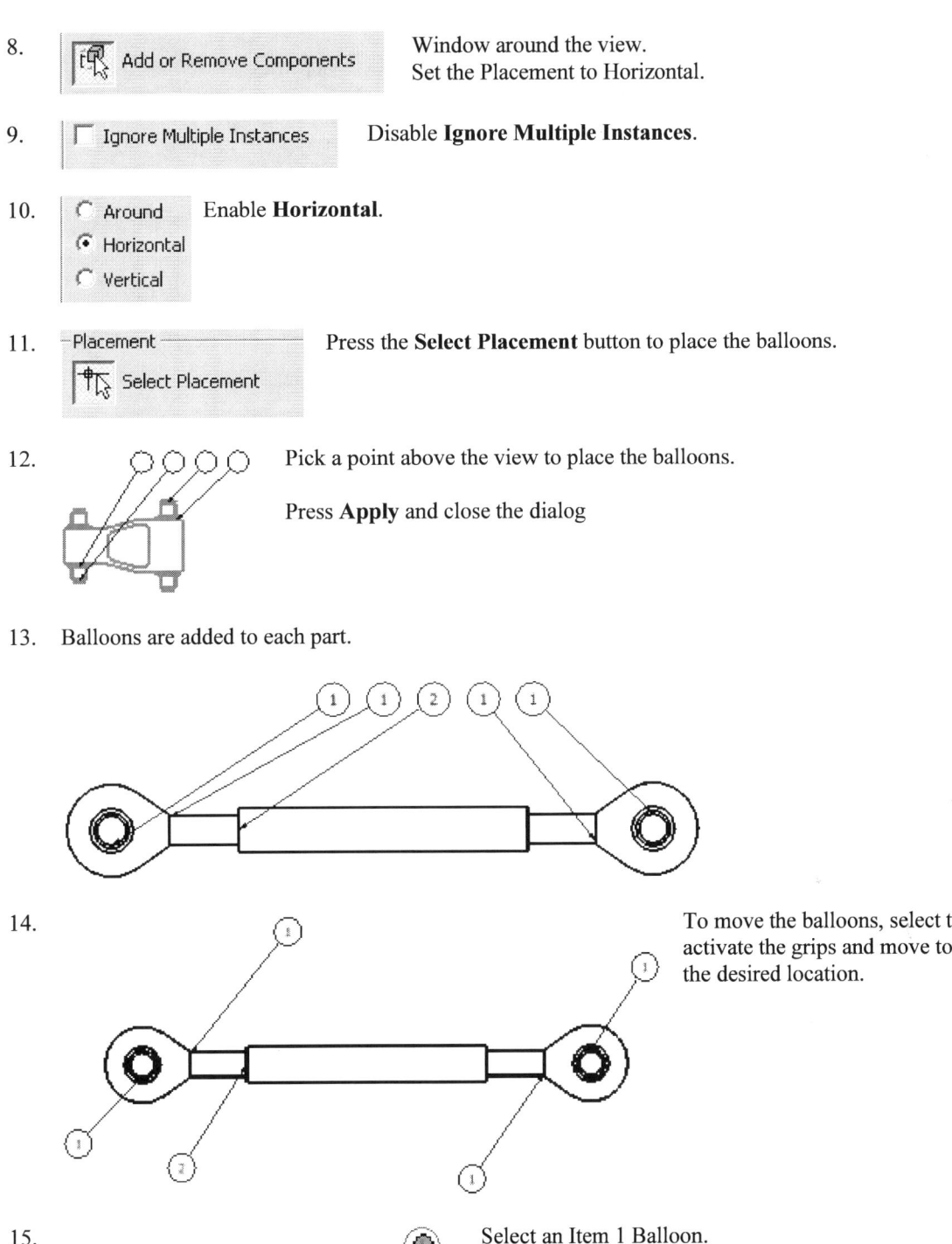

13-53

16. 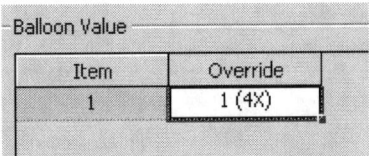 Under **Balloon Value**:
    In the **Override** column, enter **1 (4X)**.

    Press **OK**.

17. The balloon expands for the additional note.

18. Select the Item 1 balloon.
    Right click and select **Add Vertex/Leader**.

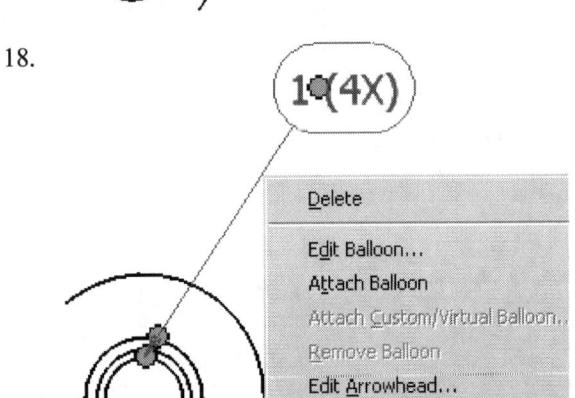

19. Pick the leader line to add a point to the line.
    A point is added to create a shoulder.

    Activate the balloon grips, then move the balloon to position the shoulder.

20. To delete a balloon, simply select, right click and select **Delete**.

21. Save the file as *ex13-9.idw*.

Parts List

Adds a parts list for the components of the assembly in the selected view.

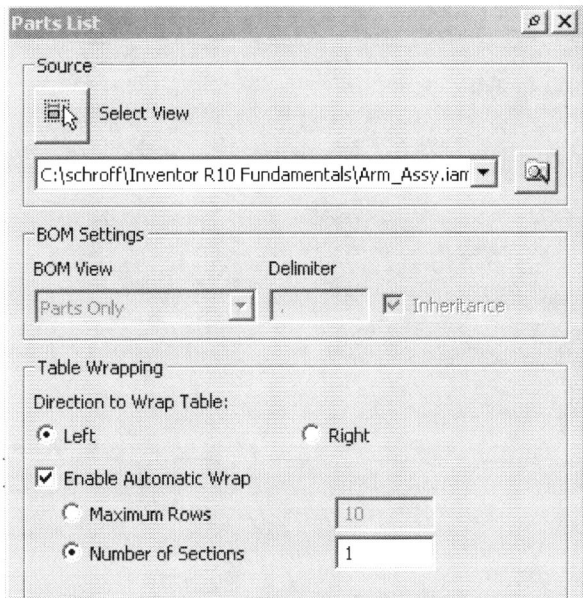

	Sets the number of levels of components to include in the parts list. Select the desired level.
First-Level Components	Creates a parts list that shows only the top level of components of the assembly in the selected view. Subassemblies and parts that are not part of a subassembly are shown, but not the parts in the subassemblies.
Only Parts	Creates a parts list that shows the parts of the assembly in the selected view. Subassemblies are not shown, but the parts in the subassemblies are shown.
	Set the scope of parts list to include all or a range of components of the assembly in the selected view when the Level is set to Only Parts. Not available when First-Level Components is selected.
All	Creates a parts list for all the components in the selected view.
Items	Creates a parts list for the specified range of parts. Select Items then enter the part numbers, separated by commas, or click the parts in the graphics window to include them in the list.

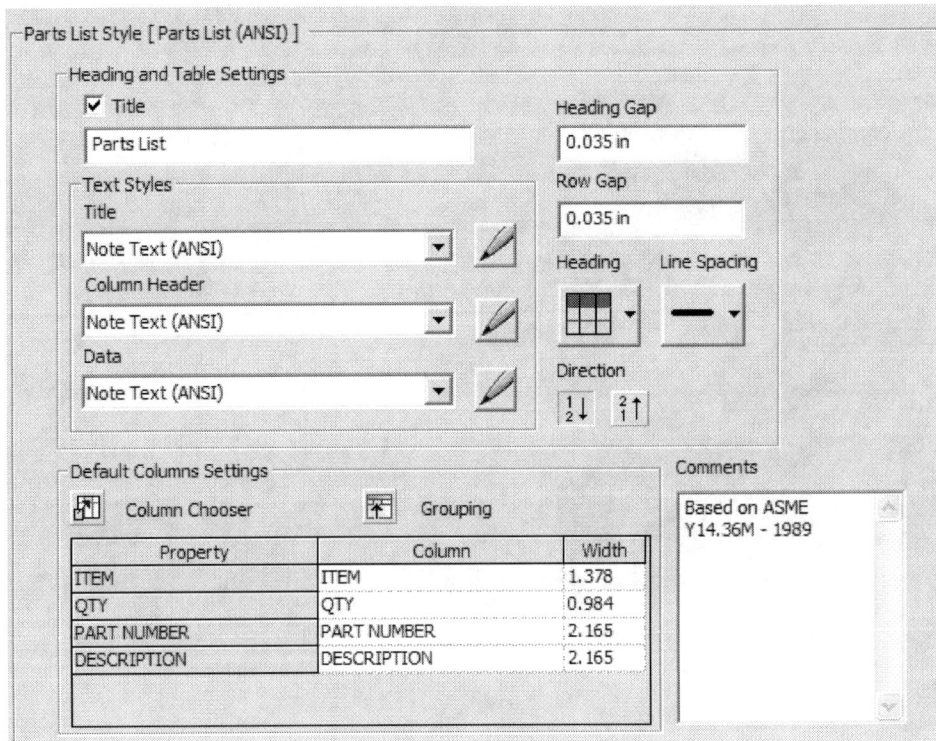

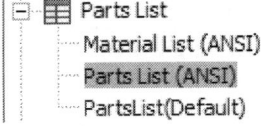

The Parts List format is set in the Styles dialog box. Access by selecting **Format→Styles Editor**.
Then, highlight the desired Parts List to be formatted from the Parts List category.

Drawing Annotation

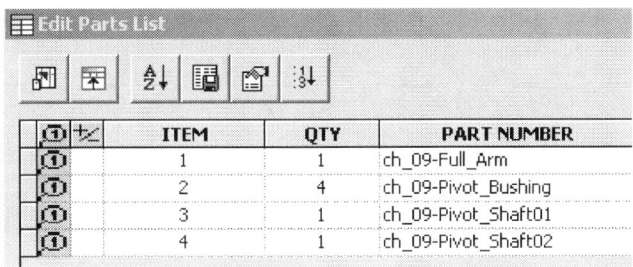

	Parts List Operations performs operations on the selected parts list. Click the appropriate button to perform the operation.
	Column Chooser opens the Column Chooser dialog box so that you can add, remove, or change the order of the columns for the selected parts list.
	Group Settings allows the user to group specific components, such as fasteners or welds.
	Sort opens the Sort Parts List dialog box so that you can change the sort order for items in the select parts list.
	Export opens the Export Parts List dialog box so that you can save the selected parts list to an external file.
	Heading opens the Heading Parts List dialog box so that you can change the title text or heading location for the selected parts list.
	Renumbers items in the Parts List.

Custom Parts	Adds non-graphic parts such as adhesives and lubricants to the parts list.

When a parts list is placed, its setup is determined by the settings in the active drafting standard. You can modify the settings for a parts list after placing it.

13-57

Autodesk Inventor Fundamentals

Exercise 13-10
Adding a Parts List

File: Ex13-9.idw
Estimated Time: 30 minutes

1. Open *ex13-9.idw* or continue working in the previous file.

2. 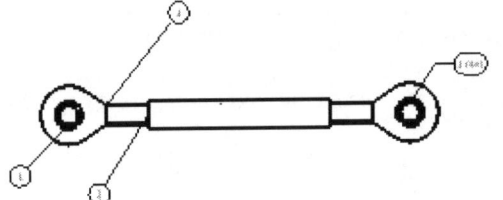 Select the **Parts List** tool.

3. Select the view.

   Press **OK**.

   Then pick below the view to place the parts list.

4. 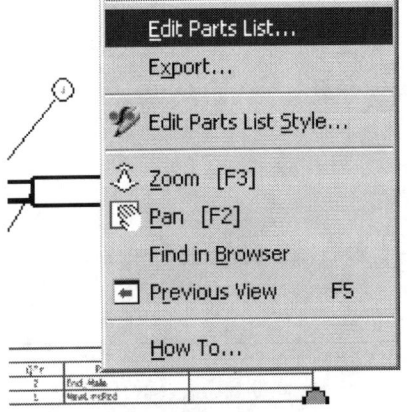 Select the Parts List.

   Right click and select **Edit Parts List**.

5.  Right click on the +/- button in the table.

   Select **Table Layout**.

13-58

Drawing Annotation

6. 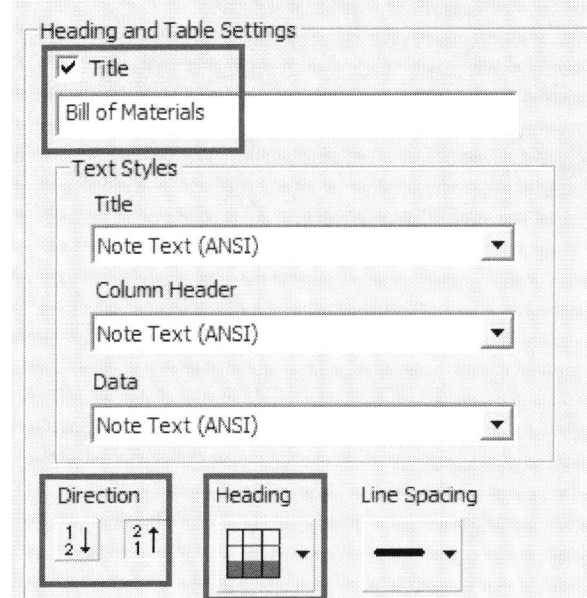 Change the title to **Bill of Materials**.

   Enable **Bottom** under **Heading**.

   Set the **Table Direction** to **Descending**.

   Press **OK**.

7. Press **OK** to exit the 'Edit Parts List' dialog.

   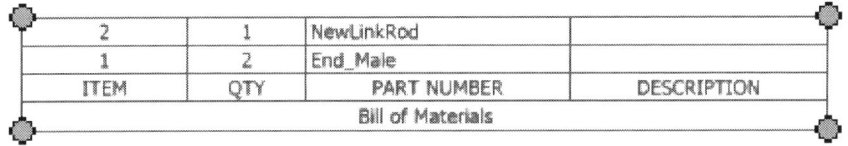

8. The Parts List changes to show the header on the bottom and the title is updated.

9. Select the parts list again.

   Right click and select **Edit Parts List**.

10. Select the **Column Chooser** tool.

11.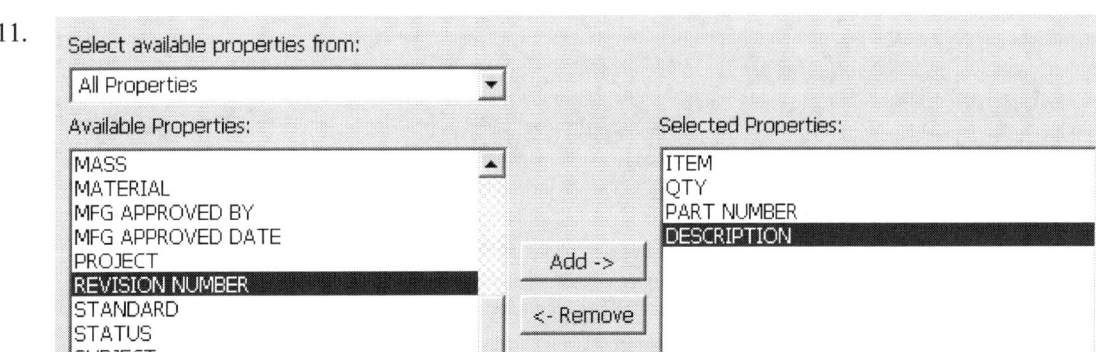

    Highlight **Description** in the right column.
    Highlight **Revision Number** in the left column.

    Select the **Add** button.

13-59

12. 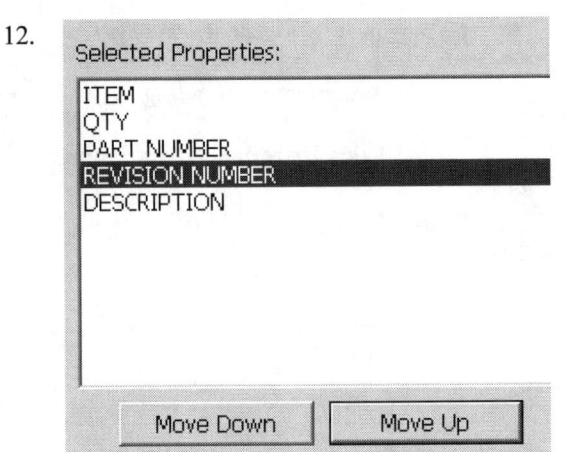 Highlight **Revision Number**.
Use the **Move Up** button to position **Revision Number** above **Description**.

    Press **OK**.

13. 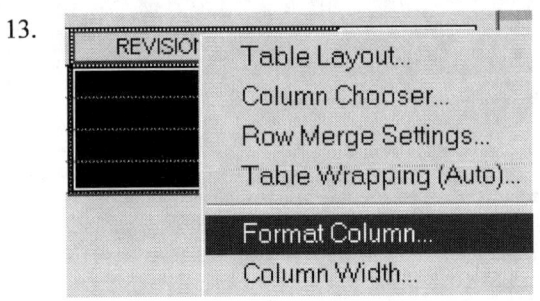 Highlight the **Revision Number** column.
Right click and select **Format Column**.

14. 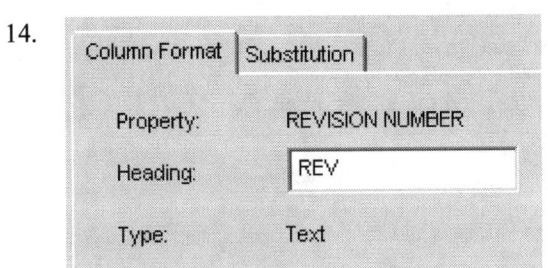 Change the name to **REV**.

15. 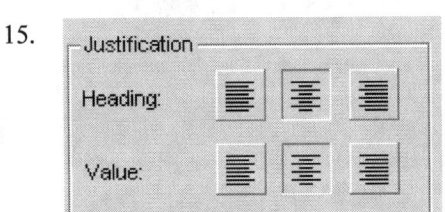 Change the **Justification** for the Value to **Centered**.

16. 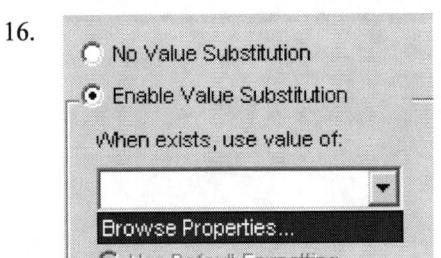 Select the **Substitution** tab.

    Select **Browse Properties** from the drop-down.

17. 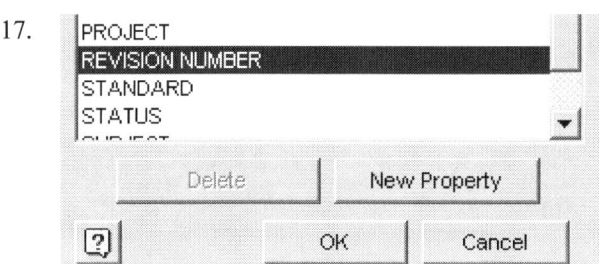 Select **Revision Number** from the drawing properties list.

    Press **OK** twice.

18. 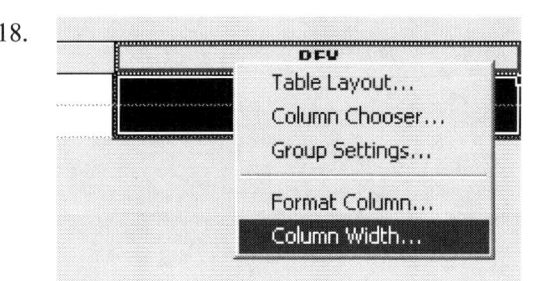 Highlight the **REV** column.

    Right click and select **Column Width**.

    Press **OK**.

19. 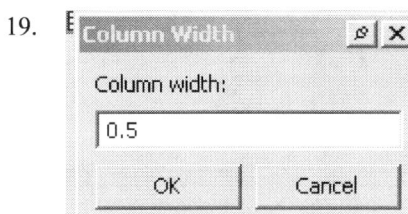 Set the **Column Width** to **0.5**.

    Press **OK** twice to close the dialog.

Autodesk Inventor Fundamentals

20.

2	1	NewLinkRod		
1	2	End_Male		
ITEM	QTY	PART NUMBER	REV	DESCRIPTION
		Bill of Materials		

The parts list updates to reflect the changes.

21. 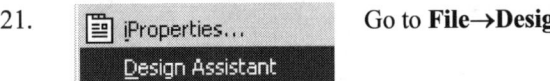   Go to **File→Design Assistant**.

22.   Expand the Parts Browser.

23.   Highlight the *End-Male.iam* part.

   Right click and select **iProperties**.

24. 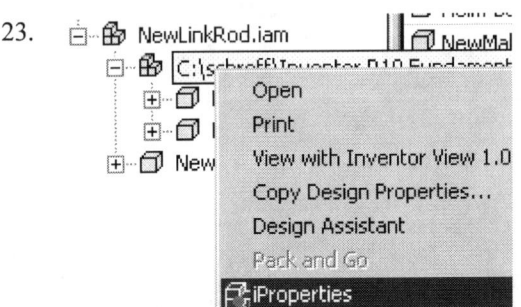   Select the **Project** tab.

   Under **Description**, enter **WASHER, MALE**.

   Under **Revision Number**, enter **A**.

   Press **Apply** and **Close**.

25.    Highlight the *NewLinkRod.ipt*.

   Right click and select **iProperties**.

13-62

Drawing Annotation

26. Select the **Project** tab.

    Under **Description**, enter **ROD, LINK**.

    Under **Revision Number**, enter **A**.

    Press **Apply** and **Close**.

27. Highlight the assembly.

28. The table updates.

    NOTE: I was not able to get the parts list to update until I selected Bill of Materials and then closed that dialog.

29. Close the Design Assistant.

2	1	NewLinkRod	A	ROD, LINK
1	2	End_Male	A	WASHER, MALE
ITEM	QTY	PART NUMBER	REV	DESCRIPTION
		Bill of Materials		

    The parts list updates.

13-63

30. 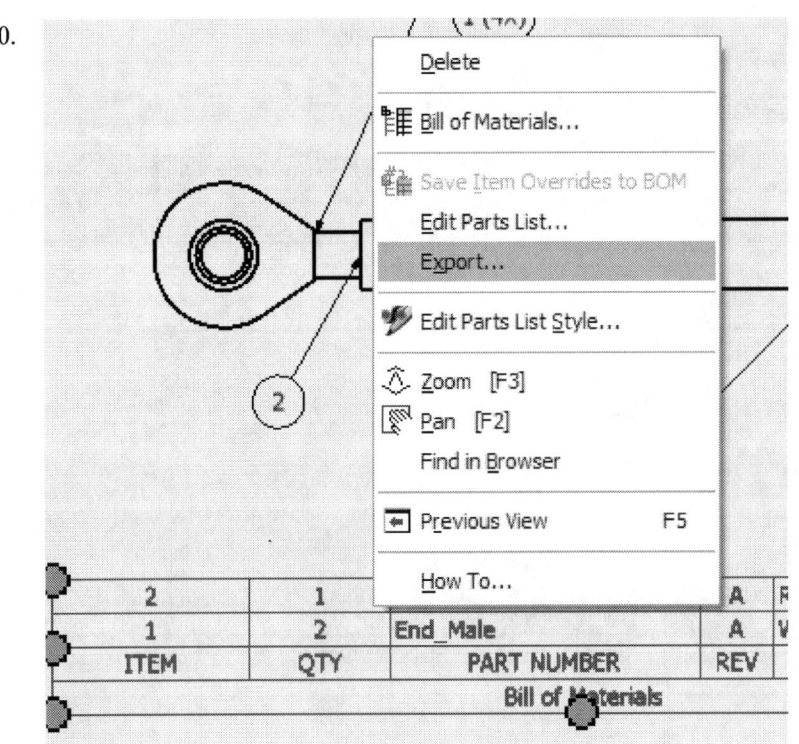 Select the **Parts List**.

Right click and select **Export**.

31. 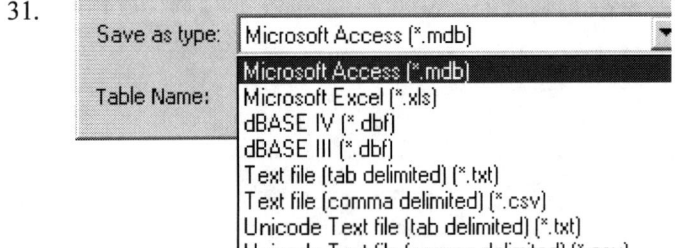 You can export your parts list into the following file types:
Mdb, xls, dbf, txt, or csv.

Drawing Annotation

32. 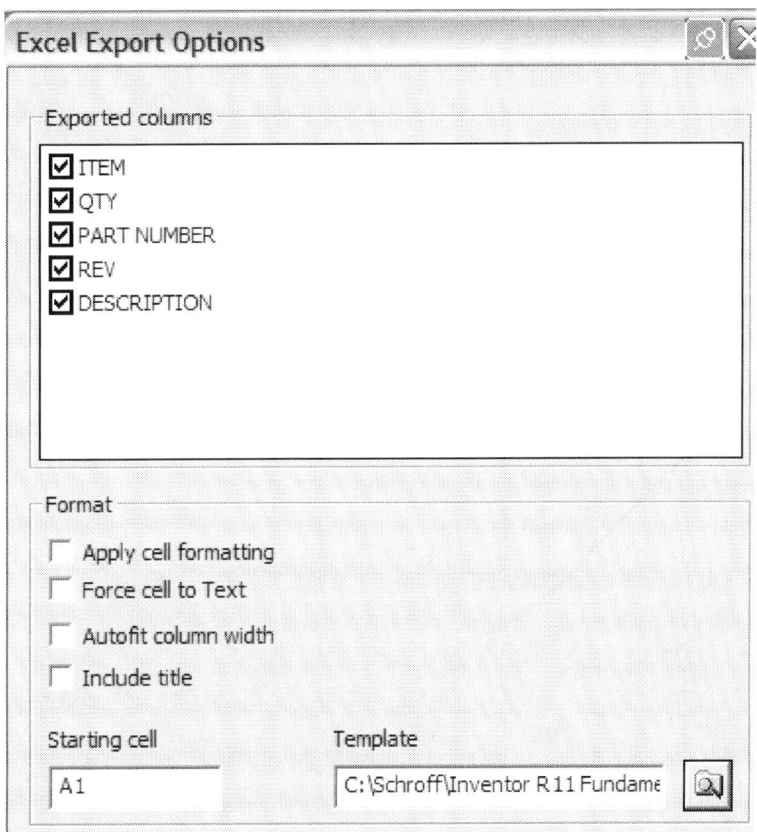 Save the file as *ex13-10.xls*.

    If you select the **Options** button, you can select which columns to include and how the columns should be formatted.

33. Locate *ex13-10.xls* using Explorer.
    Double click on the file name to launch **Excel**.

34.

	A	B	C	D	E
1	ITEM	QTY	PART NUMBER	REV	DESCRIPTION
2	1	2	End_Male	A	WASHER, MALE
3	2	1	NewLinkRod	A	ROD, LINK
4					

Your parts list opens as an Excel spreadsheet.

NOTE: You must have Excel installed for this part of the exercise to work.

Close the Excel spreadsheet without saving.

35. Save the Inventor file as *ex13-10.idw*.

13-65

Hole Chart

You can use hole tables to automatically display the location and size of either all holes or selected holes in a drawing view.

Hole tables display the X,Y coordinate locations of hole features with respect to a hole datum target point. You can place only one hole datum target point in each drawing view. This point is constrained to the geometry it is attached to.

Inventor assigns alphanumeric names to holes included in hole tables. These hole names appear as hole tags next to the holes in the drawing view. You can right-click hole tags to edit their properties.

- If you add a new hole to a part and a hole table exists for the view, the new hole is added to the hole table.
- If you move a hole or hole origin, the X,Y location of the hole in an existing hole table updates automatically.
- If you delete a hole from a part, the hole is removed from the hole table and the hole tag is removed from the view. The remaining holes in the hole table are renumbered.

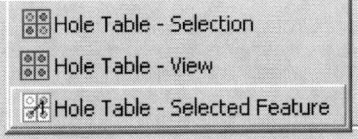 You have three tools dealing with hole tables.

Drawing Annotation

Exercise 13-11
Adding a Hole Chart

File: Ex13-7.idw
Estimated Time: 15 minutes

1. Open *ex13-7.idw*.

2. 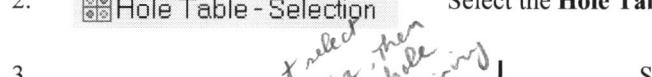 Select the **Hole Table-Selection** tool.

3.

   Select the center of the lower left hole for the hole table origin.

   Then select all four holes.

   Right click and select '**Create**'.

   Pick to place the hole chart.

4. The hole chart is added and the holes are each labeled.

5.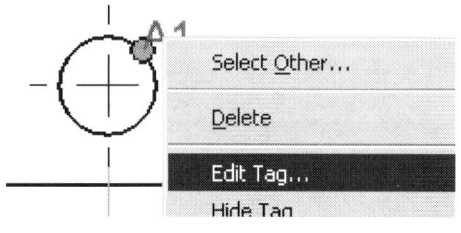

   You can control the visibility of the origin and the hole labels.

   Select the **Hole Chart**, right click, and disable the **Visibility** of the **Origin**.

   We no longer see the Origin symbol.

6.

   Select the **A1** hole label.

   Right click and select **Edit Tag**.

7. Type the word '**DRILL-**' in front of the default label.

   Press **OK**.

   This only changes a single hole label.

13-67

Autodesk Inventor Fundamentals

8. 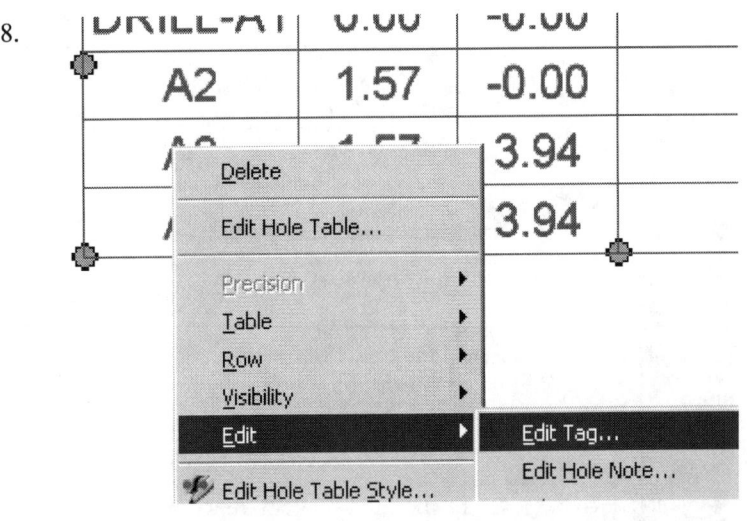 Select the **hole table**.

   Right click over the A-2 cell and select **Edit→Edit Tag**.

   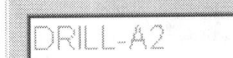 Type the word '**DRILL-**' in front of the default label.

   Press **OK**.

9. Save the file as *ex13-11.idw*.

 TIP: You can use the grips on the Hole Tag to reposition or add a leader to the Hole Tag.

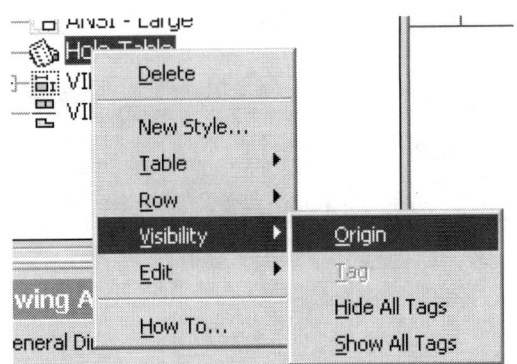 You can also access the shortcut menu by highlighting the Hole Table in the Browser and right clicking.

13-68

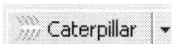

Caterpillar

This tool allows you to create the appearance of a weld on an assembly.

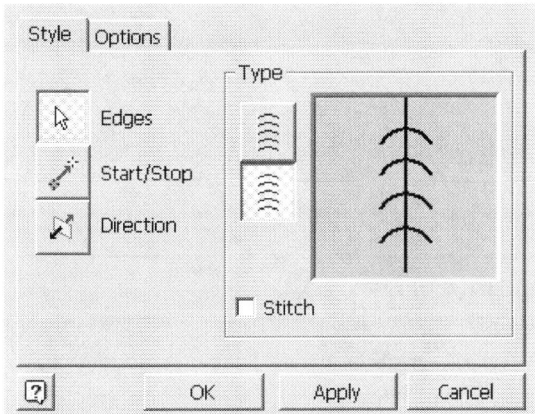

Edges	Select the Edge to apply the caterpillar.
Start/Stop	If the caterpillar is not to be shown the length of the entire selected edge, set a start and stop point for the caterpillar
Direction	Orient the direction of the caterpillar
Type	Set the caterpillar centered or to one side
Stitch	Enabling stitch adds spaces to the caterpillar.

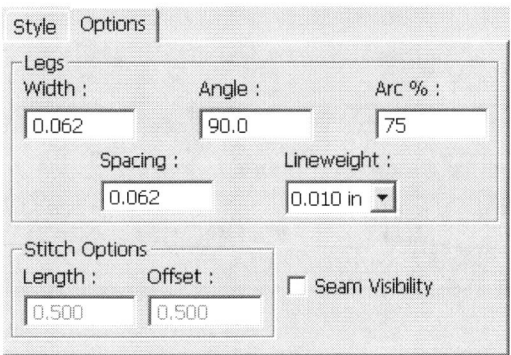

Width	Sets the distance between the endpoint of the leg and the selected edge on a partial caterpillar (when stop/start points are applied). If a full caterpillar is applied, sets the distance between the endpoints of the legs.
Angle	Applies an angle between the edge and the arc endpoints.
Arc %	Sets the radius of the arc.
Spacing	Sets the distance between the arcs.
Lineweight	Sets the thickness of the arcs.
Seam Visibility	Enables the visibility of the edge on which the weld is applied.

End Treatment

Adds a weld end treatment annotation to a view. This tool is located under the Caterpillar flyout.

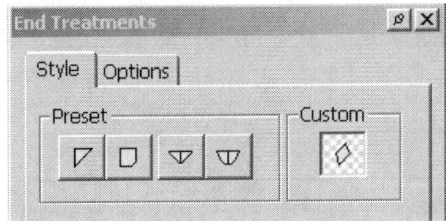

Preset	You can select among Bevel, J-Type, V-Type or U-Type.
Custom	Define a custom shape

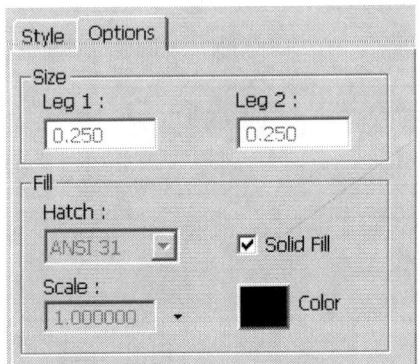

Leg 1:	Length of one side
Leg 2:	Length of second side
Fill	Hatch Pattern
Solid Fill	If enabled, a solid color instead of a hatch
Scale	Set the scale of the hatch pattern
Color	Sets the color of the hatch/fill

The leg lengths are set depending on the style selected.

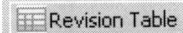
Revision Table

REVISION HISTORY				
ZONE	REV	DESCRIPTION	DATE	APPROVED
1	1	Value	11/25/2002	Elise Moss

Places a revision table on your drawing sheet.

Drawing Annotation

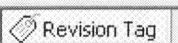
Revision Tag

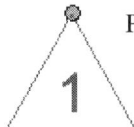
Places a tag in your drawing to identify the location of the engineering change.

Exercise 13-12
Adding Revision Tags and Tables

File: Ex13-11.idw
Estimated Time: 15 minutes

1. Open *ex13-11.idw*.

2. Select the **Revision Tag** tool.
   (This is under the Revision Table flyout.)

3. Place the tag next the hole.

   Right click and select **Continue**.
   Right click and select **Done**.

4. Select the tag.
   Right click and select **Edit Tag**.

5. Change the tag to **E-1**.

   Press **OK**.

13-71

Autodesk Inventor Fundamentals

6.  The tag updates.

7. Select the **Revision Table** tool.

8. Place the **Revision Table** in the upper right corner of the sheet.

9. Window into the revision table.
   Double click on the cell where **Value** is.

10. Edit value to read **HOLE SIZE CHANGED**.

    Press **OK**.

11. Edit the values in the **Zone**, **Date**, and **Approved** columns.
    The revision table updates.

12. 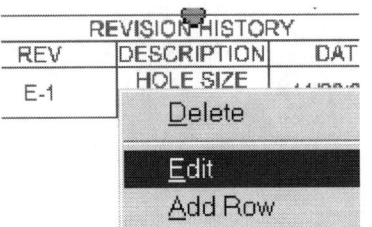 Select the **Revision Table**.
    Right click and select **Edit**.

13. Highlight **Description** in the **Selected Properties** list.
    Right click and select '**Edit Column Properties**'.

14. Change the **Data Alignment** to **Left Justified**.
    Press **OK** twice.

15. Save as *ex13-12.idw*.

13-72

Drawing Annotation

Symbols

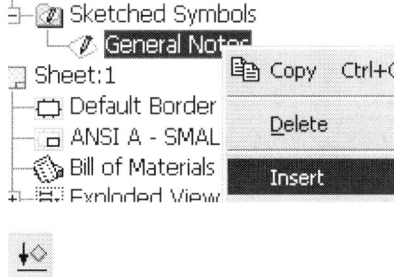

There are two ways to insert Symbols in a drawing.

You can highlight the desired symbol under Sketched Symbol, right click and select 'Insert'.

You can select the Symbols tool on the Drawing Annotation toolbar.

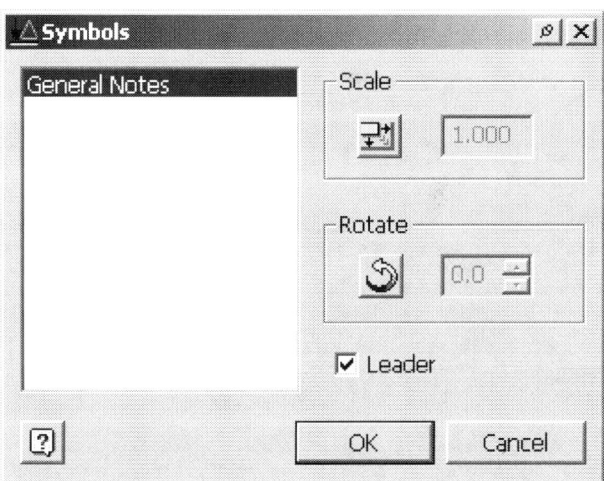

If you select the Symbols tool, a dialog appears that includes a list of symbols available.

You can set the scale for the symbol to be inserted, the rotation angle and determine whether or not a leader should be attached to the symbol.

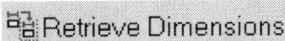

Retrieve Dimensions

This tool allows you to add model dimensions to your drawing view.

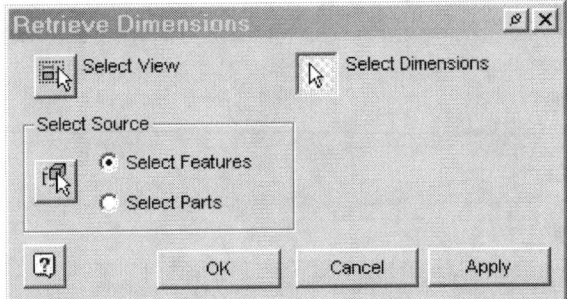

Select View	Select the view to place the dimensions	
Select Dimensions	Dimensions are displayed on the view. Pick to select the dimensions you would like added to the view.	
Select Source	**Select Features:**	Select the features to use for dimensions. You may also select sketched geometry.
	Select Parts:	Select specific parts in an assembly to be dimensioned

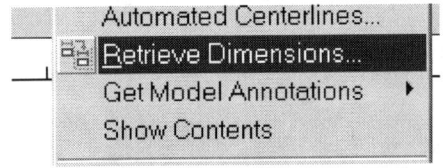

You can also select a view, right click and select **Retrieve Dimensions.**

☑ Retrieve all model dimensions on view placement

You can also automatically place dimensions when you create a view by enabling the Retrieve all model dimensions option in the Drawing tab of the Options dialog.

Drawing Annotation Tools

Button	Tool	Function	Special Instructions
	General Dimension	Creates a dimension between two points, lines, or curves.	Double click on a dimension to edit.
	Baseline Dimension	Creates a set of baseline dimensions	
	Ordinate Dimension Set	Places a set of ordinate dimensions along an axis. The first dimension placed is automatically assumed to be the origin dimension.	To edit, select then right click to access the edit options. To activate, type 'O'
	Ordinate Dimension	Creates an ordinate dimension	Requires the user to select an Origin prior to placing the dimension
	Hole/Thread Notes	Adds a hole or thread note with leader	Hole notes can only be added to features created with the Hole tool.
	Chamfer Note	Add a chamfer note	Select the chamfer edge and then an adjacent edge.
	Center Mark	Creates a center mark	Style of center mark is set up in Drafting Standards under Format
	Center Line	Creates a center line	Right click to get assistance in placing the line
	Center line Bisector	Creates an angle bisector	Select two lines to bisect
	Centered Pattern	Creates a centerline for a circular pattern	
	Surface Texture Symbol	Creates a surface texture symbol	Some options not available in ANSI mode.
	Weld Symbol	Creates a weld symbol	Drafting Standards control the linetype, color, and gap
	Feature Control Frame	Creates a feature control symbol	Typing F can be used to initiate this command
	Feature Identifier Symbol	Creates a feature identifier symbol	NOT available in ANSI standard mode
	Datum Identifier Symbol	Creates a datum identifier symbol	
	Datum Targets	Datum Target with leader	
		Datum Target with circle	
		Datum Target with line	
		Datum Target with point	
		Datum Target with rectangle	
	Text	Creates a text block	Similar to MTEXT
	Leader text	Creates text with a leader attached.	

Button	Tool	Function	Special Instructions
	Add balloon	Adds a reference balloon	Inventor assigns the reference numbers to parts automatically
	Auto Balloon	Adds balloons to all the parts in a view	
	Table	Adds a table	
	Parts List	Creates a parts list	Customize the parts list using property fields
	Hole Table-Selection	Creates a hole table based on selected holes	
	Hole Table-View	Creates a hole table for all holes in a view	
	Hole Table – Selected Type	Creates a hole table based on all holes of a selected type	
	Caterpillar	Creates a caterpillar pattern along an edge to indicate a weld	
	End Treatment	Creates a sketch of a weld end treatment using a hatch pattern or solid fill	
	Revision Table	Places a revision table in your drawing	
	Revision Tag	Places a revision tag in your drawing to indicate the location where the change occurred	
	Symbols	Inserts a sketched symbol into your drawing	A sketched symbol can be any annotation you create
	Retrieve Dimensions	Adds dimensions to a selected view	

Review Questions

1. You can select and delete only one ordinate dimension created using the Ordinate Dimension Set tool.

    A. True
    B. False

2. When placing ordinate dimensions using the Ordinate Dimension Set tool, you can switch between the two axes – that is, place a dimension along the x-axis, place a dimension along the y-axis, then place a dimension along the x-axis.

    A. True
    B. False

3. You can not hide the Origin Indicator placed with the Ordinate Dimension tool.

    A. True
    B. False

4. Inventor only has one style of Datum Target.

    A. True
    B. False

5. To modify a linear dimension:

    A. Select the dimension, right click and select 'Edit Dimension'
    B. Select Edit from the Menu.
    C. Double click on the dimension
    D. Right click in the graphics window and select 'Edit Dimension'.

6. When you place dimensions on elements in a custom title block, you need to hide the dimensions when you insert the title block into a drawing.

    A. True
    B. False

7. The Projected View tool can create an isometric view.

    A. True
    B. False

8. Isometric views are aligned with the base/primary view.

    A. True
    B. False

9. You can use the Design View window dialog to create a base view.

    A. True
    B. False

10. To have model dimensions automatically appear when creating a view:

    A. Enable 'Get Model Dimensions' in the Create View dialog box.
    B. Enable 'Get Model Dimensions' in the Drawing Options dialog.
    C. Enable 'Get Model Dimensions' in the Browser
    D. A & B, but NOT C

11. To place views so they are not aligned, hold down this key as you move and place views:

    A. CONTROL
    B. TAB
    C. SHIFT
    D. ALT

12. The draft view is used to:

    A. Create draft views
    B. Redline a drawing
    C. Create model geometry
    D. All of the above

13. Drawing Resources include all of the items listed below EXCEPT:

    A. Sheet Formats
    B. Borders
    C. Sketch Tools
    D. Title Blocks

14. You can insert the following image type file into a title block:
    A. JPEG
    B. BMP
    C. PCX
    D. GIF

15. You can define property fields for a title block. A Model Properties property fields uses the data stored here:

    A. Under File Properties
    B. Under Model Properties
    C. Under Sheet Properties
    D. Under Design Properties

ANSWERS: 1) A; 2) B; 3) B; 4) B; 5) A; 6) B; 7) A; 8) B: 9) A; 10) D; 11) A; 12) B: 13) C; 14) B; 15) A

Lesson 14
Using the Styles Library

Learning Objectives

At the conclusion of this lesson, the user will be able to:

- Create templates defining custom drafting standards
- Copy one drafting standard to another
- Modify drafting standards

Drafting Styles allow the user to set up standards for parts, assemblies and drawings. Styles control linetype, units, colors, etc.

The Styles are set up under the Format Menu.

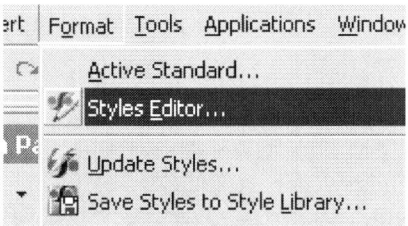

The Styles set up the drafting standards for the active drawing or template. You can apply a drafting standard to the drawing, modify an existing standard, or create a custom standard based on an existing standard.

When you create a new drawing, it is automatically assigned an active drafting standard that controls the format for dimensions, text, line weights, terminators, and other elements that are dictated. You can use the default standard, select from another named standard (ANSI, BSI, DIN, GT, ISO, or JIS), or customize a standard to meet your own requirements.

In an Autodesk Inventor drawing, drafting standard settings also include elements that are not part of the named standard, such as naming conventions for sheets and views, display color for elements on the drawing sheet, format for the parts list, and selection of the special characters and symbols for annotations.

You can change any of the elements controlled by the drafting standard. By customizing an existing standard or developing your own standard, you can enforce company conventions or optimize your working environment.

If you do not use the Style Library to store your standard settings, you can define styles within single documents. However, this method consumes additional memory and increases file size.

>
> **TIP:** The Styles Library acts as an external reference. All documents within a project are assigned to a specific style library which controls dimension styles, layers, linetypes, and colors.

> **TIP:** If you prefer to work without style libraries, set the **Use Style Library** option in your project to **No**. You will still be able to set styles for your individual documents.

By default, all pre-R9 projects are set to NOT use a style library.

Update Styles, Save Styles to Style Library and Purge Styles are all used within a document.

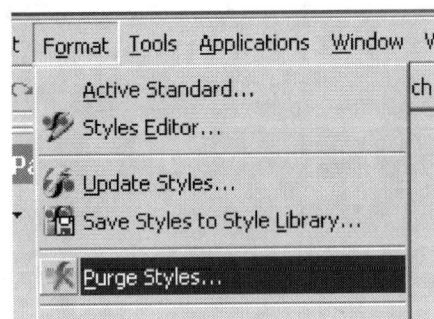

The Update Styles updates the styles inside a document to match the default style set in the Style Library.

The Save Styles to Style Library copies the styles used inside a document to the Style Library.

The Purge Styles deletes any unused styles inside a document from the document. It does not purge the Style Library.

Once your styles have been set up, you can set up template files to use the desired styles.

The template files that are included in Inventor have all the styles stored locally. Templates now specify which styles are to be used by default when a new document is created. To store a style in a document, highlight the style in the Styles Editor, right click and select **Cache in Document**. This is useful if you are sending a document to an outside vendor or team member to ensure they use the correct style without having to transmit the entire Style Library.

When you modify a style, it is not updated automatically. You can update the styles used in a document by accessing **Format→Update Styles** or inside the Styles Editor.

Templates define default styles for new documents. The default styles are listed in the Document Settings.

This is set under **Tools→Document Settings**.

Using the Styles Library

Exercise 14-1:
Creating a Part Template

File: New using Standard (inches)
Estimated Time: 30 minutes

All new part files are created with a template. You can create your own templates and add them to the templates provided by Autodesk Inventor.

1. 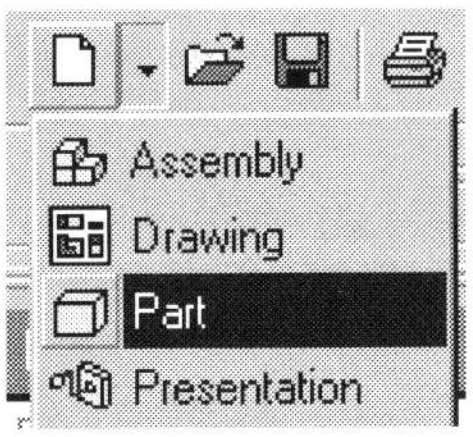 Start a new part file.

2.  Right click in the graphics window. Select **Finish Sketch**.

3. 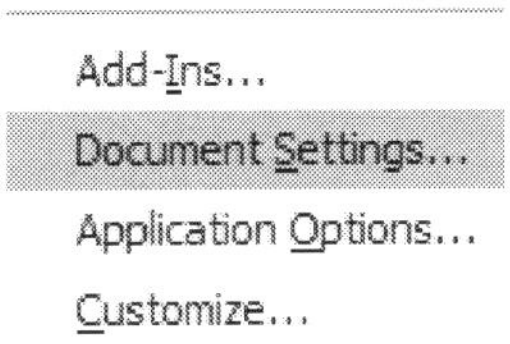 Set the default units of measurement.

   Go to **Tools→Document Settings**.

4. 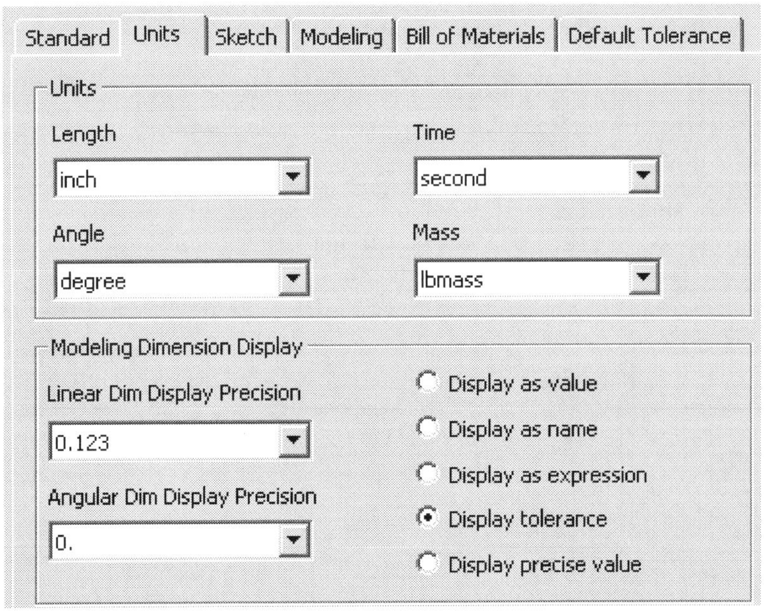 Select the **Units** tab.

   Change the **Angular Dim Display Precision** to **0**.

   This sets the decimal places for angular dimensions.

   Press **Apply** and **OK**.

5. 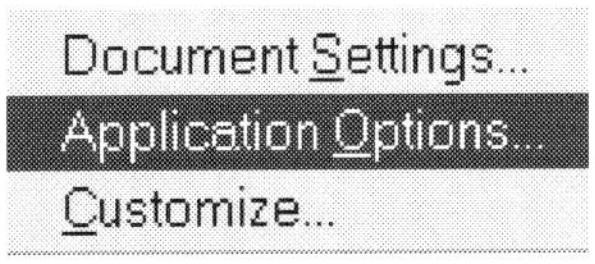 Go to **Tools→Application Options**.

14-3

6. 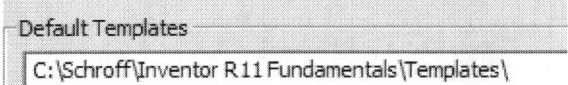 In the **General** tab, change the **Username** to how you want your name to appear in your title blocks.

7. Select the **File** tab.
We can store our custom templates in a custom directory, set the path for our projects, etc.

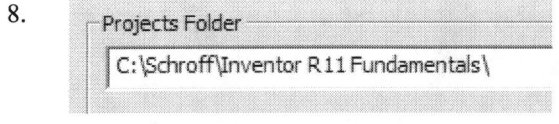 Change the Default Templates folder to your project folder.

NOTE: If you want to use a custom templates folder and store templates away from Inventor, copy the standard Inventor template files over to your templates directory.

8. Change the Projects Folder path to the directory where you will be storing your drawings.

9. Select the **Sketch** tab.

    Enable **Edit dimension when created**.

    Disable **Autoproject edges for sketch creation and edit**.

    Enable **Parallel view on sketch creation**.

    If you prefer to start your sketches on a different plane, enable your preferred plane.

    Or if you would prefer not to start a part model immediately in sketch mode so you can select the desired work plane, enable 'No New Sketch'.

10. 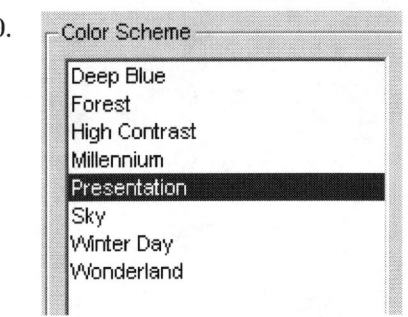 Select the **Colors** tab.

    Select the background color you would like to use for your part modeling environment.

11.
    Select the **Drawing** tab.
    Enable **Retrieve all model dimensions on view placement**.
    Enable **Center dimension text on creation**.
    Press **Apply**.

12. 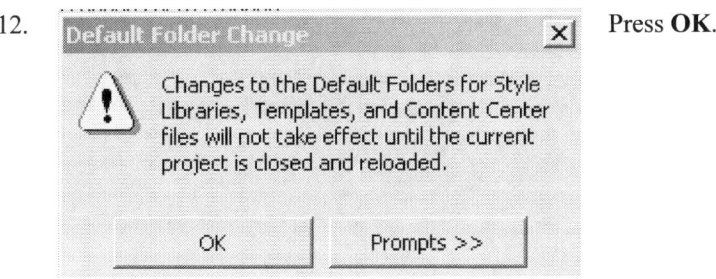 Press **OK**.

13. Press **Close** to close the dialog.

14.  Highlight the sketch in the browser.
    Right click and select **Delete**.

15. 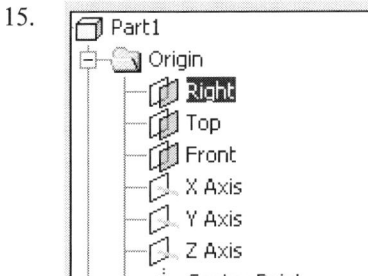 Rename the work planes: **Right**, **Top**, and **Front**.
    To rename simply left click on the name in the browser.

    Using these names is sometimes easier for beginning users.

16. 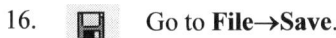 Go to **File→Save**.

17. 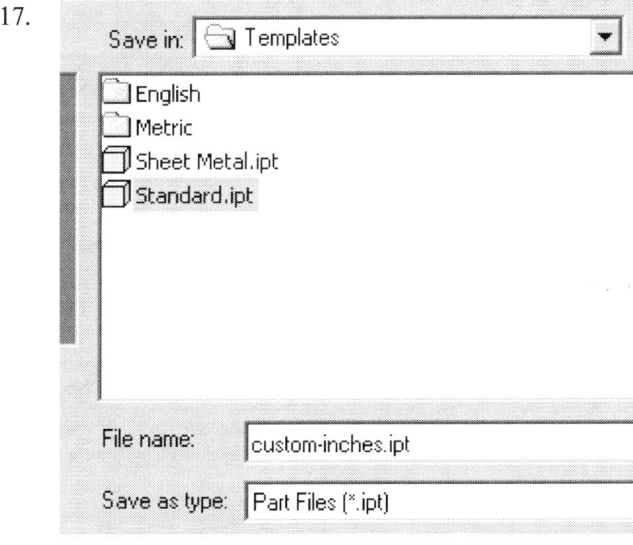 Save the file in the Templates folder of your project (use the same path you specified in Step 7). A part file automatically becomes a template when it is saved to the Templates folder.

    Name your template *custom-inches* and press **Save**.

TIP: The file *standard.ipt* in the Templates folder is the default part template. To replace the default template, remove *standard.ipt* and replace it with a template that has the same name. Then the next time you use File→New, you can access your custom part template.

18. Close and re-launch Inventor.

19. 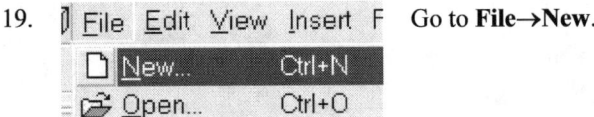 Go to **File→New**.

20. You will see your template in the menu window.

TIP: You can only access templates (and the start-up dialog) using the File→New menu. If you bypass the start-up dialog by using the pull-down tool, the new file uses the templates that begin with *standard*. If you wish to by-pass the start-up dialog, but use custom settings, save your templates as *standard.ipt*, *standard.idw* or *standard.ipn* in the templates directory. It is a good idea to save the original standard.* files in a back-up directory in case you are unhappy with the results.

TIP: To add tabs to the New dialog box, create new subfolders in the Templates folder and add template files to them. The New dialog displays a tab for each subfolder in the Templates folder.

Using the Styles Library

Exercise 14-2:
Creating a Drawing File Template

File: New idw file using Standard
Estimated Time: 60 minutes

Check List for Creating a Custom Drawing Template:

- Title Block for Sheet 1
- Title Block for Sheet 2
- Border for Sheet 1
- Border for Sheet 2
- Special Symbols defined under Sketched Symbols
- File Properties set for Drawing (Format→Active Standard) ; set Author, Company Name, etc.
- Colors set for Drawing (Format→Styles Editor)
- Dimension Values set for Drawing (Format→Styles Editor)

1. 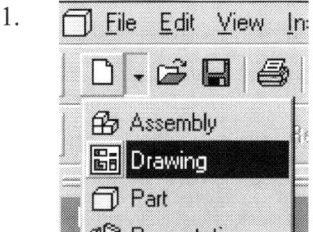 Start a new drawing file.

2. 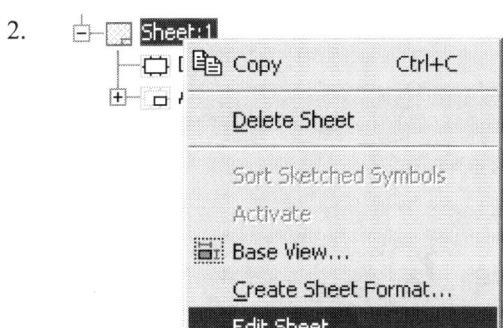 In the Browser, highlight **Sheet:1**.

   Right click and select **Edit Sheet.**

14-7

3. 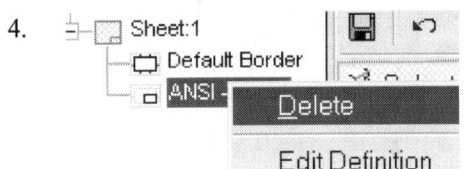 Change the size to **A**.
   Press **OK**.

4. 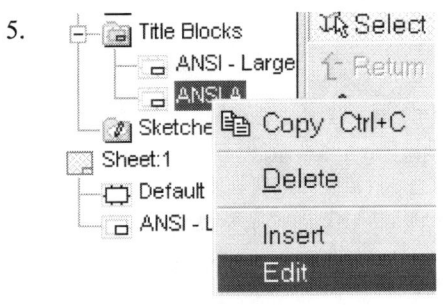 Locate the **ANSI-Large** title block in the browser.

   Right click and select **Delete**.

5. 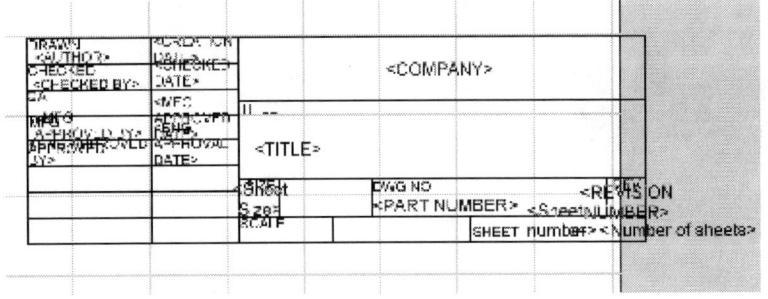 In the browser, under Drawing Resources, locate the **ANSI A** title block.

   Highlight, right click and select **Edit**.

We need to modify this title block so it is smaller.

Using the Styles Library

6. 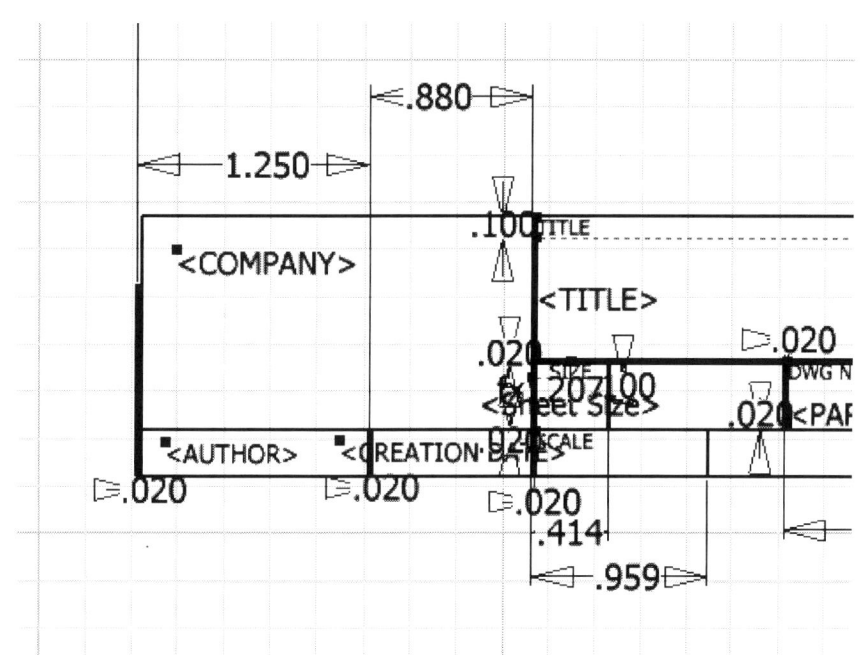   Delete all the signatures except for the **Author** and **Creation Date**.

7. Move the **Company** label above the **Author** and **Creation Date** labels.
   Move the top horizontal line down.
   Delete the unnecessary lines.
   Extend and trim lines as needed.

8. 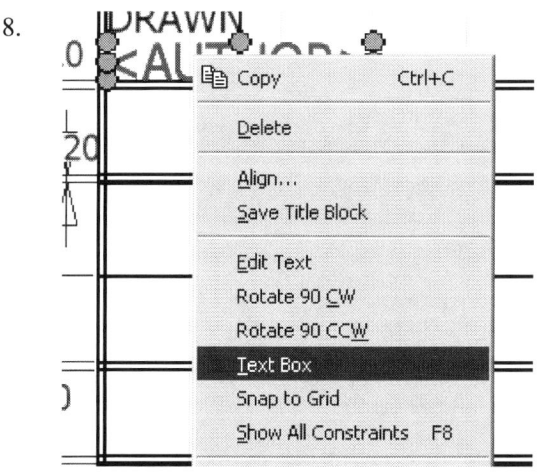   To move a label, select it so that the green grips are visible.

   Right click and select **Text Box**.

9.    A rectangle will appear around the label.

10. Select the **Move** tool.

14-9

11. 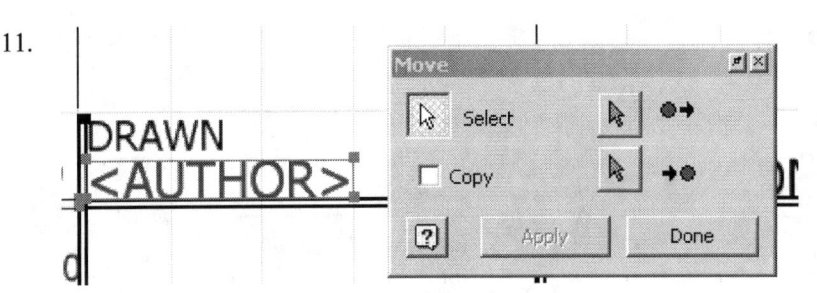 Window around the label and the rectangle to select it.

12. 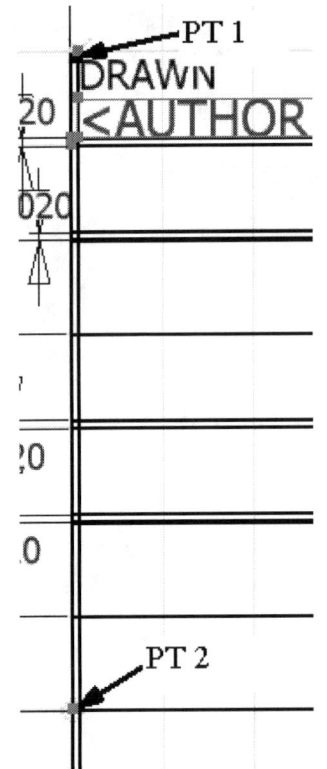 Select the **From** and **To** points to move the label.

    Enable **Copy**.

    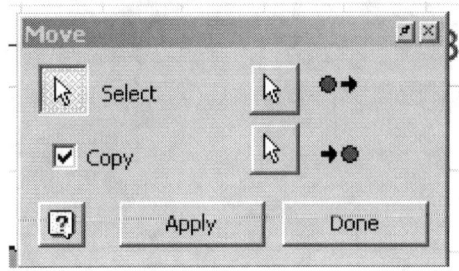

    Press **Apply**.

    If a dialog appears asking to remove constraints, press **No**.

13. The **Author** label is copied into the correct location.

14. Repeat this method to move the **Company** and **Creation Date** labels.

15. Right click and select **Save Title Block**.

Using the Styles Library

16. 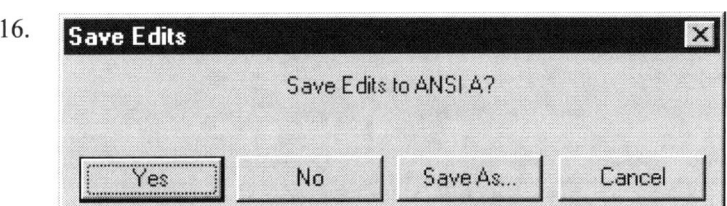 Select the **Save As** option so you don't overwrite the default title block.

17. 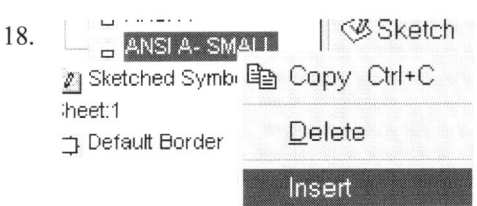 Type **ANSI A- SMALL** as the name. Press **Save**.

18. The new title block appears in your title block.

    Highlight, right click and select **Insert**.

19. 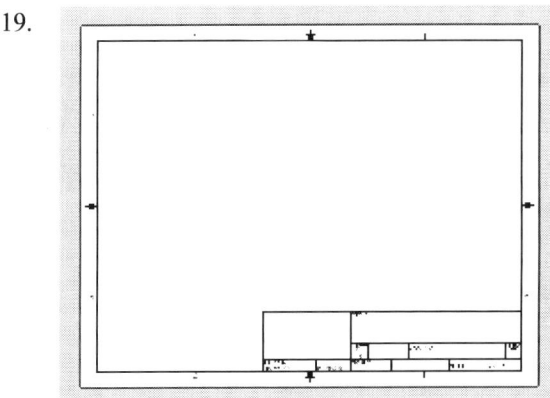 Our new title block provides a lot more space for placing views.

TIP: If you edit the title block and re-save it, the title block inserted in the drawing will automatically update to reflect the changes. You do not have to re-insert it!

14-11

20. 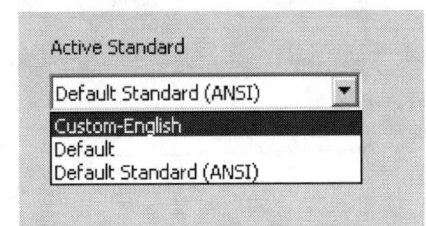 Go to **Format→Active Standard**.

21.  Set the **Active Standard** to *Custom English*.

    NOTE: This was the standard created in the previous lesson. If you skipped that exercise, you will have to go back and do it before you can proceed.

22. Select the **Drawing** tab.

23.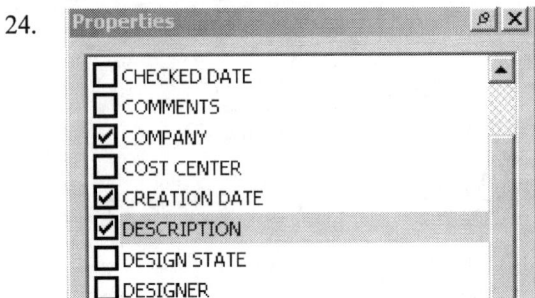

    Under **Properties in Drawing**:
    Enable **Copy Model Properties**.
    Press the **Properties** button.

24. Select the file properties that can be copied from the ipt/iam file to the idw file.

    Enable **AUTHOR, COMPANY, CREATION DATE, DESCRIPTION, PART NUMBER, REVISION NUMBER,** and **TITLE**.

    If you enable **All Properties**, then all the properties will be copied over to the idw file.

25. Press **OK**.

26. Press **Apply** and **OK** to close the dialog.

27. Go to **Format→Styles Editor**.

28.  Select **Import**.

Using the Styles Library

29.  Import the *sdc-style.styxml* file.

    NOTE: This file was created in Lesson 13 or may be downloaded from the publisher's website.

    NOTE: If you get an error when you attempt to perform the import you may be missing some files. Locate the DesignData.exe file in the Design Data folder under Inventor. Double click on the DesignData.exe file to extract all the styles files into the Styles folder under your project.

30. 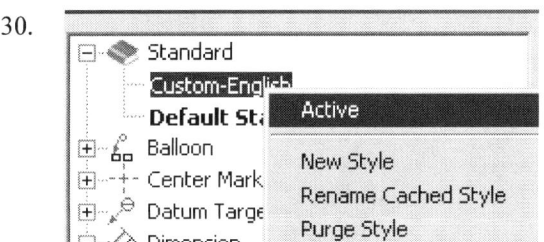 Set the **Custom-English** style as **Active**.

31. Highlight **Objects Default (ANSI)** under **Object Defaults**.

32. Set the **Filter** to **Dimension Objects**.

33. 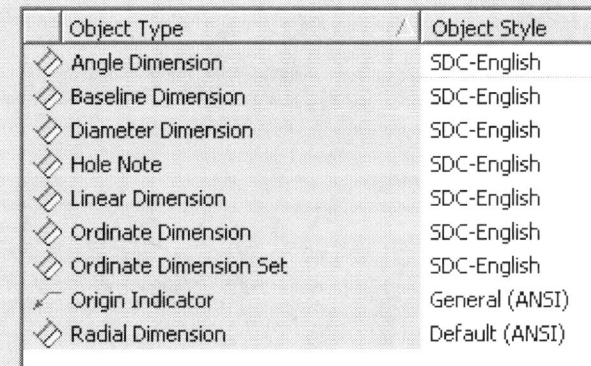 Set the **Object Style** to **SDC-English** for all the dimension types.

34.  Highlight **Objects Default (Default)** under **Object Defaults**.

14-13

Autodesk Inventor Fundamentals

35. 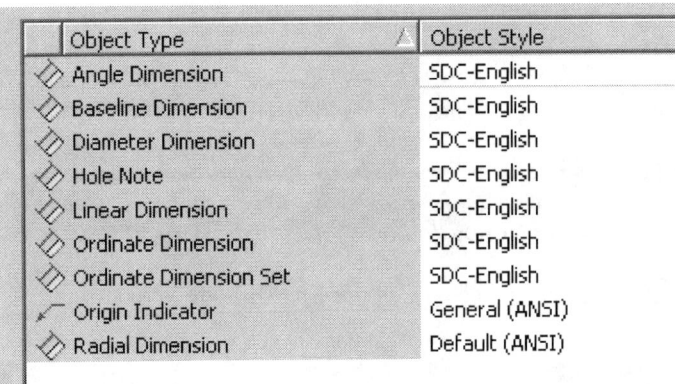 Set the **Object Style** to **SDC-English** for all the dimension types.

36. Expand the **Layers** category.

37.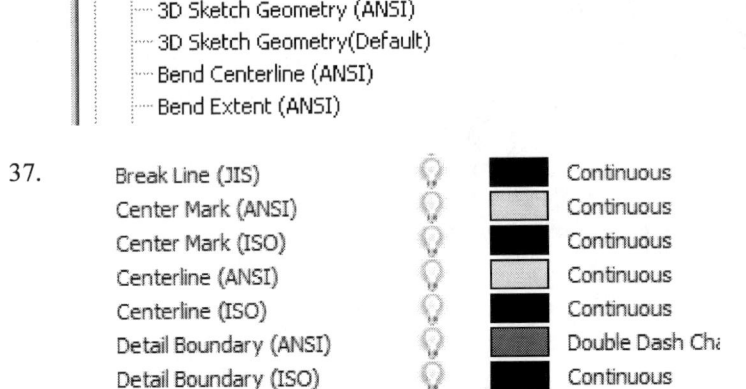

Press the **Color** button to set the colors as shown:

Hidden Edges	Yellow
Section View Lines	Dark Blue
Detail Circle Lines	Magenta
Dimensions	Red
Center Mark	Cyan
Center Line	Cyan
Hatch	Green

38. Press **Save**.

39. Expand the **Parts List** category. Select **PartsList(Default)**.

14-14

40.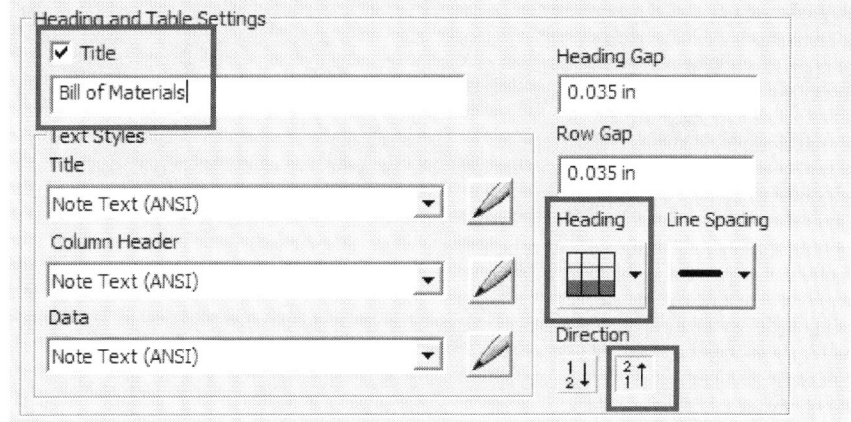

Change the **Direction** to **Ascending**.
Change the **Heading** settings to **Bottom**.
Change the **Title** to **Bill of Materials**.

41.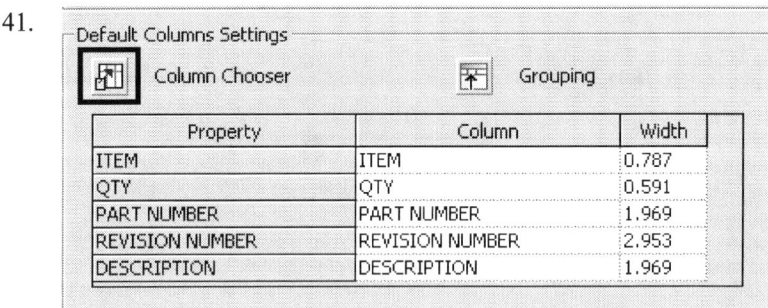

Select the **Column Chooser** button.
Add **Revision Number** to the columns.
Locate the **Revision Number** below **Part Number**.

42. Press **Save** to save the settings for this category.

43. 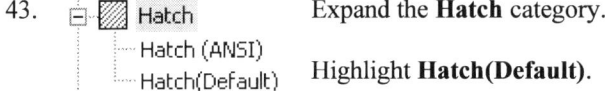 Expand the **Hatch** category.

Highlight **Hatch(Default)**.

44.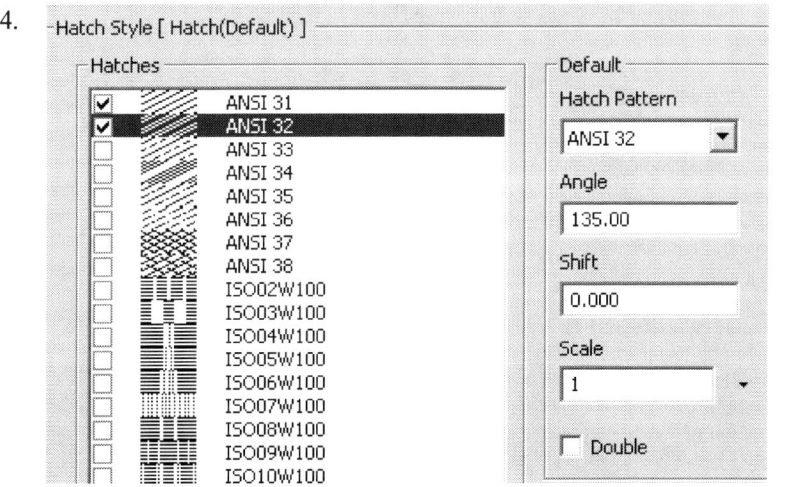

Disable all the hatch patterns except for ANSI 31 and ANSI 32.

Set **ANSI 31** to an angle of **45**.

Set **ANSI 32** to an angle of **135**.

Press **Save**.

14-15

Autodesk Inventor Fundamentals

45. Press **Done** to close the dialog box.

46. 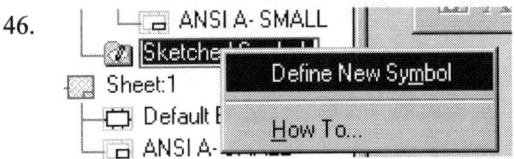 In the **Drawing Resources**, locate **Sketched Symbols**.

    Right click and select **Define New Symbol**.

47. Select the **Text** tool from the Sketch toolbar.

48.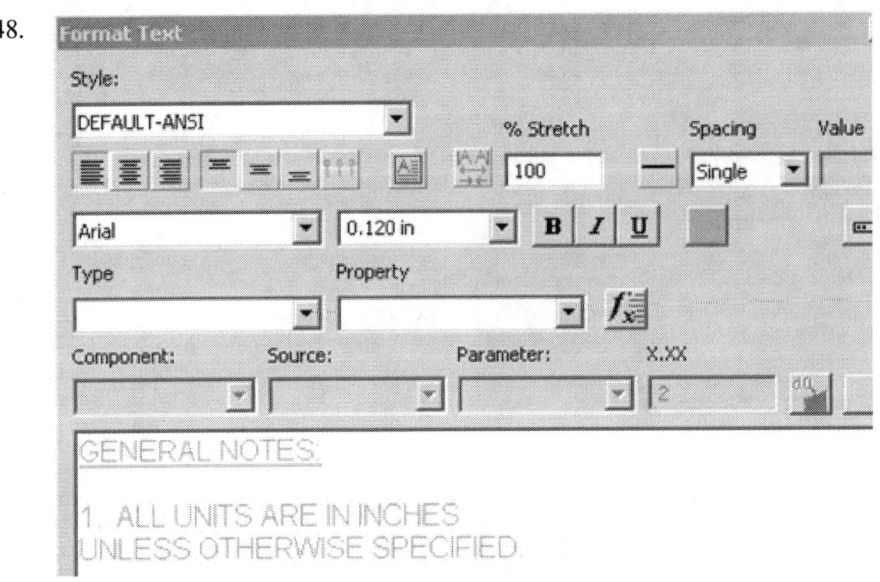

    Create a **General Note** you can apply to your drawings.
    The first note should read:

    1. ALL UNITS ARE IN INCHES, UNLESS OTHERWISE SPECIFIED.

    You may also include a note on part tolerance.

49. Press **OK**.
    Right click and select **Done**.

50.  Right click and select **Save Sketched Symbol**.

51. 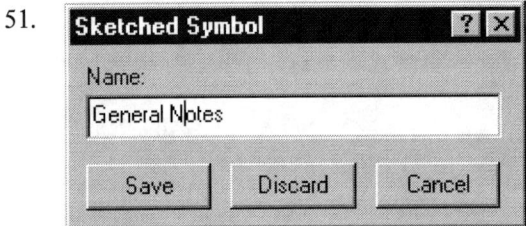 Name your Sketched Symbol **General Notes**.
    Press **Save**.

14-16

Using the Styles Library

52. Your General Notes are now available for insertion into any drawing.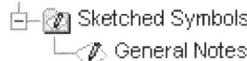

53. 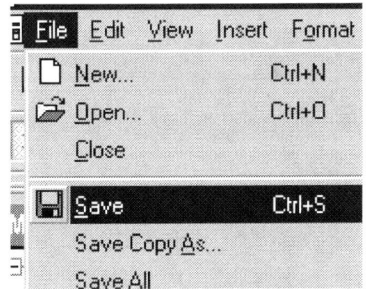 Go to **File→Save**.

54.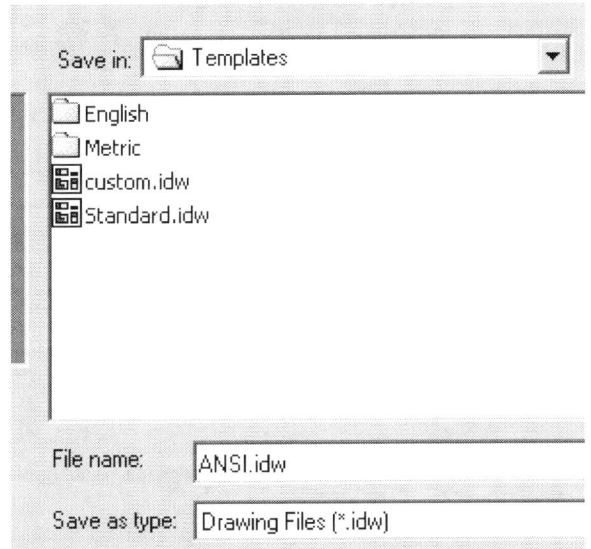

    Locate the **Templates** directory under your project.

    Name your drawing template **ANSI**.
    Press **Save**.

55. Close the drawing file.

56. Go to **File→New** and note that your drawing template is now available.

14-17

Go to **Tools→Document Settings**.

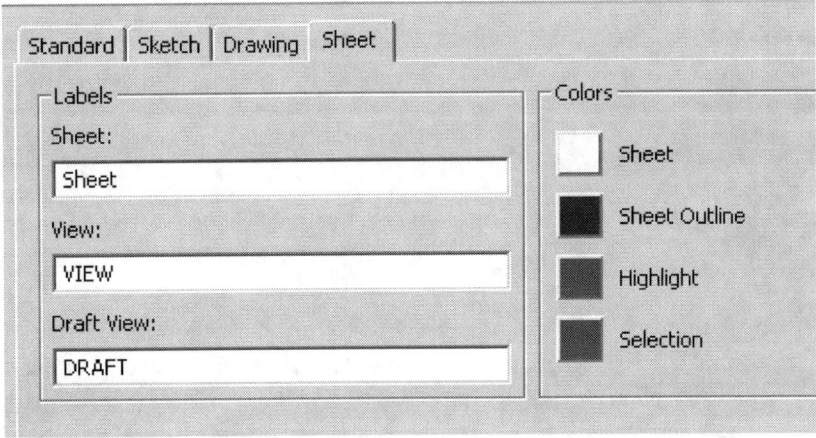

Sheet

Sets the default labels for sheets and views, and sets the colors for elements on sheets in a drawing or template.

Labels	Sets the default labels assigned to new sheets and views in the drawing browser. As a new sheet or view is added, the label is used with an incremented number (for example, Sheet1, Sheet 2, Sheet3). Click in the box and enter the label.
Colors	Sets the display colors for elements of the sheet. Click a color button to open the Color dialog box and select the color for the associated element.
Sheet Color	Sets the background color for the sheet. The color of views, symbols, and other elements does not change; so set a background color that will provide good contrast.
Sheet Outline Color	Sets the outline color for the sheet.
Highlight Color	Sets the color of highlighted elements (when the cursor passes over them).
Selection Color	Sets the color of selected elements.

Exercise 14-3:
Using a Drawing Template

File: New *.idw using ANSI.idw and Ex10-1.ipt
Estimated Time: 30 minutes

1. Go to **File→New**.

2. Select the *ANSI.idw* template we created.

3. Select the **Base View** tool.

4. Locate *ex10-1.ipt* and press **Open**.

5. Enable the **Folded Model** view.

6. Select the **Front** view.

   Set the **Scale** to **1:1**.

   Press **OK**.

14-19

Autodesk Inventor Fundamentals

7. 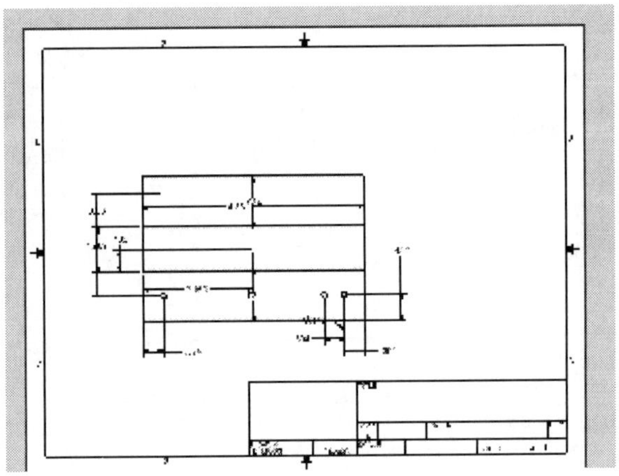 Pick the location for the view.

8. 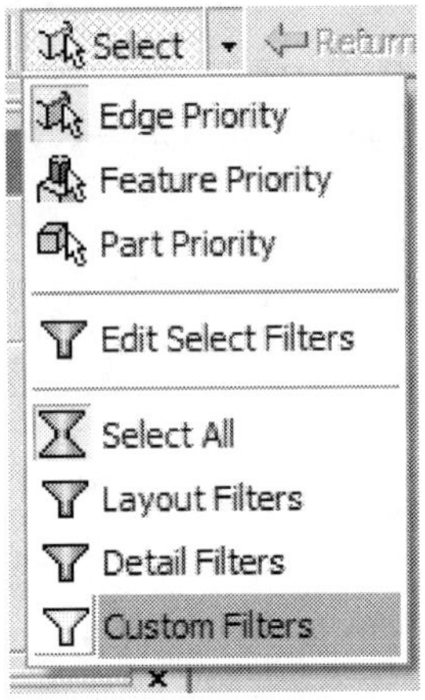 On the **Select** drop-down, select **Edit Select Filters**.

14-20

Using the Styles Library

9.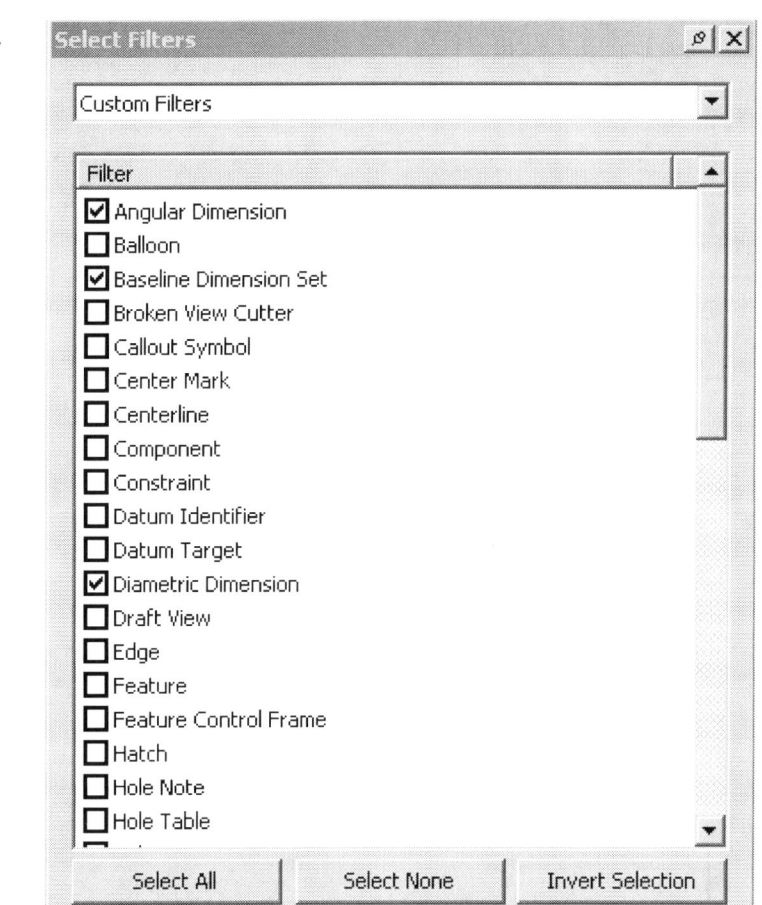

Select **Custom Filters** from the category drop-down list.

Enable all the dimensions in the list.

Press **OK**.

10. Window around the view.

11.  Select **SDC-English** from the **Style** drop-down to set all the dimensions to the SDC-English style.

12. Select the **Broken View** tool.

14-21

13. 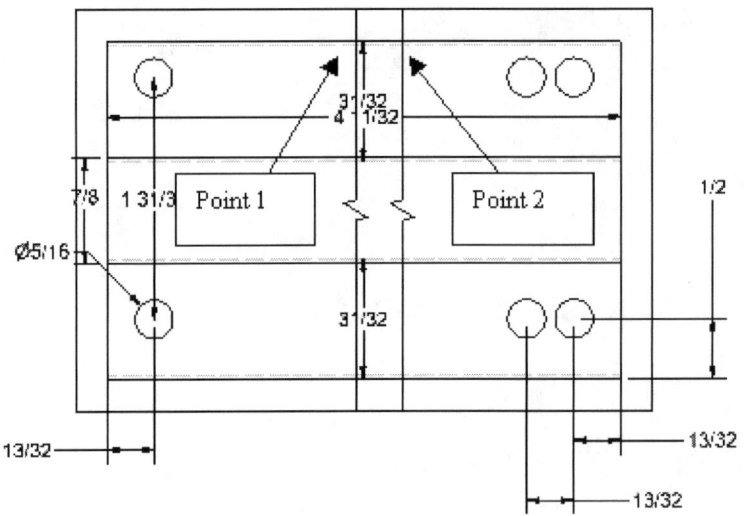 Select the front view you placed for the broken view.

Pick one point to indicate one side of the break point.

Pick a second point to indicate the other side of the break point.

14. 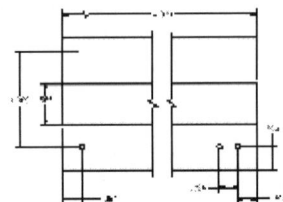 The **Base** view is updated to a **Broken View**.

15. Create a **Projected View** for the right view.

16. 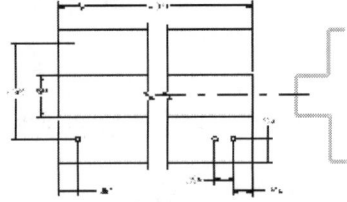 Place a view to the right.

17. 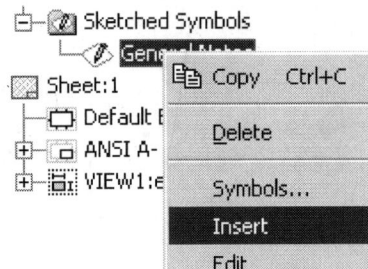 Locate the **General Notes** you created under **Sketched Symbols**.

Highlight, right click and select **Insert**.

18. 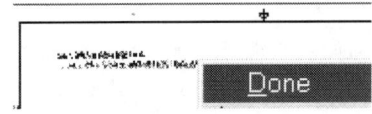 Place the notes in the drawing. Right click and select **Done**.

19. 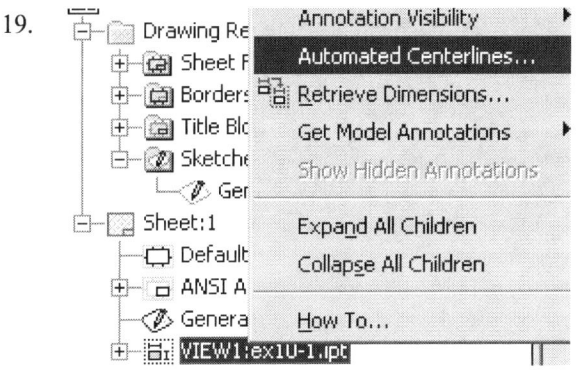 Add a **Center Mark** to all the holes.

20. 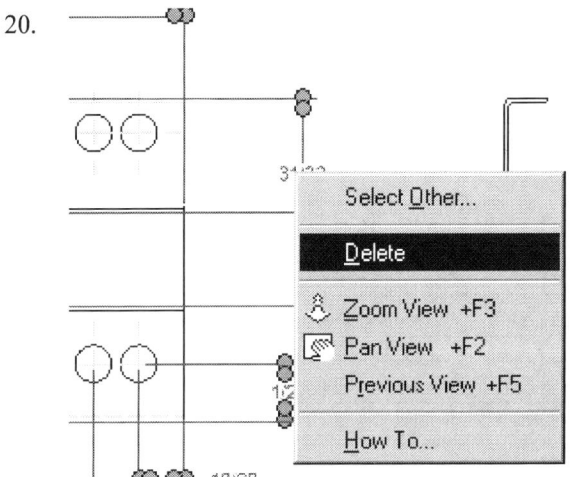 The filter you created is still active. Window around the view to select the dimensions.

    Right click and select **Delete**.

21. Select the **Ordinate Dimension** tool to place Ordinate Dimensions on the front view.

22. 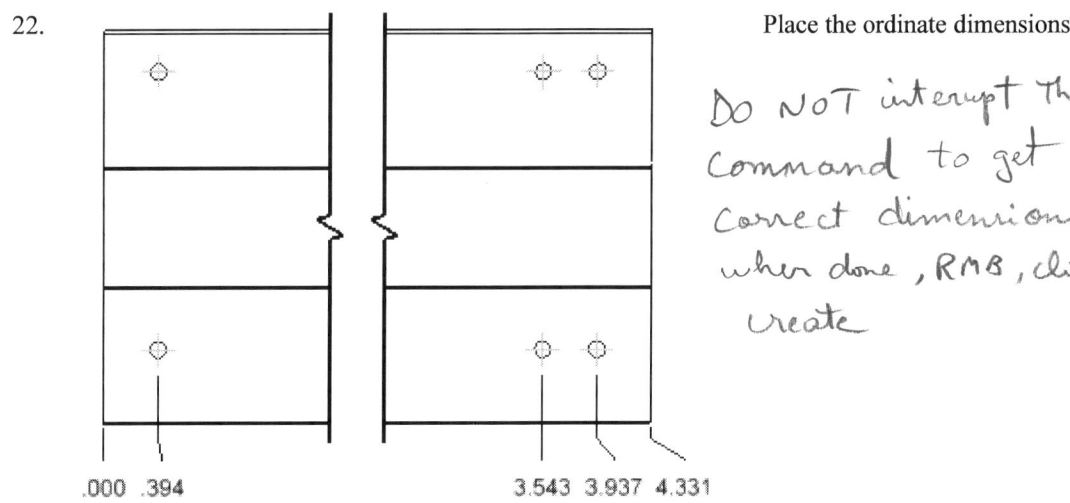 Place the ordinate dimensions.

    Do NOT interupt the command to get the correct dimensions. when done, RMB, click create

23. Select the lower left corner and pick to the left side.
Select the holes and edges going up the vertical side.

Right click and select **Create**.

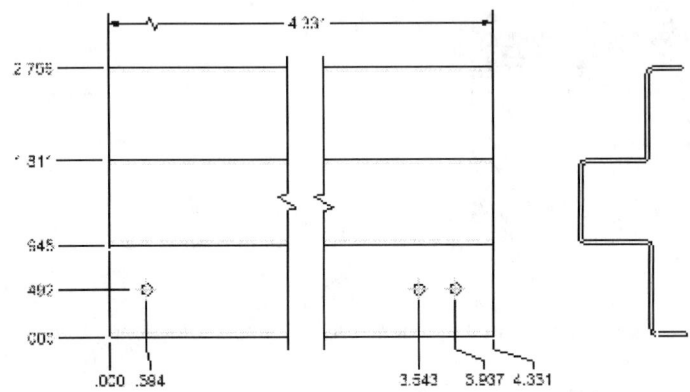

24. Select the **Baseline Dimension** tool.

25. You can place the dimensions and then use the Grips to re-arrange the dimensions the way you want.

26. Go to **File→iProperties**.

27. 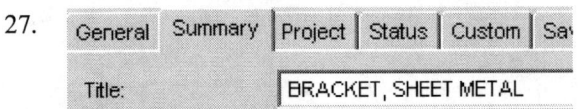 Under the **Summary** tab, enter **BRACKET, SHEET METAL** as the **Title**.

Using the Styles Library

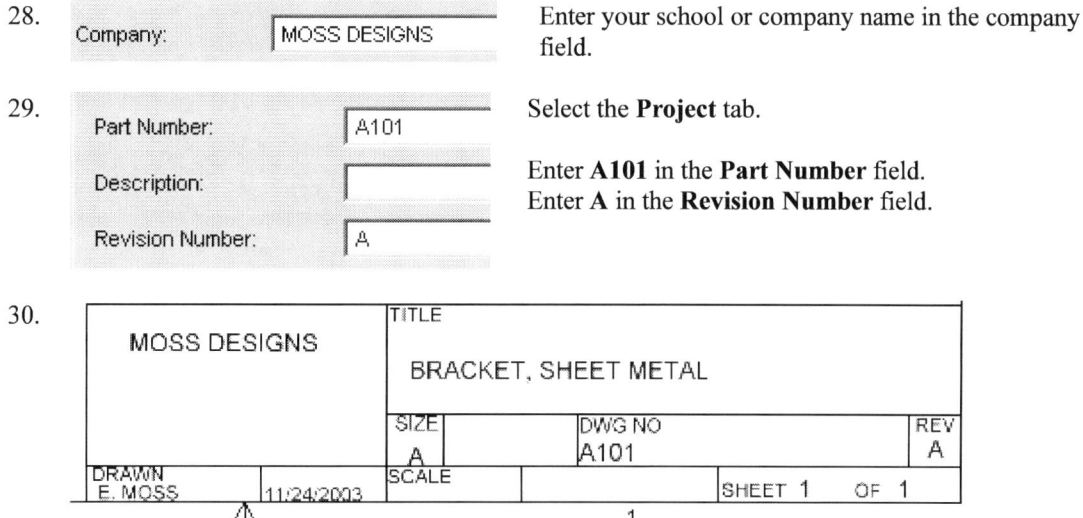

28. Enter your school or company name in the company field.

29. Select the **Project** tab.

    Enter **A101** in the **Part Number** field.
    Enter **A** in the **Revision Number** field.

30. Press **Apply** and **OK**.

    Your title block updates.

31. Save as *ex14-3.idw*.

Review Questions

1. Styles set up standards for parts, assemblies and drawings.

    A. True
    B. False

2. If you have more than one file open at a time, they all must use the same Drafting Standard.

    A. True
    B. False

3. To set up the background color for a drawing sheet:

    A. File→Properties→Custom
    B. Tools→Document Settings
    C. Format→Standards→Sheet
    D. Sheet→Edit

4. To set up the number of decimals to be used in dimensions:

    A. File→Properties→Units
    B. Tools→Options→Design Elements
    C. Format→Styles Editor→Dimension
    D. Format→Dimension Style→Units

5. To set up the color of dimension text to be used in drawings:

    A. File→Properties→Units
    B. Tools→Options→Colors
    C. Format→Styles Editor→Dimension
    D. Format→Dimension Style→Display

6. Custom templates must be saved in this directory:

    A. Support
    B. Projects
    C. Templates
    D. Root

ANSWERS: 1) A; 2) B; 3) B; 4) C; 5) C; 6) C

Lesson 15
Textures and Colors

You can create color styles and use them to change the colors of individual parts for better visibility in an assembly or for design presentations. Color styles interact with lighting styles to control the online display of color in a part or assembly model.

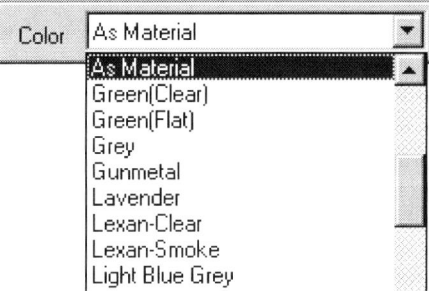

To apply a color style to a part, select the part in the Browser or graphics window, click the arrow next to the Color box on the Command toolbar, and select the color style from the list.

TIP: To use the same color styles in all parts and assemblies, save the color style to the Style Library.

Exercise 15-1
Importing a Colors Style Library

File: Ex5-14.ipt
Estimated Time: 10 minutes

1. Open *ex5-14.ipt*.

2. 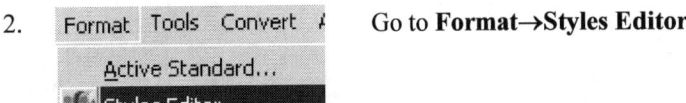 Go to **Format→Styles Editor**.

3.  Expand the **Color** category.
   The colors available depend on what styles library you have set.

   In **Projects** you can set to use your own style library.

4.  Select the **Import** button.

5. 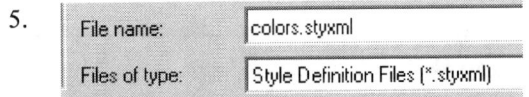 Locate the *colors.styxml* file available from the publisher's website.

   Press **Open**.

6. 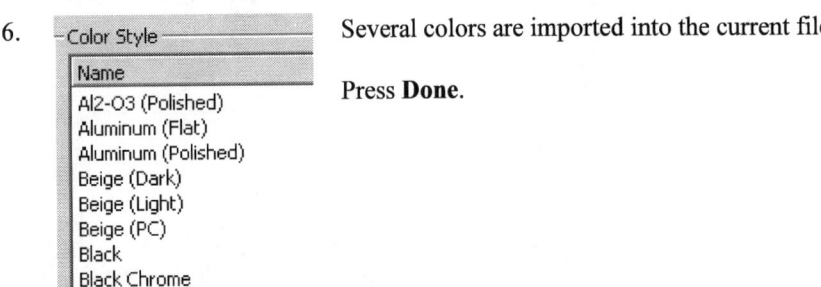 Several colors are imported into the current file.

   Press **Done**.

7. 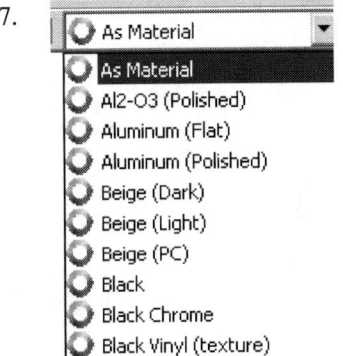 The colors are available from the drop-down list on the **Standard** toolbar.

8. Save as *ex15-1.ipt*.

Exercise 15-2
Adding Color to a Feature

File: Ex15-1.ipt
Estimated Time: 5 minutes

1. 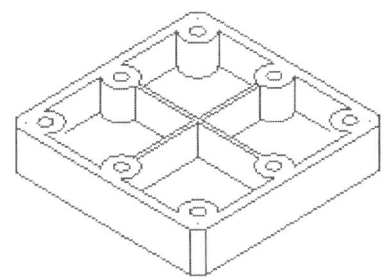 Open *ex15-1.ipt*.

    We see several features.

    Sometimes we want to apply a color to just a single feature or feature type to make it easier to locate or identify.

2. 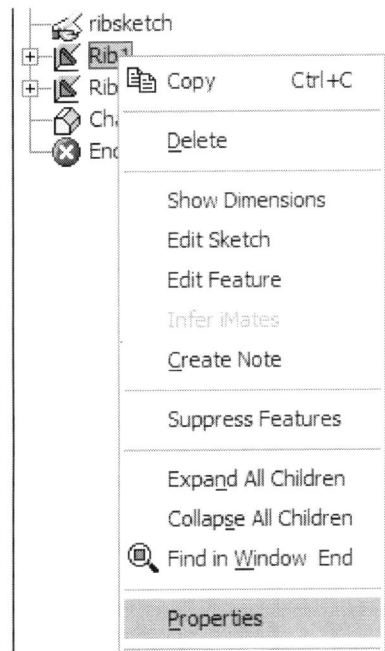 Locate the first **Rib** in the Browser.

    Highlight, right click and select **Properties**.

3.  Change the name of the rib to **Rib-Horizontal**.

4. Locate the **Feature Color** of **Metal-Bronze** under the drop down list.

    Press **OK**.

5. Save the file as *ex15-2.ipt*.

Autodesk Inventor Fundamentals

Exercise 15-3
Changing the Color of a Part

File: Ex15-2.ipt
Estimated Time: 15 minutes

1. 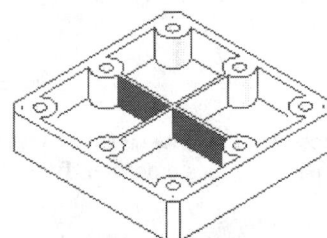 Open *ex15-2.ipt*.

   We see several features.

   We can also apply a color to an entire part.

2. 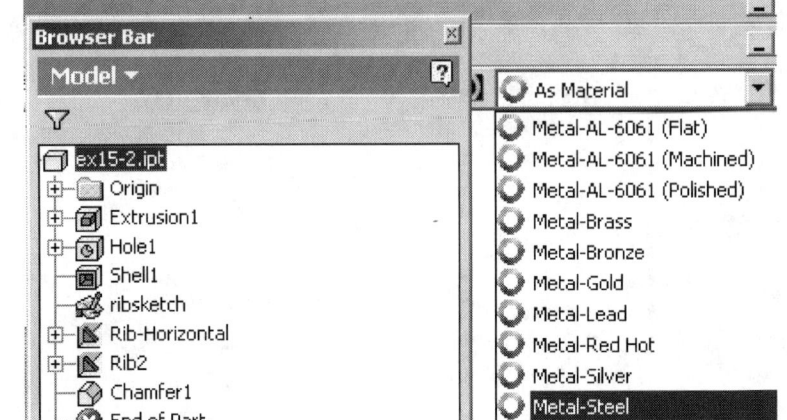 Highlight the part name in the Browser so that the entire part is selected.

   Locate the color '**Metal-Steel**' in the drop down list.

   Highlight the desired color and release.

   Left pick anywhere in the graphics window to complete the operation.

   You will see the entire part change color, except for any features where you have applied specific color/materials.

3. Save as *ex15-3.ipt*.

 TIP: You must be in shaded display mode in order to see any applied colors or materials.

Creating a Texture

One of the big features on many users' Wish Lists was the ability to apply textures to features and parts in Inventor.

Autodesk Inventor comes with a library of surface textures. You can also define, store and apply your own textures. A texture that you define can be especially useful when, for example, you'd like to represent a series of holes or slots in a part but there is no need to model each hole or slot for manufacturing or analysis purposes. For example, for a part such as a grill or a screen, it may be more convenient or economical to represent each hole or slot as a texture rather than a modeled feature. To that end, you can create textures with transparent pixels so that a part will display as transparent wherever those transparent pixels appear.

15-4

Textures and Colors

Defining your own textures gives you the flexibility to precisely customize the texture color, pattern and size. You can also adjust the texture map pixel measurement and dpi resolution to create texture patterns that match the scale of the parts to which the textures are applied.

When applying a texture to a part, keep in mind that the texture color combined with the part color displays as the product of those colors. For example, a texture with a color that is primarily a dark gray will, when combined with a part color, which is a light blue, display as a darker blue, relative to those two shades. Combining a texture with a color that is primarily red with a part color of green will display as black, just as if you viewed a red object through a green filter.

If you want the texture color to display on the part with exact fidelity to the texture color as it appears in the preview window of the Texture Chooser dialog box, you must set the part color to white. Set each color property in the Colors field, on the Color tab, of the Colors dialog box to White. Then apply a texture.

Exercise 15-4:
Creating a Texture

File: New using Standard.ipt (inches)
Estimated Time: 15 minutes

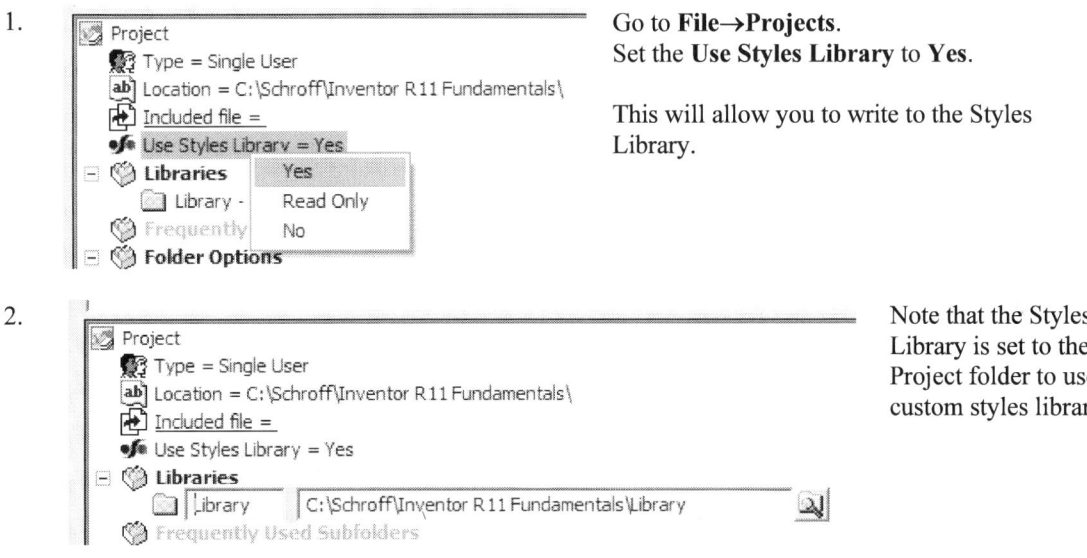

1. Go to **File→Projects**.
   Set the **Use Styles Library** to **Yes**.

   This will allow you to write to the Styles Library.

2. Note that the Styles Library is set to the Project folder to use a custom styles library.

3. Save the changes to the project.

4. Open a **New** part file.

5. 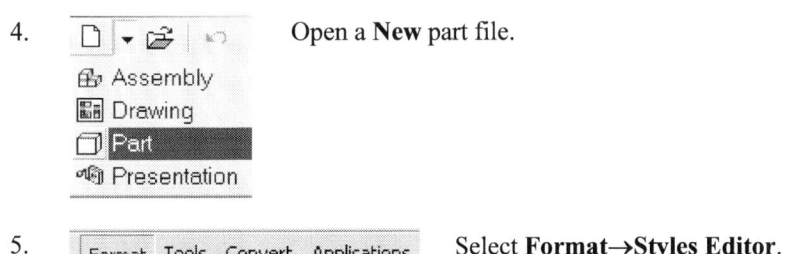 Select **Format→Styles Editor**.

15-5

Autodesk Inventor Fundamentals

   Note that the library is set to **Read/Write**.

6.    Set the Filter to **All Styles**.

If you only see one color, you will have to import the styles library file.

Select **Import**.

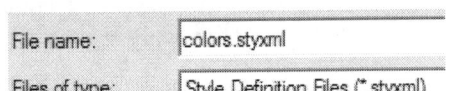   Select the *colors.styxml* file available for download from the publisher's website.

7. Highlight **White** under the **Colors** category.

8. Press the **New** button.

9. Under **Style Name**, type **Knurl**.

10.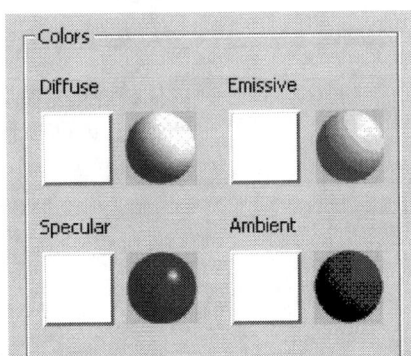

Pick the rectangle titled **Diffuse**.

The Color dialog will appear.
Select **White**.

Set all the rectangles (Specular, Emissive, Ambient, and Diffuse) to White.

11.    Under **Texture** click the **Choose** button.

15-6

Textures and Colors

12. Under **Texture Library**, select **Knurl_1.bmp**. Notice that you can control the density of the Knurl pattern by moving the slider on the left.

Set the slider so it is around **2**.

Knurl_1   Click **OK**.

13. Press **Save**.

14. The **Knurl** definition has now been added to your **Style Name** list.

   - Gunmetal (Antique/Polished)
   - Gunmetal (New/Polished)
   - Knurl

15.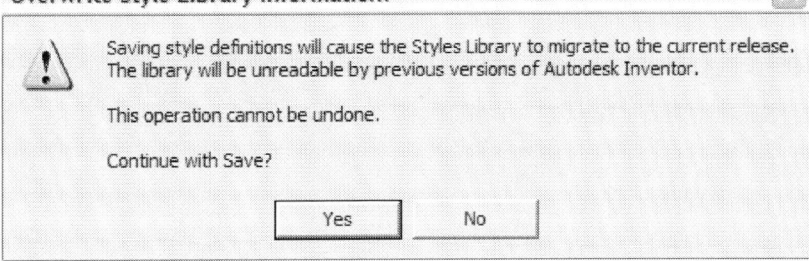

Highlight the Knurl material.

Right click and select **Save to Styles Library**.

16. Press **Yes**.

17. Click **Done**.

18. Exit sketch mode.
    Save as *ex15-4.ipt*.

Once you save the new material to the Styles Library, it will be available for any drawing you open or create.

15-7

Review Questions

1. When you create a color style, it is only available in the file where it is created.

    A. True
    B. False

2. To copy a color style from one file to another, use Insert.

    A. True
    B. False

3. Textures have the best appearance when you set the color to white.

    A. True
    B. False

4. Colors can be applied to the entire part only, not to individual features.

    A. True
    B. False

5. It is possible to apply a different color to each face on a part. features.

    A. True
    B. False

6. To modify Color Styles, access the color/material from the Command bar.

    A. True
    B. False

7. You can apply a color style to a part or feature by selecting the part or feature in the graphics window.

    A. True
    B. False

ANSWERS: 1) A; 2) B; 3) A; 4) B; 5) B; 6) B; 7) A

Lesson 16
Assembly Tools

Learning Objective

At the conclusion of this lesson, the user will have a good overall understanding of the tools used in constraining and managing assemblies.

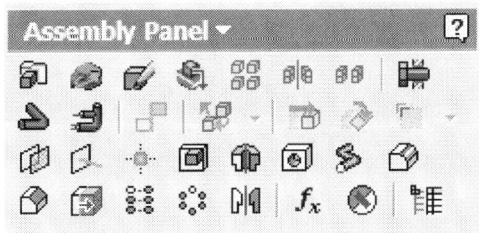

The Assembly tool bar is divided into three sections: Component, Assembly Constraints, Component Management, and Component Viewing.

NOTE: The toolbar shown is for Inventor Professional. Some of the tools are not available in Inventor Standard.

Assembly components can be individual parts or subassemblies that behave as a single unit. For example, a single-part base plate and a multi-part air cylinder subassembly are both components when placed in an assembly.

To make sure that they are always available when you open the assembly, add the paths for all components to the project file for the assembly.

The behavior and characteristics of a component depend on its origin.

A Mechanical Desktop part placed as a component in an Autodesk Inventor assembly acts much like any assembly component. You can add assembly constraints, set its visibility, and perform other assembly operations. However, you cannot edit the part in Autodesk Inventor.

Each Mechanical Desktop part is linked to the assembly through a special file called a proxy file. The proxy file contains the linking information so that the assembly component updates when you edit the part in Mechanical Desktop.

TIP: If you make extensive changes to any component in an assembly, some assembly constraints may not compute correctly when the Autodesk Inventor assembly file is updated. These constraints must be recreated.

Parts or subassemblies created using another CAD system can be inserted as components in the active assembly. You cannot change the size or shape of external components, but you can customize them by adding features.

Adaptive parts can change size and shape to satisfy assembly design requirements. When an adaptive part is constrained to other assembly components, underconstrained geometry in the adaptive part resizes.

When a part is first placed in an assembly, it is not defined as adaptive in the assembly context. You can create fixed-size geometry, and then place the part in an assembly. Select one occurrence in the assembly and designate it as adaptive.

Most assemblies contain a combination of existing components and components (parts and subassemblies) created in the assembly environment.

When you create components in place, you can use geometry from other parts (such as edges and hole centers) in feature sketches. Parts based on existing geometry are sized and positioned in relation to that geometry. Parts created in place have an automatic mate constraint applied between the part XY sketch plane and the part face you sketch on. You can define a part created in place as adaptive so that its size and shape can adjust as assembly requirements change.

Any part in an assembly may have all of its degrees of freedom removed and be fixed in position, relative to the assembly coordinate system. The origin of a grounded part will not move when you place assembly constraints, but a grounded part can still be designated as adaptive. The features on a grounded, adaptive part can change size or shape although its position is fixed.

TIP: The first component placed in an assembly is automatically grounded, so that subsequent parts may be placed and constrained in relation to it. If necessary, you can remove the grounded status of a part.

Placing the First Component

The first component placed in an assembly should be a fundamental part or subassembly, such as a frame or base plate, on which the rest of the assembly is built.

The first component in an assembly file sets the orientation of all subsequent parts and subassemblies. The part origin is coincident with the origin of the assembly coordinates and the part is grounded (all degrees of freedom are removed).

If necessary, you can restore degrees of freedom to the grounded part (the base component) and reposition it. Any components you have constrained to it will also move.

Although there is no distinction in an assembly between components, you can think of the first component you place as the base component because it is usually a fundamental component to which others are constrained. If you place a first component and then want to change to a different base component, you can place a new component, specify it as grounded, and then reconstrain any components you placed earlier, including the first component. Right-click on the first component, clear the Grounded check box, and then constrain it to the new base component.

There is no limit to how many components can be grounded, but most assemblies have only one grounded component. Grounded components are appropriate for fixed objects in assemblies because their position is absolute (relative to the assembly coordinate origin) and all degrees of freedom are removed. Grounded components have no dependencies on other components.

You can use the Move button on the Assembly toolbar to relocate the grounded component. You can drag the component to its new position. The component is grounded in the new location.

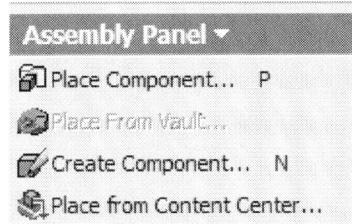

Assemblies can also be managed from the Menu. You can place an Existing Component, Create a New Component, Access the Standard Parts Library using Shared Content, Pattern a Component, or Add a Constraint under the Insert Menu.

Place Component

The first component in an assembly is automatically positioned with its origin coincident with the assembly coordinate origin. Additional components are positioned with the cursor, attached at the component center of gravity.

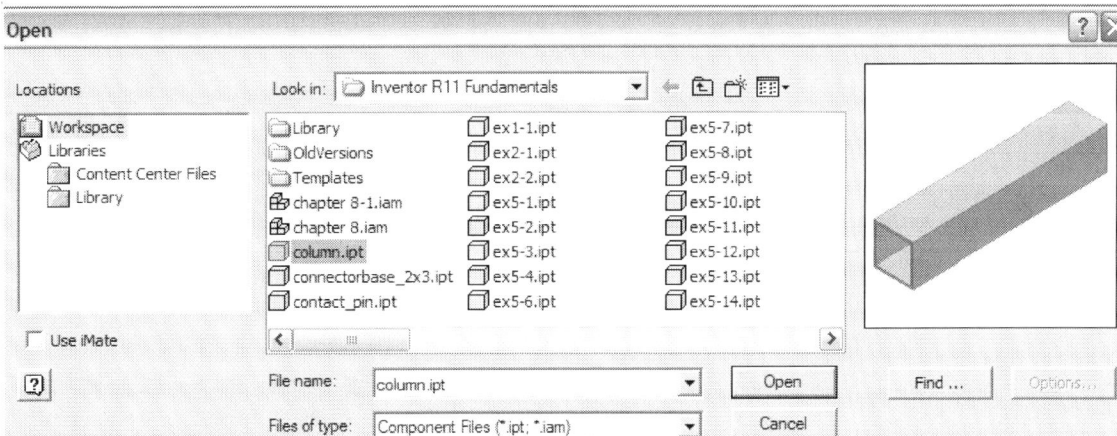

Autodesk Inventor Fundamentals

Menu	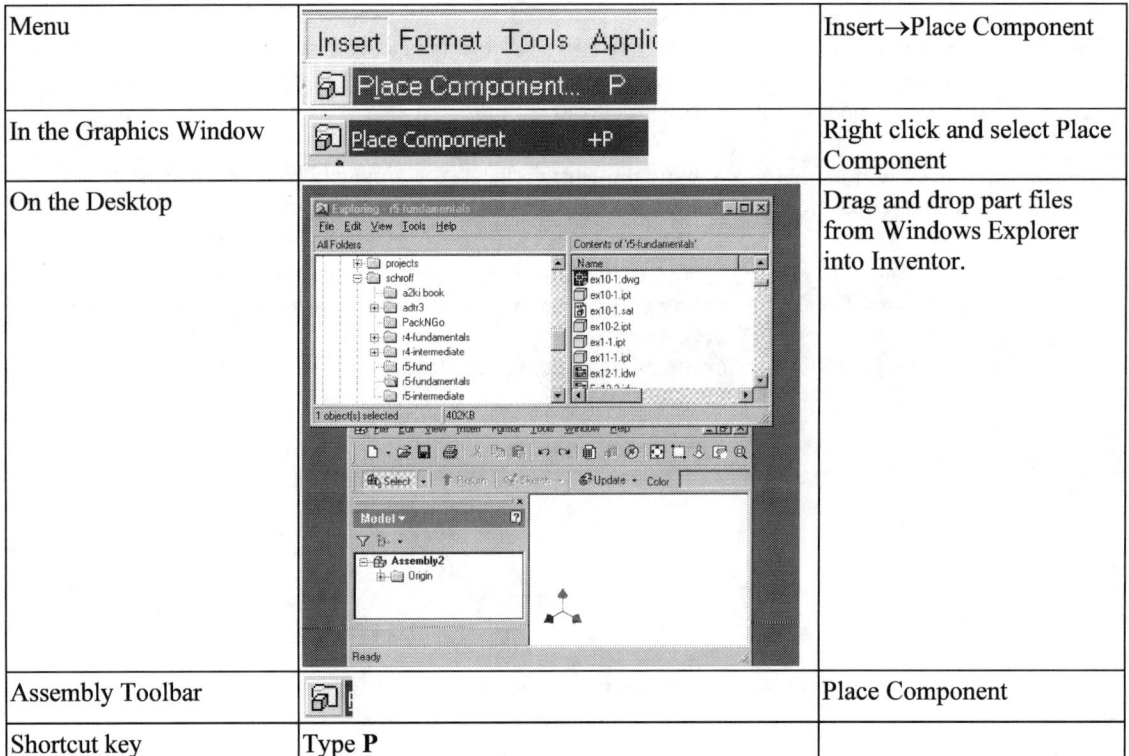	Insert→Place Component
In the Graphics Window		Right click and select Place Component
On the Desktop		Drag and drop part files from Windows Explorer into Inventor.
Assembly Toolbar		Place Component
Shortcut key	Type **P**	

TIP: A mate constraint is automatically placed between the new sketch and the face or work plane. To omit this constraint, clear the check box in the Create Part In-Place dialog box when you create the part file.
You can set options on the Adaptive tab of the Options dialog box to control feature termination.

Assembly Tools

Exercise 16-1
Place Component

File: New Assembly file using Standard (mm)
Estimated Time: 10 minutes

1. 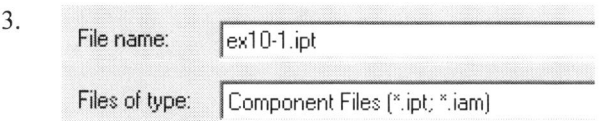 Use **File→New** so you can access the metric tab.
   Start a New Assembly file.

2. Select the **Place Component** tool from the Panel Bar.

3. Locate the file *ex10-1.ipt* created in Lesson 10.

   Press **Open**.

4. 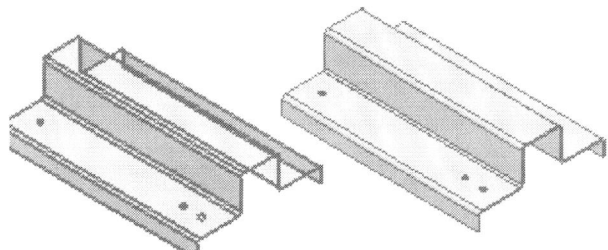 Inventor automatically assumes you want to place more than one instance of each selected part.

   Notice how the cursor image changes to indicate that you are in Part Placement mode.

   Left click the mouse to insert the part into the assembly.

5. After placing one instance of the part, right click and select **Done**.

6. 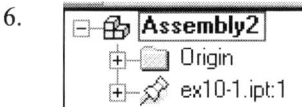 The first part placed is automatically grounded. This is indicated by the pushpin next to the part name in the Browser.

   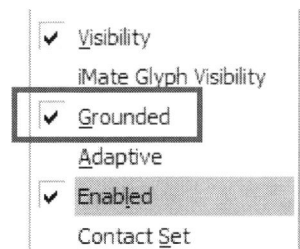 Grounding means the part is fixed in place and cannot be moved.

   Remove the ground.

   Highlight the part name, right click and disable **Grounded**.

7. Save the assembly file as *Ex16-1.iam*.

 TIP: It is a good idea to place parts and subassemblies in the order in which they would be assembled in manufacturing.

16-5

Autodesk Inventor Fundamentals

Exercise 16-2
Place Component

File: New Assembly file using Standard (inches)
Estimated Time: 10 minutes

1.  Start a New Assembly file.

2. Open Windows Explorer.

   Locate the path:
   Program Files/Autodesk/Inventor 11/Samples/Models/Assemblies/Tuner/Tuner Components

3. 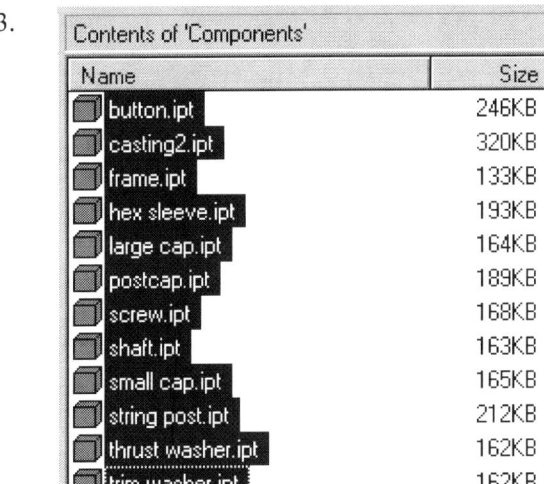 Select all the components in the **Tuner Components** folder.

You can use **Edit→Select All**, Shift-Control, or hold down the Control key and pick to select all the files.

4.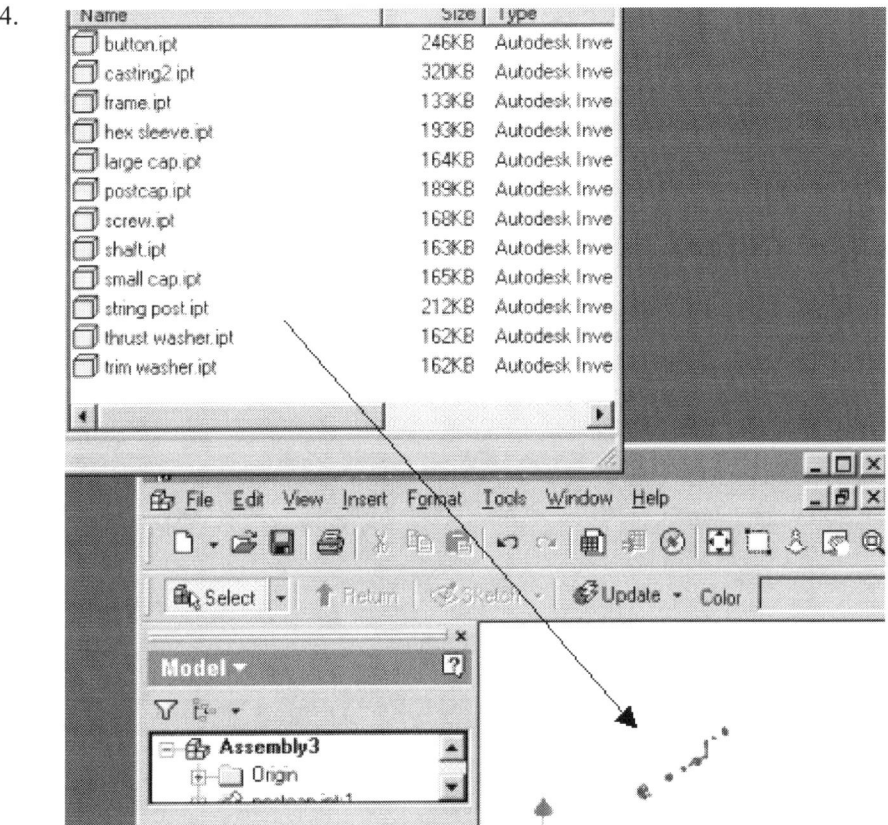

Hold down your left mouse button, drag and drop all the files into your assembly file.

5.  All the parts are automatically inserted.

6. Save your file as *Ex16-2.iam*.

7. 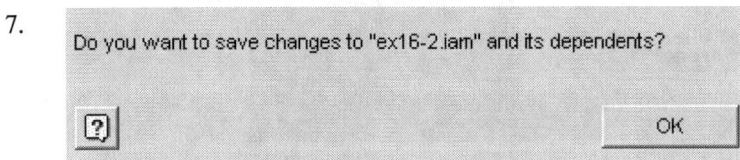 You may be asked if you want to save all the dependent files along with the assembly.

   Press **OK**.

8. You may be asked if you want to migrate the files from the previous release. Press **OK**.

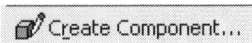

Create Component

You can create a new part in the context of an assembly file. Creating an in-place part has the same result as opening a part file, with the additional option of sketching on the face of an assembly component or an assembly work plane. To allow the size of the new part to change with assembly requirements, you can designate the part as adaptive and constrain it to fixed geometry in the assembly.

(create part in the assembly (attached like a child))

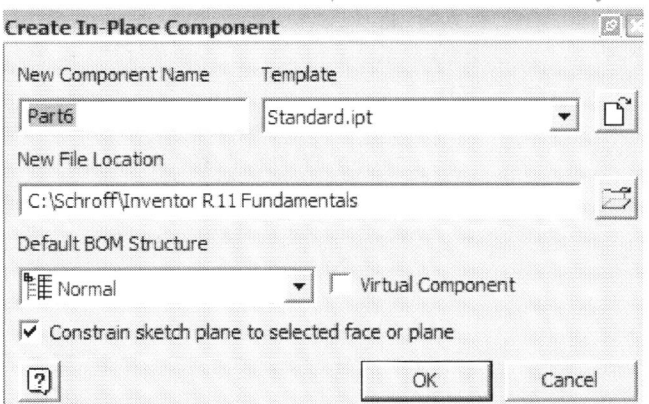

New File Name	Name of the file for the new component
File Type (Part / Assembly / Part)	Part – defines part as a single component Assembly – defines part as a subassembly
New File Location	Set the subdirectory where the file is to be saved.
Standard.ipt (Sheet Metal.ipt / Standard.iam / Standard.ipt)	Template sets the template to be used for the new part file.
Default BOM Structure (Normal / Inseparable / Purchased / Phantom / Reference)	Determines how the new component will be listed in the parts list. Normal – treats the component as a standard part Inseparable – assemblies that are welded or glued together Purchased – vendor supplied parts Phantom – parts that are not listed as line items on the bill of materials. This could be silkscreens, labels, manuals, or small hardware components Reference – components that are placed in an assembly for visualization purposes, such as a table or a wall.
Virtual Component	A virtual component has no geometry and no part file. An example of a virtual component might be an adhesive or a grease. Use this option if you wish to have the component listed on the parts list, but do not need it to be modeled.
Constrain sketch plane to selected face	Enabling will automatically add a mate constraint for the new part/ subassembly Disabling means that the part will be "free-floating" and will need to be constrained later

16-9

Autodesk Inventor Fundamentals

Menu	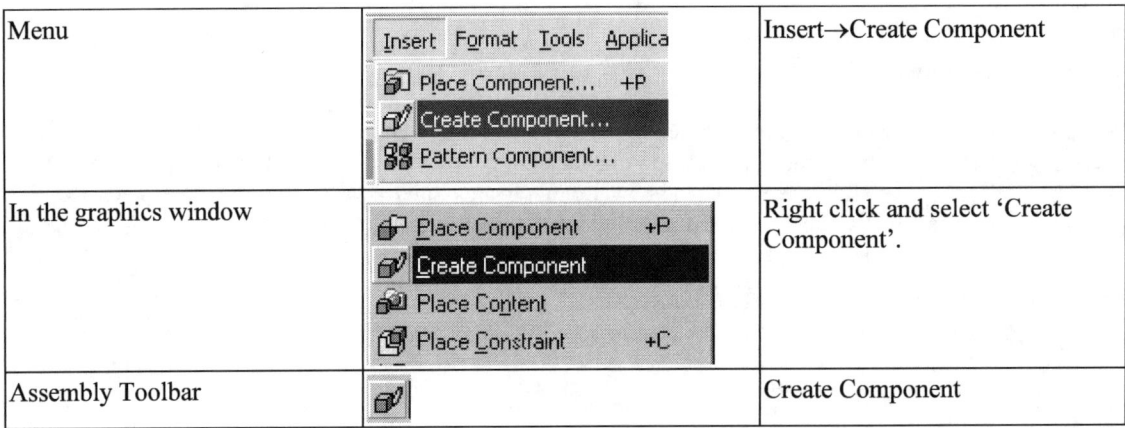	Insert→Create Component
In the graphics window		Right click and select 'Create Component'.
Assembly Toolbar		Create Component

Exercise 16-3
Create Component

File: Ex16-1.iam
Estimated Time: 20 minutes

1. Open *ex16-1.iam* file.

2. 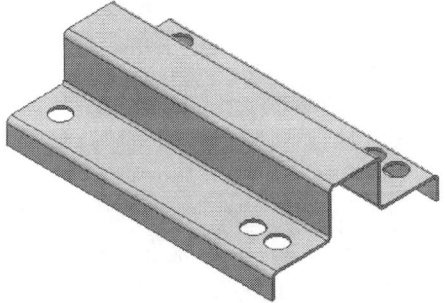   We'll create a rail to mount the bracket on.

3. Select the **Create Component** tool.

4.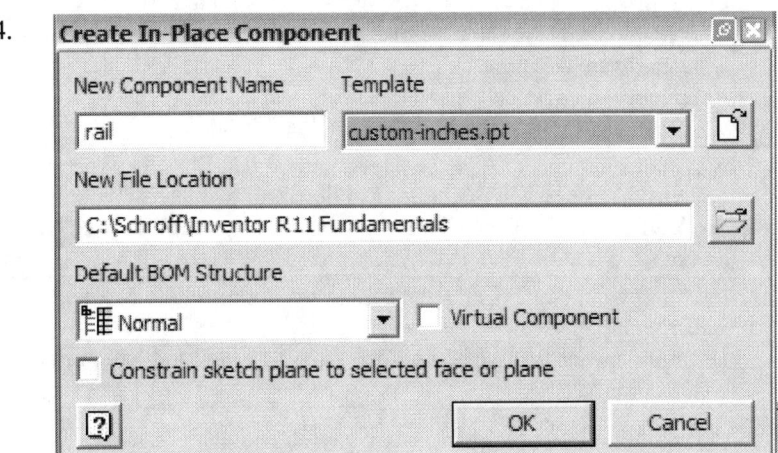

Enter *rail* for the new file name.

Set the template to use ***custom-inches.ipt***. It should be available from the drop-down list.

Disable the **Constrain** sketch plane option.

Press **OK**.

16-10

Assembly Tools

5. 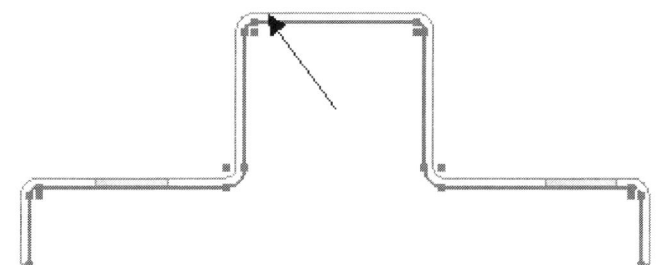 Select the end face as the sketch plane.

6. Use the **Project Geometry** tool to project the bracket edges onto the current sketch.

   Project the inner edge of the bracket.

7. Draw a horizontal line below the projected geometry sketch.

   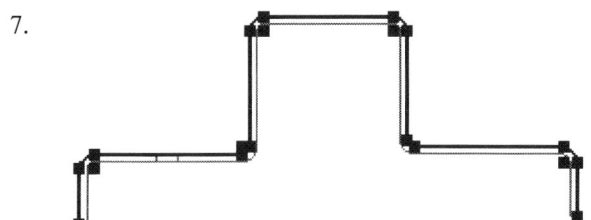

8. Draw two vertical line segments to close the profile.

   Apply the 0.750 dimension.

9.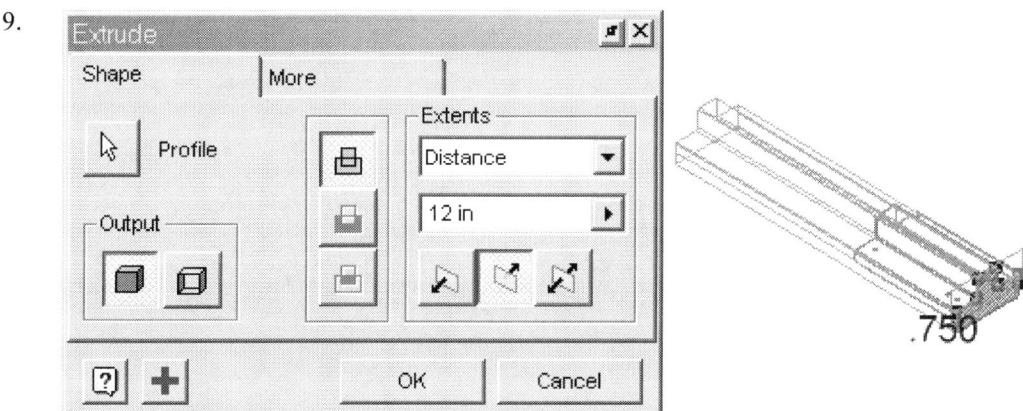

   Extrude the sketch 12 in.

10. ⇐ Return   To exit out of Part Edit mode, press the **Return**.

11. Move Component   Select the **Move Component** tool.

16-11

12.

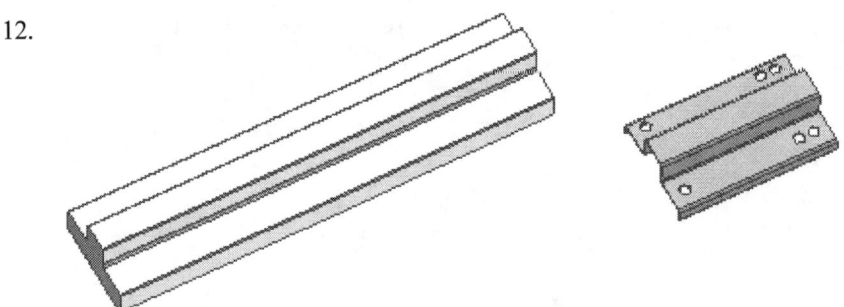

Select the rail by pressing down the left mouse button and drag it away from the bracket.

13. 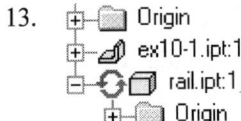 When we created an in-place component, Inventor automatically presumed that we wanted it to be Adaptive. This is indicated by the red and green arrow symbol next to the part name.

14. 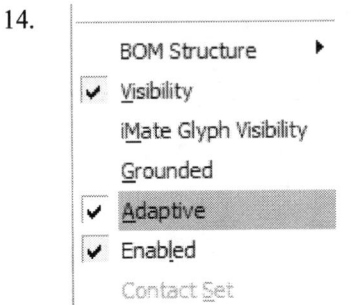 To remove the adaptivity, highlight the part name in the Browser and disable **Adaptive**.

15. 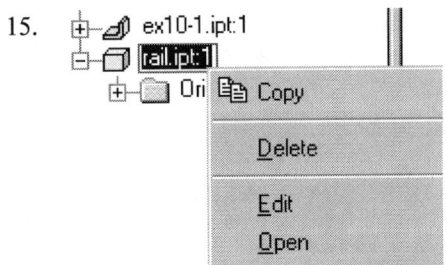 Highlight the *rail.ipt* in the Browser. Right click and select '**Edit**'.

 TIP: You can also double click on top of the part you wish to edit to activate Part Edit mode in an assembly file.

Assembly Tools

16. 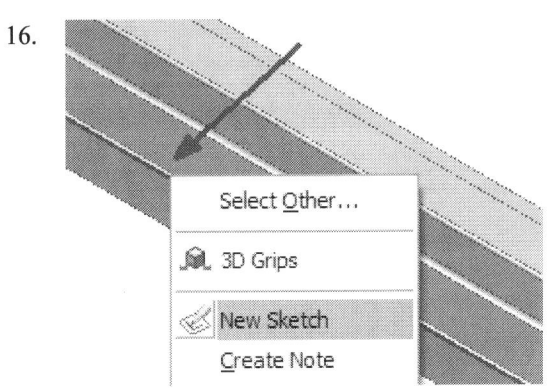 Select the face indicated for a **New Sketch**.

17.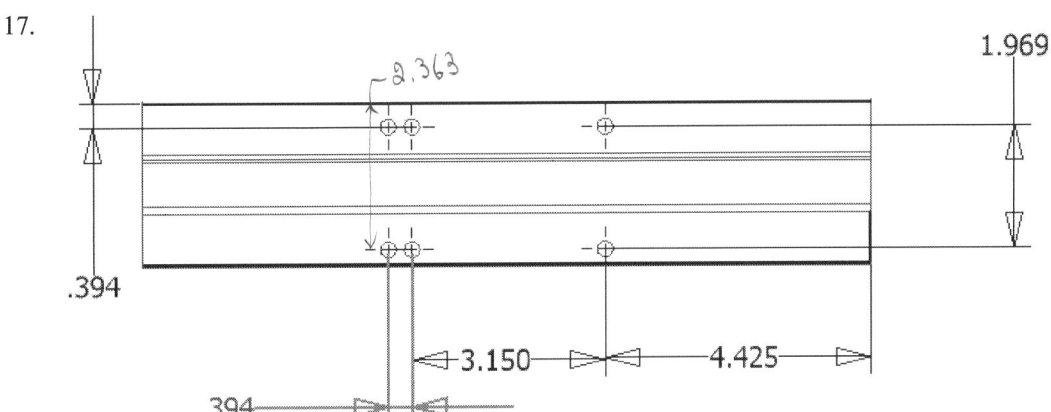

Place six hole points as shown.

Autodesk Inventor Fundamentals

18.

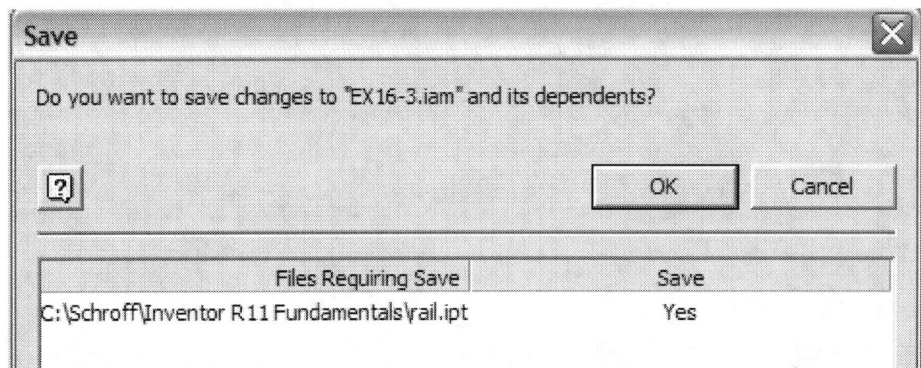

Holes should be **Tapped Drilled M8x1.25 x .50 in Deep**.

19. ⇐ Return ▼   Exit out of Part Edit mode.

20. Save the file as *ex16-3.iam*.

21. Press **OK** when you are prompted to save changes to the components.
    Close the file without saving.

Assembly Tools

TIP: You can drag and drop a component from Windows Explorer or a Browser window, but the component includes undisplayed default work planes that may offset the part from the cursor. Drag and drop places a single instance, unlike multiple occurrences as described above. Whether you place components through the dialog box or drag and drop, use assembly constraints to position components and remove degrees of freedom.

Typing a 'P' will initiate the Place Component command.

Place Component from Content Center

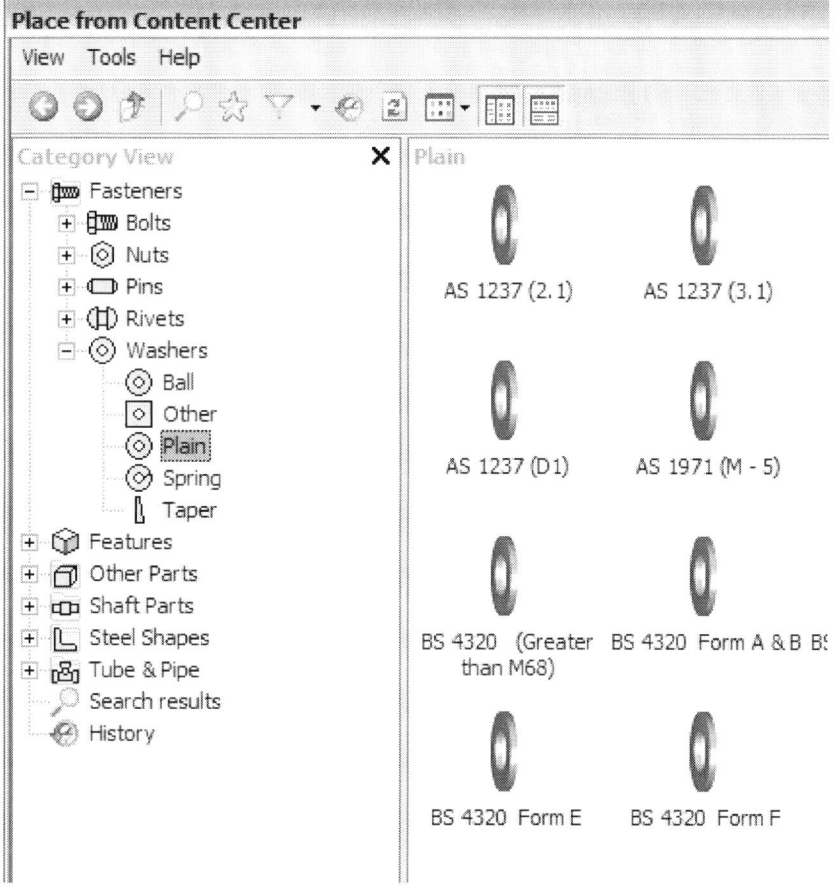

16-15

Autodesk Inventor Fundamentals

Users can download additional content using the Autodesk Inventor link.

Go to Web→Supplier Content Center.

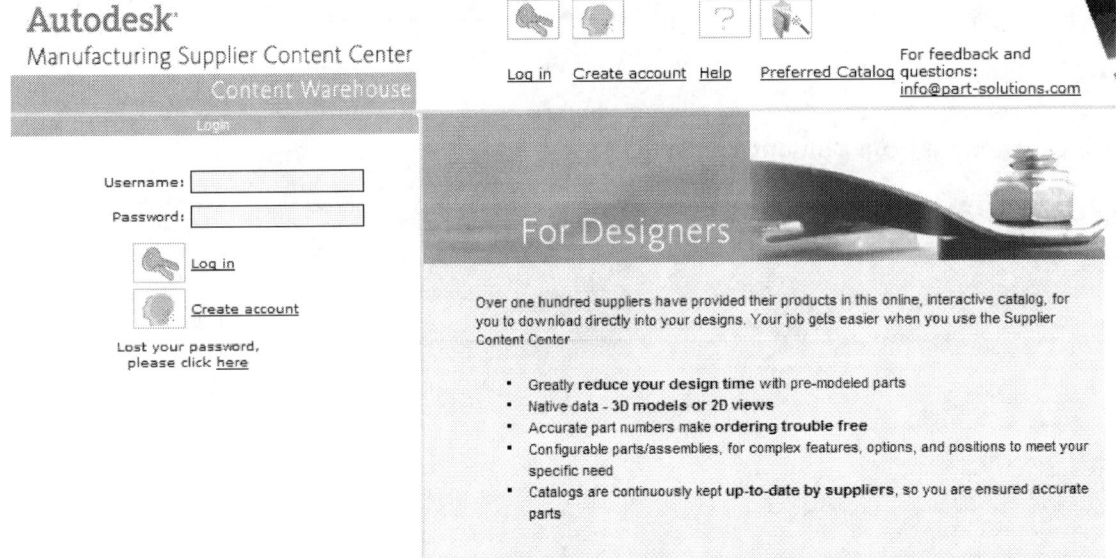

More content can be found at *http://cad.thomasnet.com/*.

Exercise 16-4
Place Inventor Content

File: Ex16-3.iam
Estimated Time: 20 minutes

1. Open *ex16-3.iam*.

2. 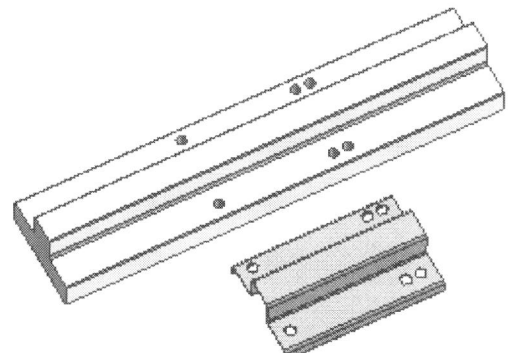 The assembly includes the bracket and the rail.

3. Select **Place from Content Center**.

4. Expand the *Fasteners* category.

   Expand the *Bolts* category.

   Highlight the **Socket Head** category.

5. You can set the Browser to use a Thumbnail, List or Detail View.
   Switch to a **Thumbnail View**.

6. Switch to a **List View**.

   Note that there are several screw types in this folder.

7. Select the **Search** tool.

16-17

8.

Type **Broached Hexagon** in the Search for: field.
Press **Search Now**.

9. 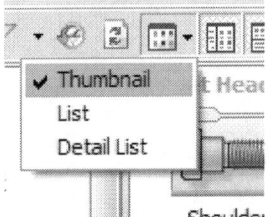   Switch back to a **Thumbnail View**.

10.    When you highlight the icons, a complete description will appear.

Double click on the picture.

11. 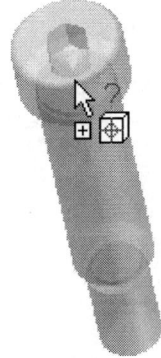   Highlight the **Broached Hexagon Socket Head Shoulder Screw- Metric** part and press **OK**.

The part will appear at the end of the cursor.

12.    Right click and select **Edit the component**.

There will be a slight pause as the component's parametric table is loaded.

13.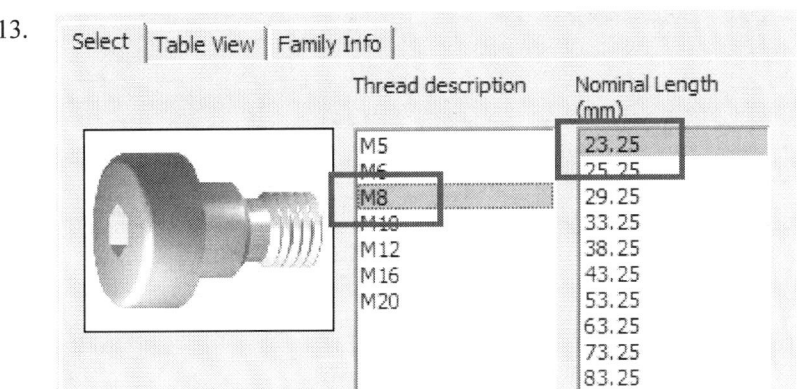

In the drop down, select **M8** for the **Thread**.
Set the **Nominal Length** to **23.25 mm**.

Press **Apply**.

14.  The preview of the fastener updates.

Press **OK** to accept the new values.

15. 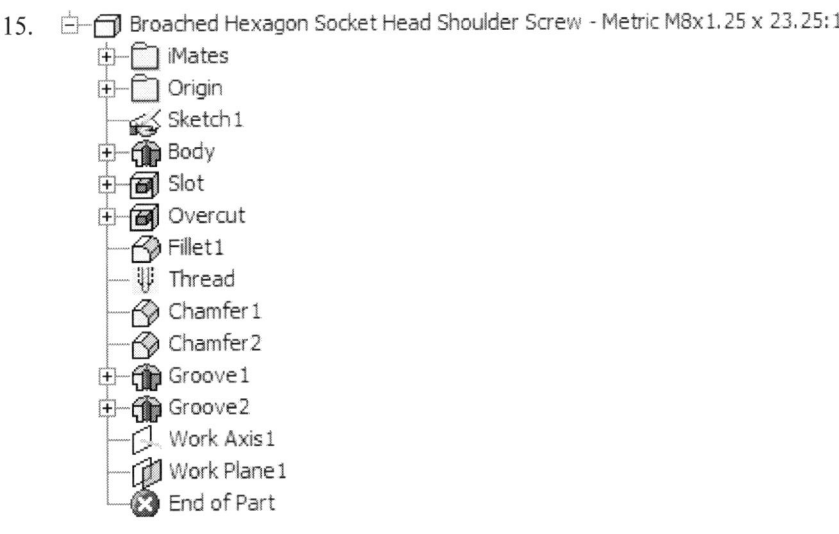 The fastener is listed in your assembly. This part is fully parametric and can be modified like any other part.

16. Save the file as *ex16-4.iam*.

Autodesk Inventor Fundamentals

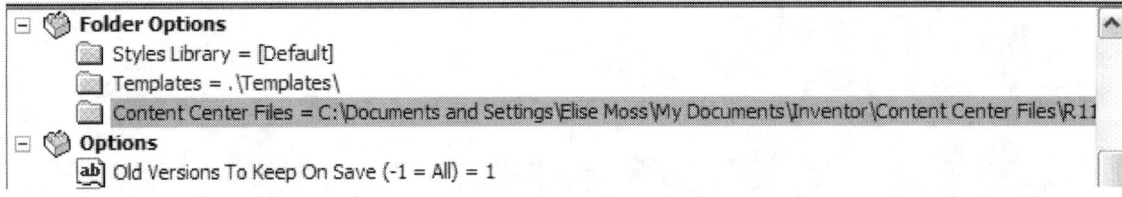

Inventor automatically saves any shared content to a Standard Parts library folder. The idea is that users tend to use the same fasteners and hardware over more than one project. Inventor saves all your hardware to the same folder (Library), so that any project may access it.

You can set where to locate your standard parts in your Projects definition.

If you are in a classroom situation or if you need to email the assembly to someone, you may want to create a part file for the fastener within your project to ensure that it does not disappear. You can also set your Library folder as a path on your server so you can access it from any computer.

Exercise 16-5
Saving a Copy of a Standard Part

File: Ex16-4.iam
Estimated Time: 10 minutes

1. Open or continue working in *ex16-4.iam*.

2.    Highlight the fastener in the Browser.

   Right click and select **Edit**.

3. Note that the fastener is fully parametric.

16-20

Assembly Tools

4. 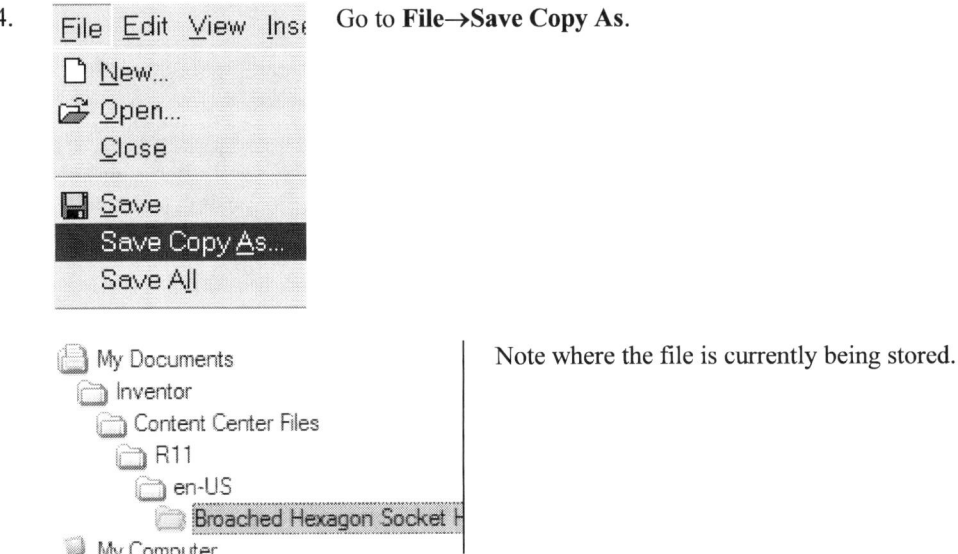 Go to **File→Save Copy As**.

Note where the file is currently being stored.

5. Browse to the Library folder under your project.

 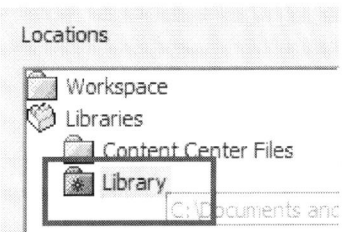

You can also simply left click on the Library folder in the dialog to go directly to that folder.

6. 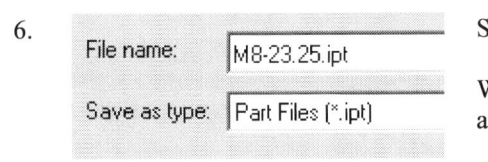 Save the part.

When you load the assembly at the new location, Inventor will ask you for that file name.

You can then set Inventor to replace all instances of that part with the file in your project directory.

7. 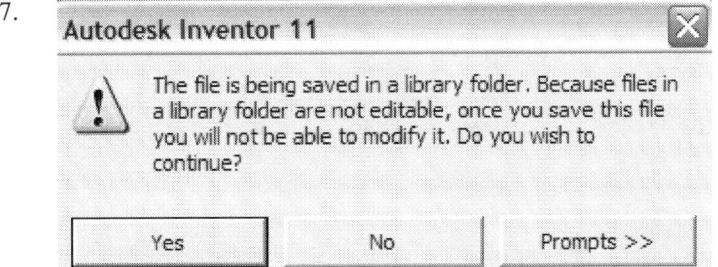 Press **Yes**.

8. ![Finish Edit] Right click and select **Finish Edit**.

9. Close the assembly file without saving.

16-21

Pattern Component

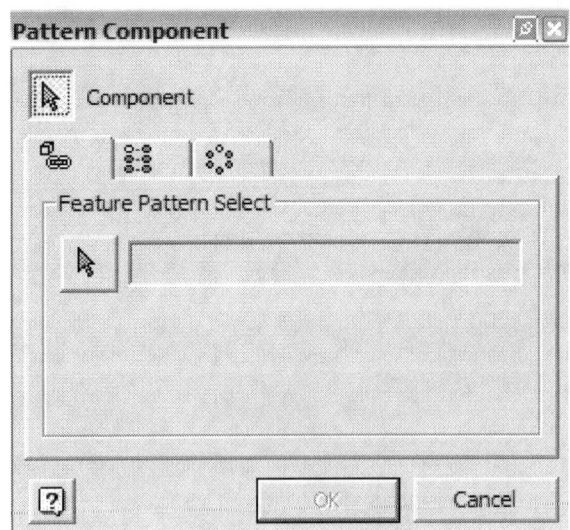

Arranging assembly components in a pattern saves time, increases your productivity, and captures design intent. For example, you may need to place multiple bolts to fasten one component to another or place multiple subassemblies into a complex assembly.

You can create a circular pattern by specifying the number of components and the angle between them. You can create a rectangular pattern by specifying column and row spacing. You can create both circular and rectangular patterns by matching features patterned on a part.

Usually, you pattern components at several points in the assembly design process after you place a component in an assembly. When you place a component:

- You position it using an existing part feature pattern.
- You select the component and copy it into a pattern.

Individual occurrences are listed in the Browser as individual parts. Individual or all occurrences can have visibility turned on or off.

You can arrange components in a circular or arc pattern by specifying the number, angle spacing, and rotation axis or by matching the spacing of features a part.

TIP: In the graphics window, you can select a part feature to pattern, but you must select an occurrence of the feature, not the original feature.

Menu		Insert→Pattern Component
	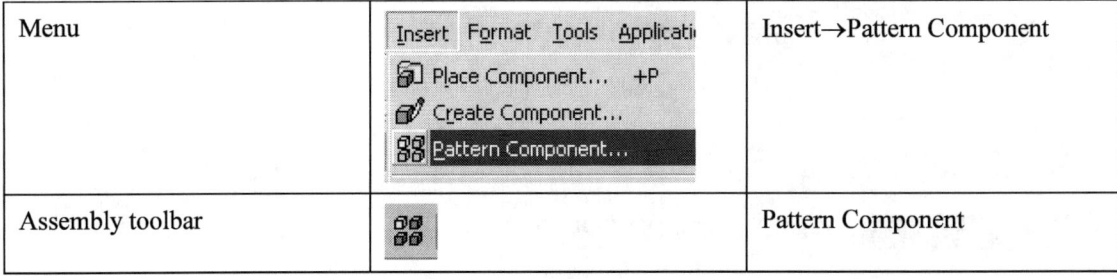	
Assembly toolbar		Pattern Component

Assembly Tools

Mirror Component

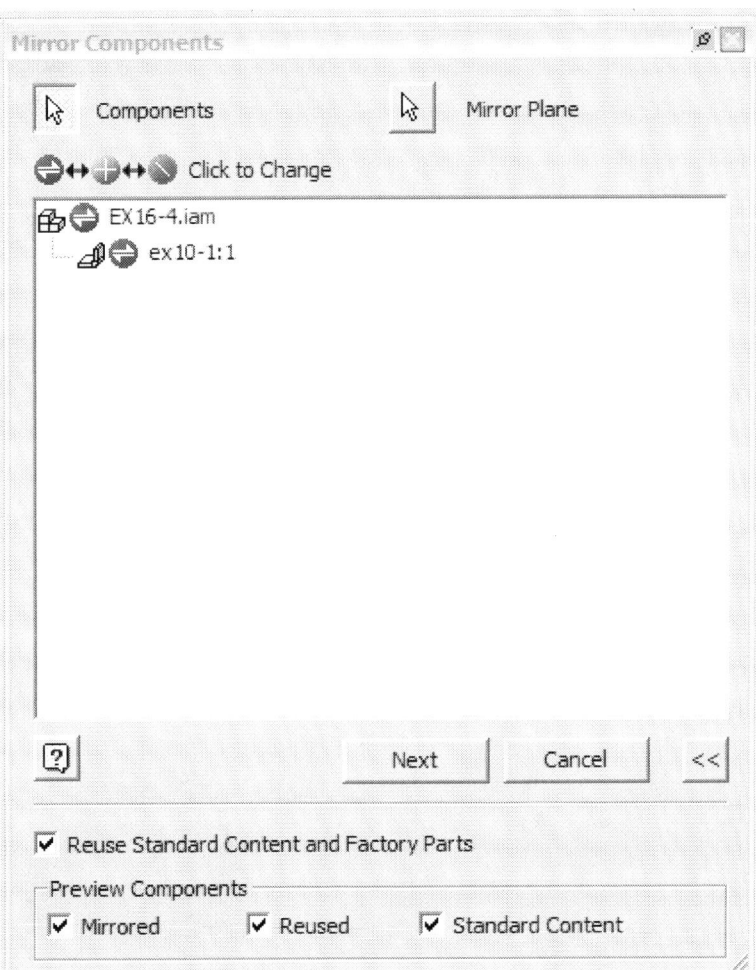

The Mirror Component tool allows the user to mirror a single component or group of components around a plane. The plane can be the face of a part or a work plane.

Copy Components

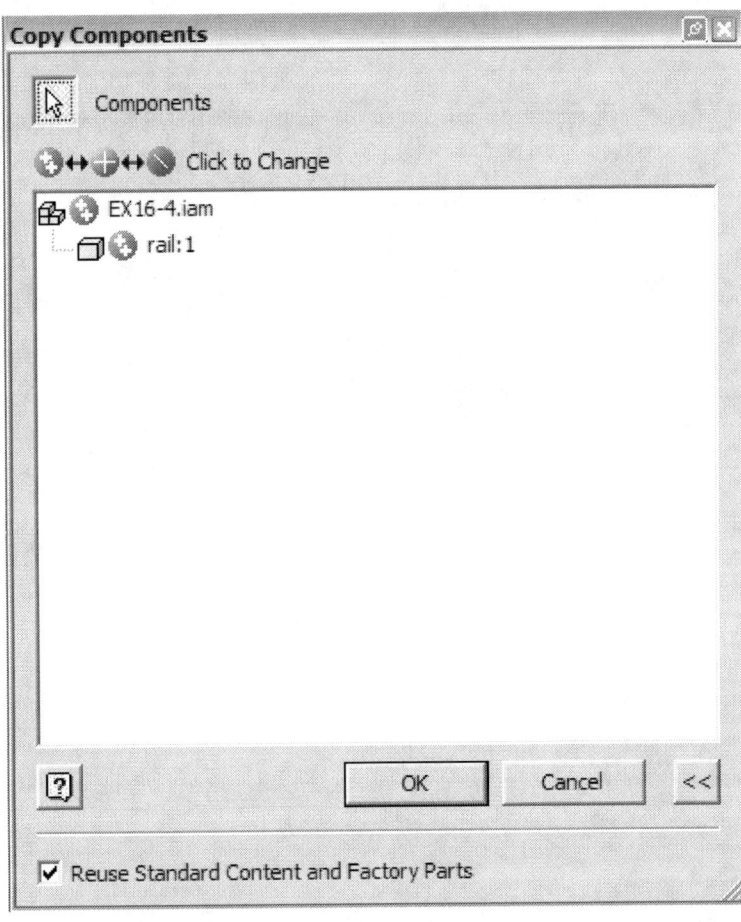

The Copy Components tool is different from simply copying a part by using the Copy/Paste method.

Use this method if you would like to assign a different part number or other information to the copied part that would be displayed in the Bill of Materials.

Assembly Tools

Place Constraint

Assembly constraints determine how components in the assembly fit together. As you apply constraints, you remove degrees of freedom, restricting the ways components can move.

To help you position components correctly, you can preview the effects of a constraint before it is applied. After you select the constraint type, the two components, and set the angle or offset, the components move into the constrained position. You can make adjustments in settings as needed, then apply it.

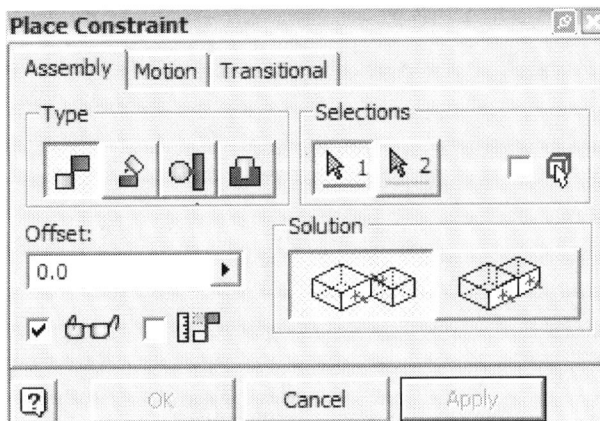

The Place Constraints dialog box creates constraints to control position and animation. Motion constraints do not affect position constraints.

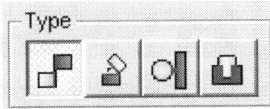

The Assembly tab has constraints to control position. The four types of assembly constraints are: Mate, Tangent, Angle, and Insert.

- A **mate** constraint positions selected faces normal to one another, with faces coincident or aligns parts adjacent to one another with faces flush. The faces may be offset from one another.
- An **angle** constraint positions linear or planar faces on two components at a specified angle.
- A **tangent** constraint between planes, cylinders, spheres, and cones causes geometry to contact at the point of tangency. Tangency may be inside or outside a curve.
- An **insert** constraint positions cylindrical features with planar faces perpendicular to the cylinder axis.

To create a complex assembly, create several small assemblies and save each one as a separate file. Combine them in larger assemblies, constraining them to other subassemblies and parts as a single unit.

Group parts in subassemblies if you want to use them in more than one assembly. Modify small subassemblies or regroup parts to change assembly configuration.

Mate Constraint

Mate constraint positions components face-to-face or adjacent to one another with faces flush. Removes one degree of linear translation and two degrees of angular rotation between planar surfaces.	
	Mate constraint positions selected faces normal to one another, with faces coincident.
	Flush constraint aligns components adjacent to one another with faces flush. Positions selected faces, curves, or points so that they are aligned with surface normals pointing in the same direction.

Angle Constraint

Angle constraint positions edges or planar faces on two components at a specified angle to define a pivot point. Removes one degree of angular rotation.	
	Directed Angle applies the Right Hand Rule.
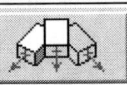	Undirected Angle is the default. This allows you to flip part directions to specify the angle.

Tangent Constraint

Tangent constraint causes faces, planes, cylinders, spheres, and cones to contact at the point of tangency. Tangency may be inside or outside a curve, depending on the direction of the selected surface normal. A tangent constraint removes one degree of linear translation.	
	Inside Positions the first selected part inside the second selected part at the tangent point.
	Outside Positions the first selected part outside the second selected part at the tangent point. Outside tangency is the default solution.

Insert Constraint

Insert constraint is a combination of a face-to-face mate constraint between planar faces and a mate constraint between the axes of the two components. The Insert constraint is used to position a bolt shank in a hole, for example, with the shank aligned with the hole and the bottom of the bolt head mated with the planar face. A rotational degree of freedom remains open.

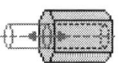	Opposed reverses the mate direction of the first selected component.
	Aligned reverses the mate direction of the second selected component.

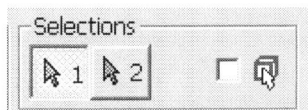

Selections select geometry on two components to constrain together. You can specify one or more curves, planes, or points to define how features fit together.

	First Selection Selects curves, planes or points on the first component. To end the first selection, click the Second Selection button.
	Second Selection Selects curves, planes, or points on the second component. To select different geometry on the first component, click the First Selection tool and reselect.
	Pick Part First Limits the selectable geometry to a single component. Use when components are in close proximity or partially obscure one another. Clear the check box to restore selection mode.

 The Offset text box specifies distance by which constrained components are offset from one another.

Use to enter a value equal to a distance or angle that exists in the assembly, but when you do not know the offset or angle. Click the down arrow to measure the angle or distance between components, show dimensions of selected component, or enter a recently used value.

Specify positive or negative values. Default setting is zero. The first picked component determines the positive direction. Enter a negative number to reverse the offset or angle direction.

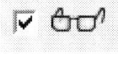 **Show Preview** shows the effect of the constraint on the selected geometry. After both selections are made, underconstrained objects automatically move into constrained positions. Default setting is on. Clear the check box to turn preview off.

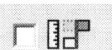 **Predict Offset and Orientation** calculates the value of the offset distance. In order to use this option, it must be enabled and the Offset field must be empty.

Motion Constraints

Motion constraints specify the intended motion between assembly components. Because they operate only on open degrees of freedom, they do not conflict with positional constraints, resize adaptive parts, or move grounded components.

Motion constraints are shown in the Browser. When clicked or the cursor hovers over the Browser entry, constrained components are highlighted in the graphics window.

Drive constraints are not available for motion constraints. However, parts that are constrained using motion constraints will drive according to the direction and ratio specified.

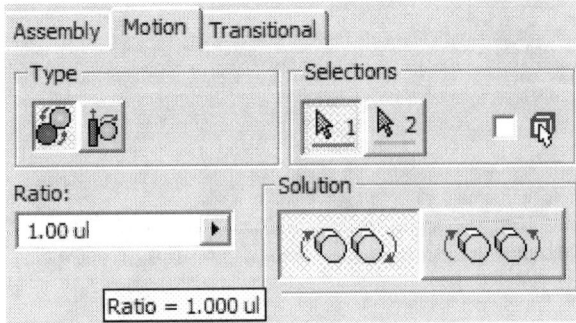

The Motion tab has constraints to specify intended motion ratios between assembly components:

- A **rotation** constraint specifies rotation of one part relative to another part using a specified ratio.
- A **rotation-translation** constraint specifies rotation of one part relative to translation of a second part.

The first part in an assembly is grounded. Its position is fixed, with the part origin coincident with the assembly origin.

When the next part is placed and constrained to the grounded part, it moves to the grounded part and fits together according to the type of constraint applied.

As you add parts, you can add constraints to position the new parts relative to the other assembled parts.

After constraints are positioned, you can use motion constraints to control rotation and translation in the remaining degrees of freedom. You specify a ratio to set movement between two components.

Drive constraints do not control motion between components, but simulate mechanical motion by driving a constraint through a sequence of steps for a single component. You can, however, animate two components by using the Equation tool to create algebraic relationships between components. A drive constraint operation is a temporary animation.

Motion constraints specify motion ratios between components, either by rotation or by rotation and translation. Such constraints are useful for specifying motion of gears and pulleys, a rack and pinion, or specifying motion between third-party components such as a gearbox and input and output shafts. Use work geometry and assembly constraints to limit the range of motion.

Assembly Tools

Type specifies the constraint type and illustrates the solution that shows the intended motion between selected components. May be applied between linear, planar, cylindrical, and conical elements.

You can change constraint type when the dialog box is open during constraint placement or editing. When the cursor hovers over a component, an arrow shows the direction of the constraint. Click Forward or Reverse to change solution.

	Rotation constraint specifies that the first selected part rotates in relation to another part using a specified ratio. Typically used for gears and pulleys.
	Rotation-Translation constraint specifies that the first selected part rotates in relation to translation of another part using a specified distance. Typically used to show planar motion, such as a rack and pinions.

Selections select geometry on two components to constrain together. You can specify one or more curves, planes, or points to define how features fit together.

	First Selection Selects curves, planes or points on the first component. To end the first selection, click the Second Selection button.
	Second Selection Selects curves, planes, or points on the second component. To select different geometry on the first component, click the First Selection tool and reselect.
	Pick Part First Limits the selectable geometry to a single component. Use when components are in close proximity or partially obscure one another. Clear the check box to restore selection mode.
Ratio	For Rotation constraints, the ratio specifies how much the second selection rotates when the first selection rotates. For example, a value of 4.0 (4:1) rotates the second selection four units for every unit the first selection rotates. A value of 0.25 (1:4) rotates the second selection one unit for every four units the first selection rotates. The default value is 1.0 (1:1). If two cylindrical surfaces are selected, Autodesk Inventor computes and displays a default ratio that is relative to the radii of the two selections.
Distance	For Rotation-Translation constraints, the distance specifies how much the second selection moves relative to one rotation of the first selection. For example, a value of 4.0 mm moves the second selection 4.0 mm for every complete rotation of the first selection. If the first selection is a cylindrical surface, Autodesk Inventor computes and displays a default distance that is the circumference of the first selection.

Autodesk Inventor Fundamentals

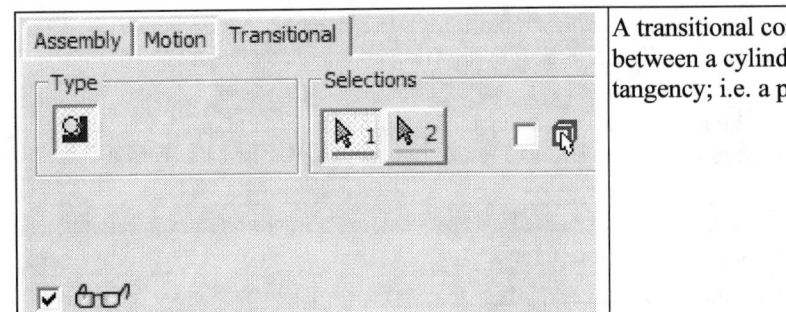		A transitional constraint applies a constraint between a cylindrical object moving through a tangency; i.e. a pulley belt on a pulley.
Menu	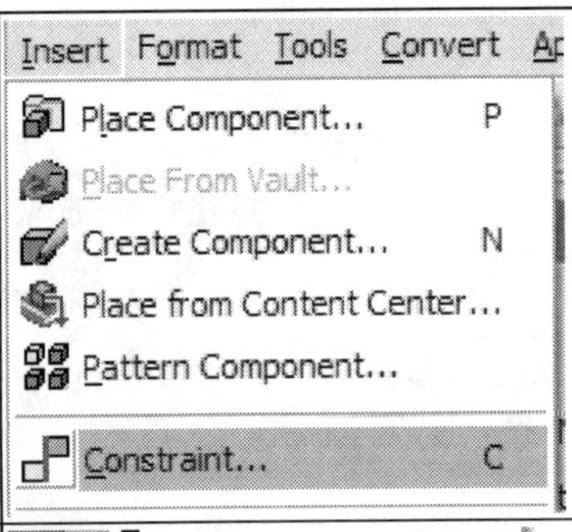 Insert → Place Component... / Place From Vault... / Create Component... / Place from Content Center... / Pattern Component... / Constraint... C	Insert→Constraint
In the graphics window	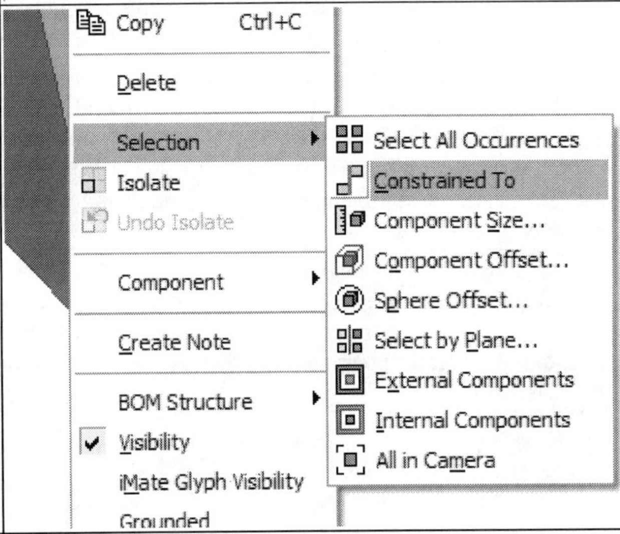 Copy / Delete / Selection → Select All Occurrences / Constrained To / Component Size... / Component Offset... / Sphere Offset... / Select by Plane... / External Components / Internal Components / All in Camera / Isolate / Undo Isolate / Component / Create Note / BOM Structure / Visibility / iMate Glyph Visibility / Grounded	Select a component. Right click and select **Selection→Constrained To**.
Assembly toolbar		Constraint
Shortcut Key	Press 'C'	

TIP: If you select a Rotation Type, then you must indicate Ratio.

If you select Rotation-Translation Type, then you must indicate Distance.

Although the ratio and distance parameters are used to specify of amount of movement for the second selection with respect to the first selection, the constraint is bi-directional so that if the second selection is moved, the first selection will move by an inverse amount of either the ratio or distance as appropriate to the constraint type.

Assembly Tools

Exercise 16-6
Place MATE Constraint

File: Ex16-4.iam
Estimated Time: 10 minutes

1. Open the *ex16-4.iam* file.

2. 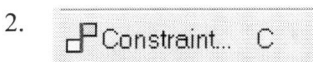 Select the **Constraint** tool.

3. The Place Constraint dialog appears.

   The default constraint is **MATE**.

4.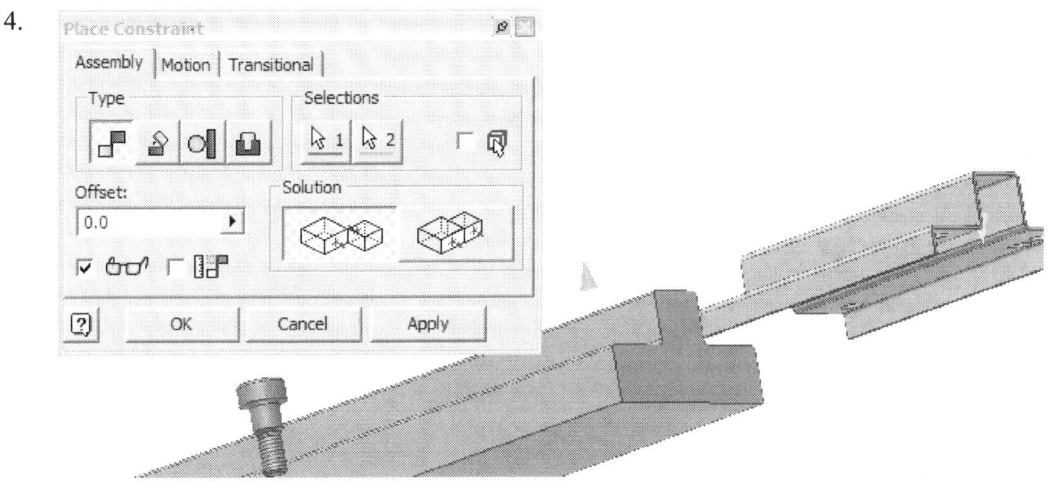

   Use a **MATE** constraint to place the bracket on top of the rail.

5. Press **Apply**.

6. 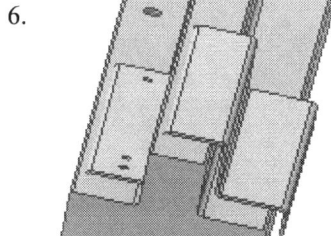 Even though the bracket sits on the top of the rail, it is not aligned properly.

7. You can use the **Degrees of Freedom** tool under View in the menu to help you see how your parts are constrained together.

   Go to **View→Degrees of Freedom**.

16-31

Autodesk Inventor Fundamentals

8.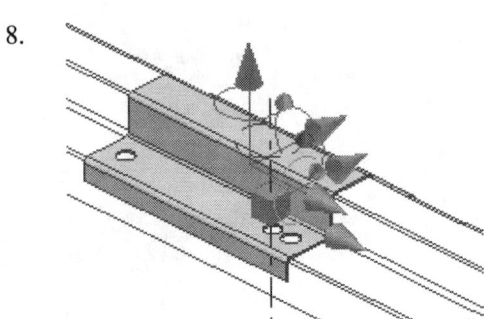

Turning on the Degrees of Freedom shows us how the rail is constrained and how the bracket is constrained.

We see that the bracket is free to move in all directions. The rail has two degrees of freedom and rotation available.

9. Enable **Grounded** for the rail part.
We can do this by selecting the part in the Browser or in the graphics window.
Right click and enable **Grounded**.

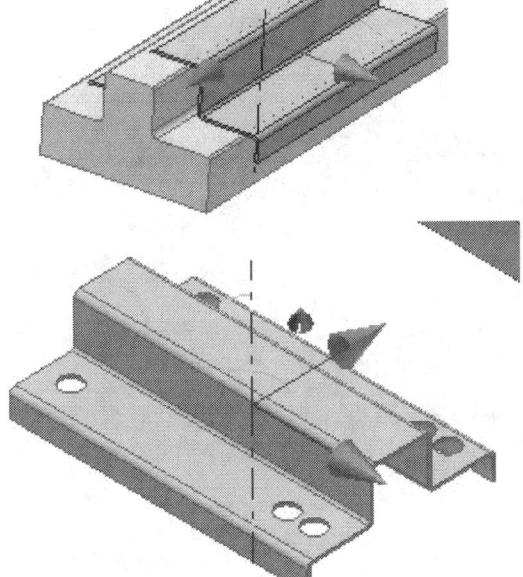

The Degrees of Freedom disappears for the rail as soon as we ground the part.

Note that the bracket has three degrees of freedom available. One for rotation indicated by the red arc and arrow and two planar/directional degrees of freedom.

10. Save the file as *ex16-6.iam*.

Assembly Tools

Exercise 16-7
Place ANGLE Constraint

File: Ex16-6.iam
Estimated Time: 10 minutes

1. Open the *Ex16-6.iam* file or continue working in the open file.

2. 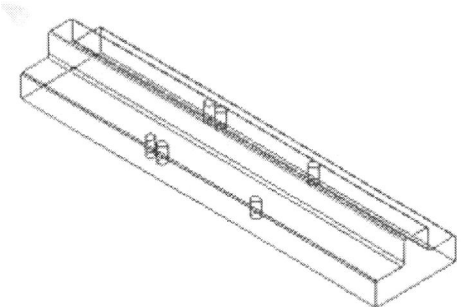 Select the **Constraint** tool.

3.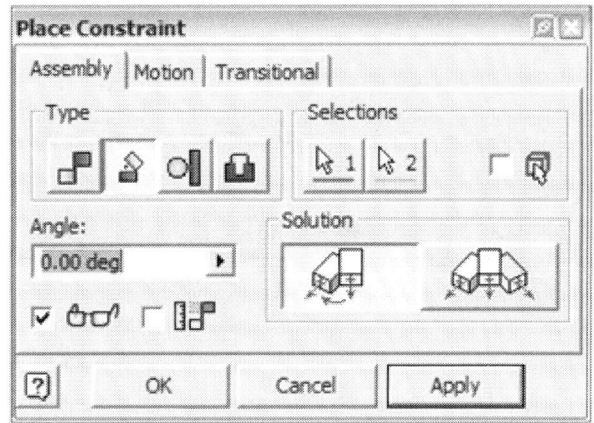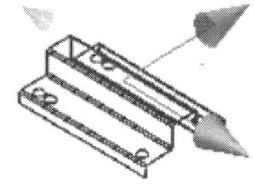

Select the **ANGLE** as **Type**.
Select one edge of the bracket.
Select a parallel edge of the rail.
The angle should be set to 0 if the angle arrows are pointing the same direction or 180 if the arrows are opposed.

Press **Apply**.

Close the dialog.

16-33

4. 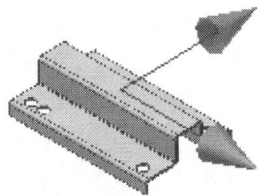 We have eliminated the rotation degree of freedom, but we see that two degrees of freedom remain.

5.
```
rail:1
   Origin
   Work Plane1
   Mate:1
   Angle:1 (180.00 deg)
```
You can see the constraints we have placed listed in the Browser.

6. Save the file as *ex16-7.iam*.

Exercise 16-8
Place INSERT Constraint

File: Ex16-7.iam
Estimated Time: 10 minutes

1. Open the *ex16-7.iam* file or continue working in the open file.

2. You can separate parts to assist you in placing additional constraints without deleting the constraints you already applied.

    Use the **Move** and **Rotate** tools on the Assembly toolbar to move and rotate the bracket up and over to make it easier to add the next constraint.

3. Select the **Constraint** tool.

Assembly Tools

4.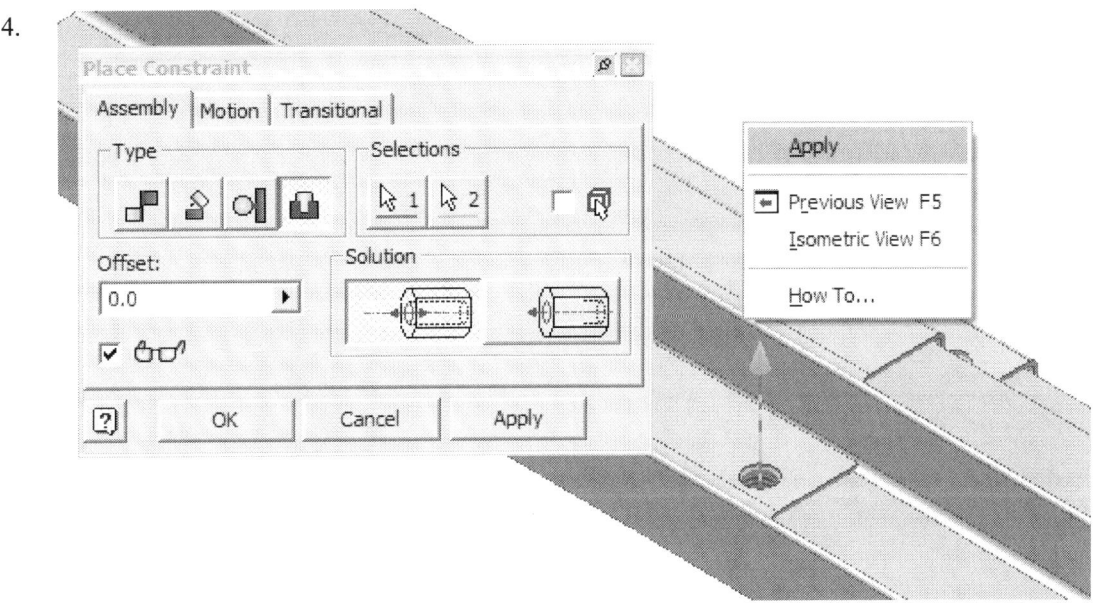

Select the **INSERT** type.

Select one of the holes in underside of the bracket and select one of the corresponding holes in the rail.

Press **Apply**.

NOTE: You can also right click and select Apply.

5. We see that the rail is completely constrained now because the Degrees of Freedom icon disappears.

Press **Shift+E** to turn off the DOF display.

6. 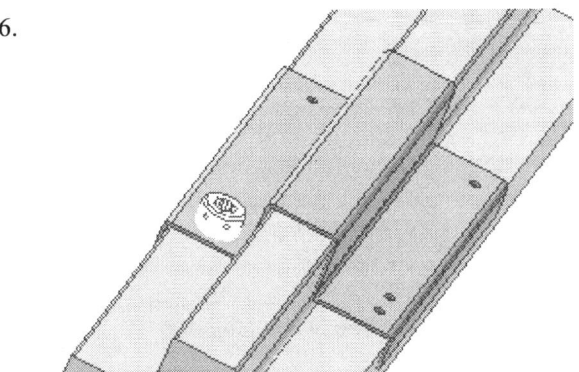 Select the **Constraint** tool again and use it to **INSERT** the screw into one of the holes.

7. Save the file as *ex16-8.iam*.

Autodesk Inventor Fundamentals

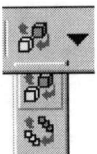

Replace Component/Replace All

The next tool in the Assembly toolbar is a flyout with two options: Replace Component and Replace All.

Replace Component

In the design process, you often need to replace one or more components in an assembly. You may design a placeholder component that you eventually replace with a standard purchased component, or replace one vendor's component with another.

You can select a part to replace an existing assembly component regardless of its location in the directory structure. In a networked environment, you need write permission to replace a component in the open assembly.

The new component is placed in the same location as the original component. The origin of the replacement component is coincident with the origin of the replaced component. Constraints can re-map. Mate and Flush constraints will usually be retained, but Angle and Insert constraints are often lost.

You can replace one assembly component with another component, but existing assembly constraints may be deleted.

TIP: You can select the component to be replaced either in the Browser or by picking a part in the drawing window.

Replace All

The Replace All tool will replace all the occurrences of a component with another component.

TIP: If you have an earlier release of Inventor, you can still use the Fastener Library that came with R3 through R5.

16-36

Assembly Tools

Move Component

When you constrain assembly components to one another, you control their position. To move a component, either temporarily or permanently, use one of these methods:

You can move a component to get a better view of its features. An unconstrained move is simply a temporary "get out of the way" move. You might want to move components to:

- See a face or feature on the selected component.
- See a face or feature on a part that is obscured by the selected component.
- Facilitate selection of a face or feature on a component by moving it to an uncluttered area of the screen.

An unconstrained move is convenient but it is temporary. The part remains in the moved location but snaps back to its constrained position when you apply a new constraint or update or refresh the assembly.

To see how a constrained component moves, you can drag it (and all components constrained to it). A constrained move honors previously applied constraints. That is, the selected component and parts constrained move together in their constrained positions.

A grounded component remains grounded at the new location. Components constrained to the grounded component remain in their constrained positions at the new location.

You can click any component in the edit target (the file that contains edits) and drag it to a new location. If you select a component that is not a child of the edit target (a part in a subassembly), the component acts as a handle and drags the whole subassembly.

You can simulate mechanical motion by driving a constraint through a sequence of steps. After you have constrained a component, you can use the Drive Constraint tool to animate it by incrementally changing the value of the constraint. For example, you can rotate a component by driving an angular constraint from zero to 360 degrees. The Drive Constraint tool is limited to one constraint, but you can drive additional constraints by using the Equations tool to create algebraic relationships between constraints.

Rotate Component

The Rotate tool on the Standard toolbar rotates the entire assembly. When you want to rotate a single component, use the Rotate Component tool on the Assembly toolbar. Operation of both Rotate tools is the same.

Keep the following behaviors in mind when you rotate components:

- You can rotate a constrained component.
- When you update the assembly, the components constrained to the rotated component snap into their constrained positions. The rotated component determines the viewing orientation of the components.
- Unconstrained components remain in their rotated positions when you update the assembly.
- Rotating a grounded component overrides the grounded position. The component is still grounded but its position is rotated.

Autodesk Inventor Fundamentals

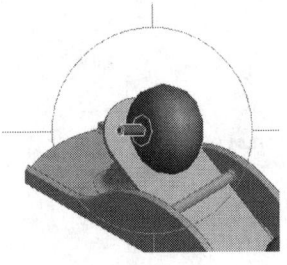

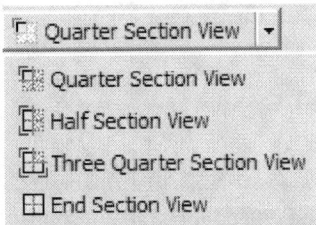

Section Views of Assemblies

You create a section view to visualize portions of an assembly within chambers or that are obscured by components. While the assembly is sectioned, you use part and assembly tools to create or modify parts-in-place.

TIP: From Quarter section and Three-quarter section views, you can right-click and select the opposite view.

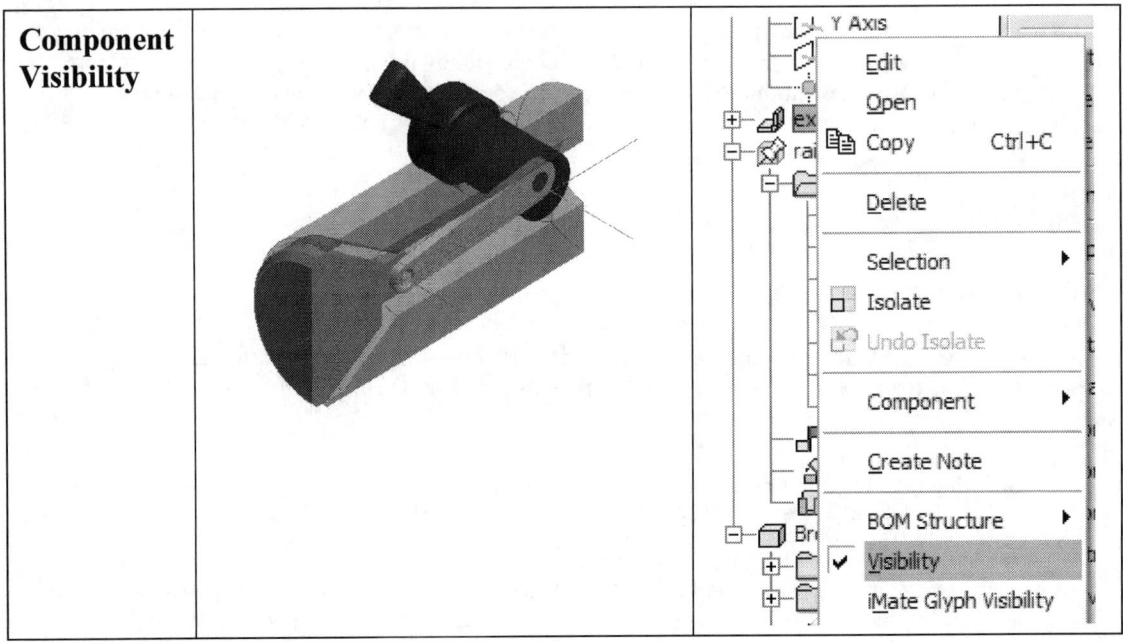

16-38

Assembly Tools

Visible and enabled components can be selected.
Visible but not enabled components cannot be selected. They are displayed in background style (wireframe).
Invisible and not enabled components cannot be selected and are not visible in the graphics window. Select an invisible and disabled components from the Browser, then right-click to change its visibility status.
Invisible and enabled components can be selected in the Browser but are not visible in the graphics window. The icon shown in the Browser looks the same as that for components that are visible and not enabled. Select an invisible and enabled component from the Browser, then right-click to change its status.

TIP: It is possible to turn off component visibility, but have the component still be enabled. This may be useful for quickly removing a needed component from view. Enabled components are fully loaded in an assembly file, while only the graphic portion of not enabled components is loaded. The assembly calculates faster because the data structure of not enabled components is not present, but its graphics are useful for a frame of reference.

Occurrence Properties

Occurrence properties control characteristics for an individual occurrence of a component in an assembly.

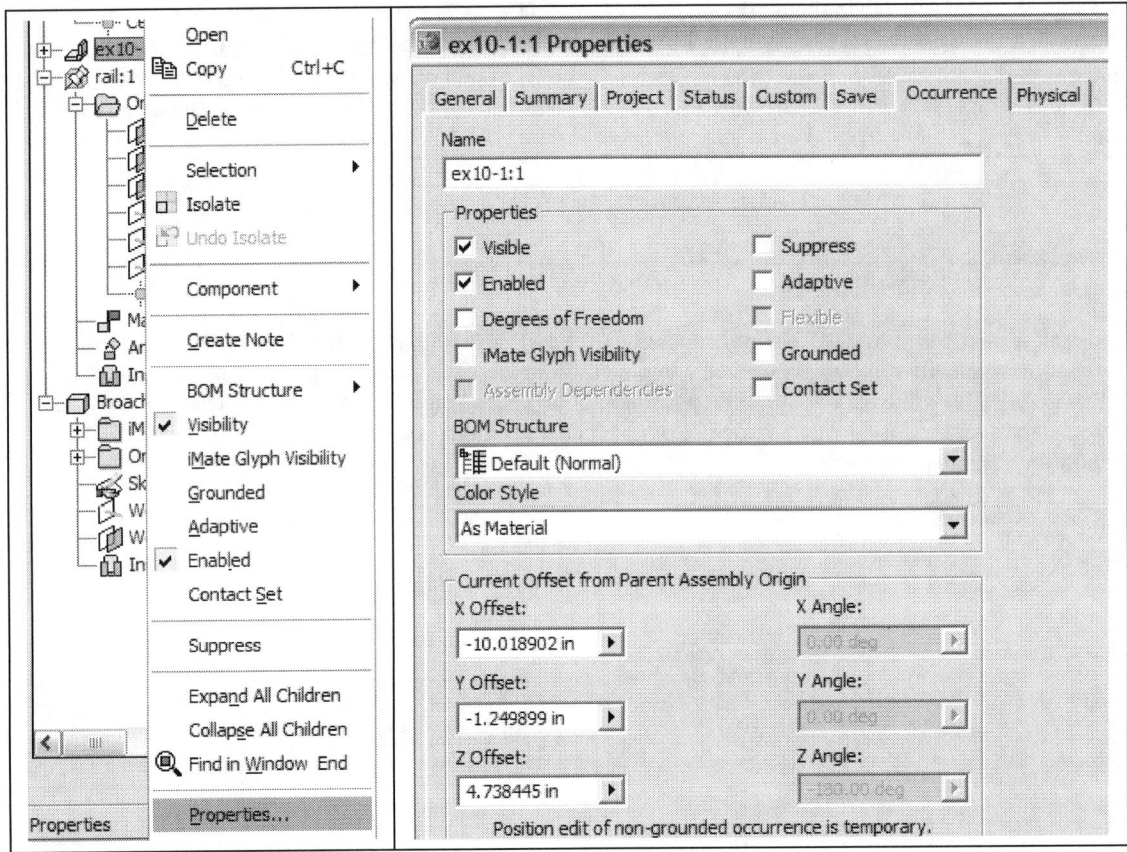

Highlight the component in the Browser, then right-click and select **Properties**. Select the **Occurrence** tab.

TIP: The adaptive status of an occurrence controls all occurrences in the assembly. When an adaptive part resizes, all occurrences of the part in other assemblies also resize.

Assembly Tools

Exercise 16-9
Pattern Component

File: Ex16-8.iam
Estimated Time: 10 minutes

1. Open the *ex16-8.iam* or continue working in the open file.

   We will use the Pattern Component tool to pattern the screw to populate the remaining three holes.

2. Select the **Pattern Component** tool.

3. 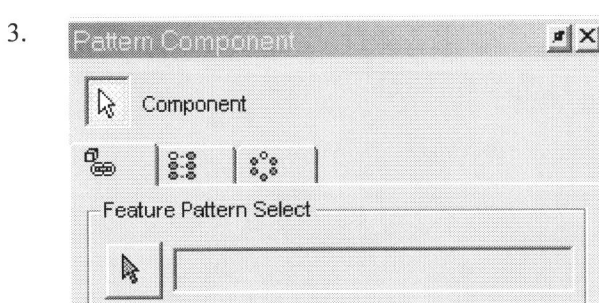 With the Component button depressed, select the fastener in the Browser.

4.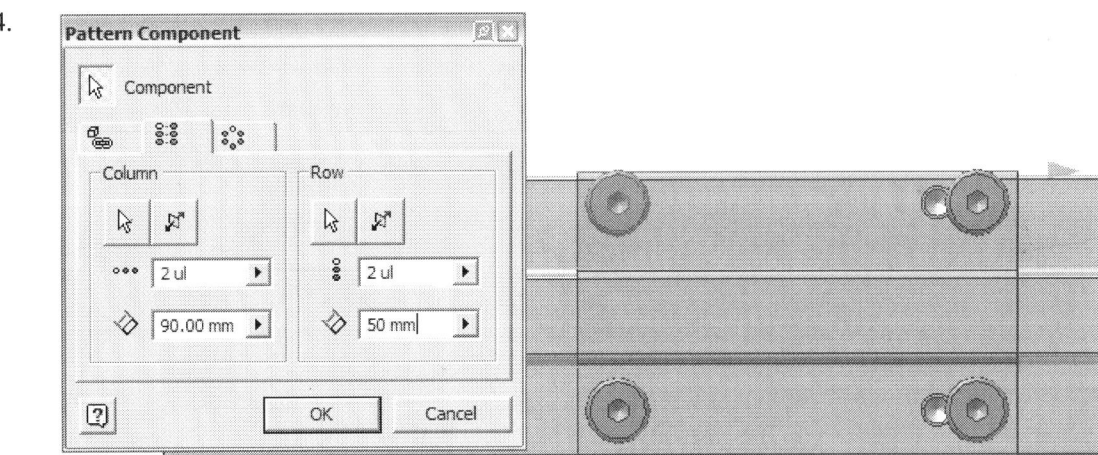

   Select the **Rectangular Pattern** tab.
   For column placement, select the side edge of the rail.
   Use the **Direction** button to flip direction if necessary.
   For the row placement, select a horizontal edge of the bracket.
   Use the **Direction** button to flip direction if necessary.
   Set the **Column Count** to **2**.
   Set the **Column Spacing** to **90**.
   Set the **Row Count** to **2**.
   Set the **Row Spacing** to **50**.
   Inventor provides a preview so you can see if your pattern is correct.
   Press '**OK**'.

5.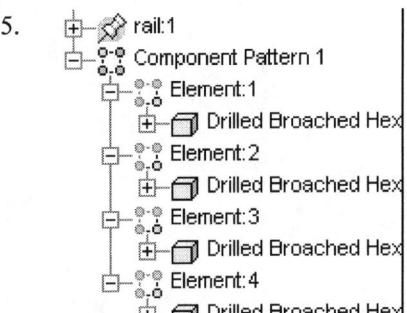

In the Browser, we see that the original screw is moved under the Component Pattern and the Insert Constraint is retained.

The other components have no assembly constraints. Their location in the assembly is based on the definition of the pattern.

6. Use the **Move** tool to move the screws out of the holes.

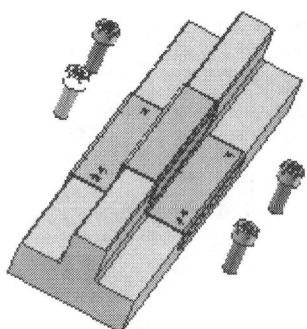

7. Press the **Update** button and see what happens.

8. Save the file as *ex16-9.iam*.

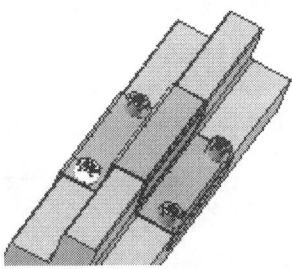

TIP: Except for the current Offset from Parent Assembly Origin, Color Style, or Name, you can set occurrence properties from the context menu. In the Browser, right-click the part in the Browser to view the context menu. A check mark beside the property indicates it is On. Clear the check mark to switch the property Off.

Assembly Tools

Exercise 16-10
Pattern Component

File: New assembly; Ex11-1.ipt
Estimated Time: 15 minutes

1.  Select the new assembly template to start a new file.

2. Select the **Place Component** tool.

3. Locate *ex11-1.ipt* and press **Open**.
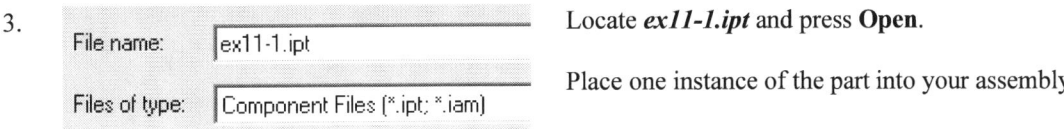
   Place one instance of the part into your assembly.

4. Save your file as *ex16-10.iam*.

5. Select the **Place from Content Center** tool.

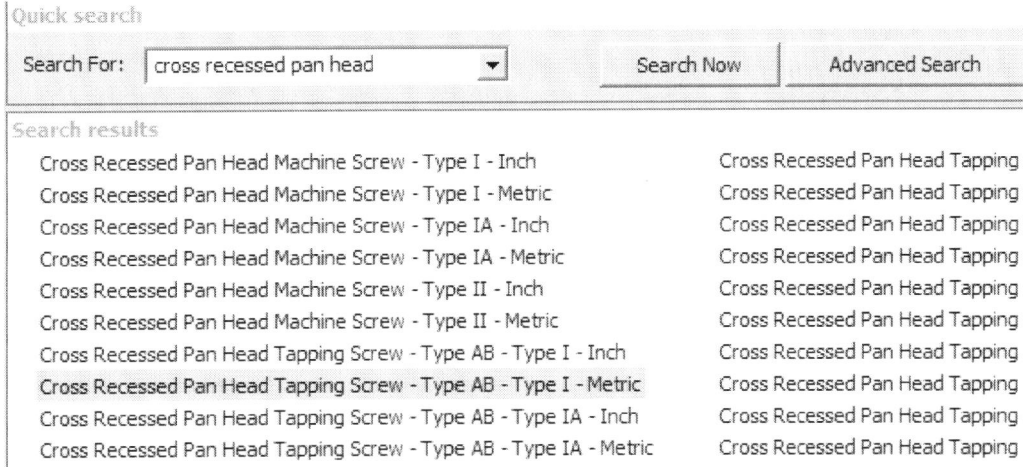

6. Use **Search** to locate a **Cross Recessed Pan Head Tapping Screw-Type AB – Type I- Metric**.

7. Highlight the fastener name. Press **OK**.

8. The component appears on the end of the cursor.

   Right click and select **Edit the component**.

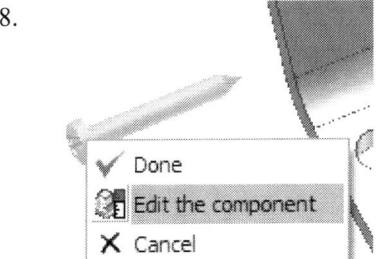

16-43

9. 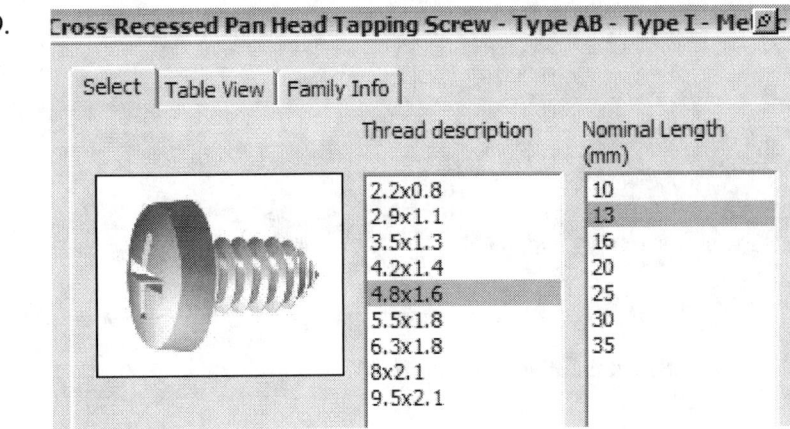 Set the **Thread** description to **M4.8 x 1.6**.

   Set the **Nominal Length** to **13 mm**.

   Press **Apply** to set the parameters.

10. Press **OK** to close the dialog.

11.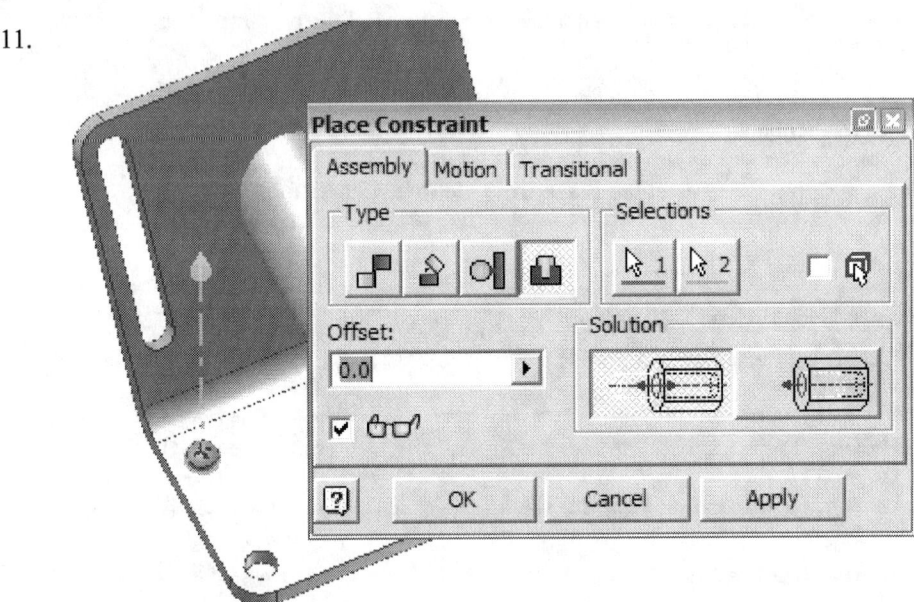

    Use an **Insert** constraint to place the screw into a hole.

12. Select the **Pattern Component** tool.

13.

Select the screw you just inserted.
Select the **Feature Pattern Select**.
Pick one of the empty holes.
A screw will be placed in each hole in the rectangular pattern.
Press **OK**.

14. 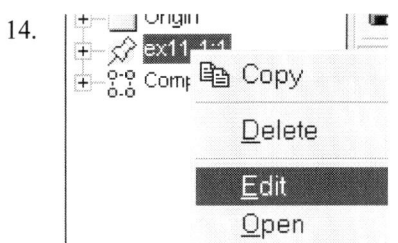 Highlight *ex11-1.ipt* in the Browser.

Right click and select **Edit**.

15. 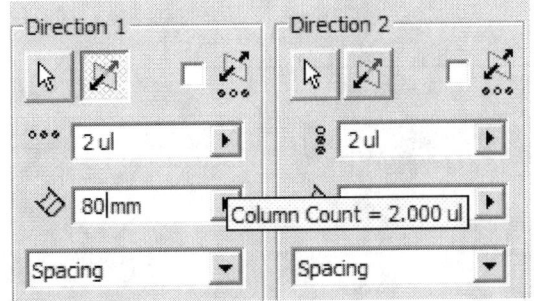 Highlight the Rectangular Pattern in the Browser. Right click and select **Edit Feature**.

16. 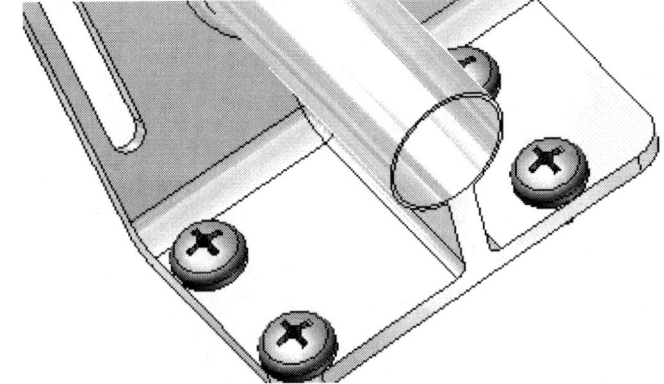 Select the 100 mm dimension.

    Change it to **80 mm**.

    Press **OK**.

17. Select the **Update** tool.

18. Select the **Return** tool to return to the assembly environment.

19. The screws are automatically repositioned.

20. Save as *ex16-10.iam*.

TIP: In order to use the Pattern Select option, the component used must have a pattern as one of the features.

Assembly Tools

Button	Tool	Function
	Place Component	Places a link to an existing part of subassembly in an assembly. A change to any instance updates all other instances of a component.
	Create Component	Creates a new part or subassembly in an assembly
	Pattern Component	Creates copies of a component in a rectangular or circular pattern
	Mirror Component	Mirrors components using a work plane or face
	Copy Component	Creates a derived copy of the selected part
	Bolted Connection Generator	Creates an assembly consisting of bolts and washers
	Place from Content Center	Access to library of standard parts
	Refresh Standard Components	Reloads the standard components in the assembly
	Constraint	Places an assembly constraint between two parts.
	Replace Component	Replaces a component in an assembly with another component
	Replace All	Replaces all occurrences of a component in an assembly
	Move Component	Enables a temporary translation of a constrained component. A constrained component returns to proper position when the user clicks Update. Enables permanent translation of a grounded component. A grounded component will remain in the placed position when Update is clicked.
	Rotate Component	Enables a temporary rotation of a constrained component. A constrained component returns to proper position when the user clicks Update. Enables permanent rotation of a grounded component. A grounded component will remain in the placed position when Update is clicked.
	Section Views	Displays a quarter section view of a model defined by hiding portions of components on one side of a defined cutting edge
		Displays a three quarter section view
		Displays a half section view
		Displays an unsectioned view of the model

Review Questions

1. The first component placed in an assembly is automatically _____.

    A. constrained
    B. placed on Plane XY
    C. grounded
    D. adaptive

2. A grounded part can not be made adaptive.

    A. True
    B. False

3. In a large assembly, you can ground _____ component(s).

    A. Only one
    B. Eight
    C. All
    D. Six

4. Use 'Place Component' to:

    A. Insert an existing part file into an assembly
    B. Create a new part in-place in an assembly
    C. Insert a Part from the Standards Parts RedSpark Plugin
    D. Move a component into position

5. You can drag and drop one or more part files from WINDOWS Explorer into an Inventor Assembly file.

    A. True
    B. False

6. When you create an in-place component in an assembly file, you must do all the items listed below EXCEPT:

    A. Specify a file name
    B. Specify a file location
    C. Specify a template
    D. Specify a material

7. ⟲🗏 Ex18-6.ipt
    The symbol shown in front of the part name indicates that the part is:

    A. Grounded
    B. Adaptive
    C. Recycled
    D. Rotated

ANSWERS: 1) C; 2) B; 3) C; 4) A; 5) A; 6) D; 7) B

Lesson 17
Bottom Up Assemblies – Yoke Assembly

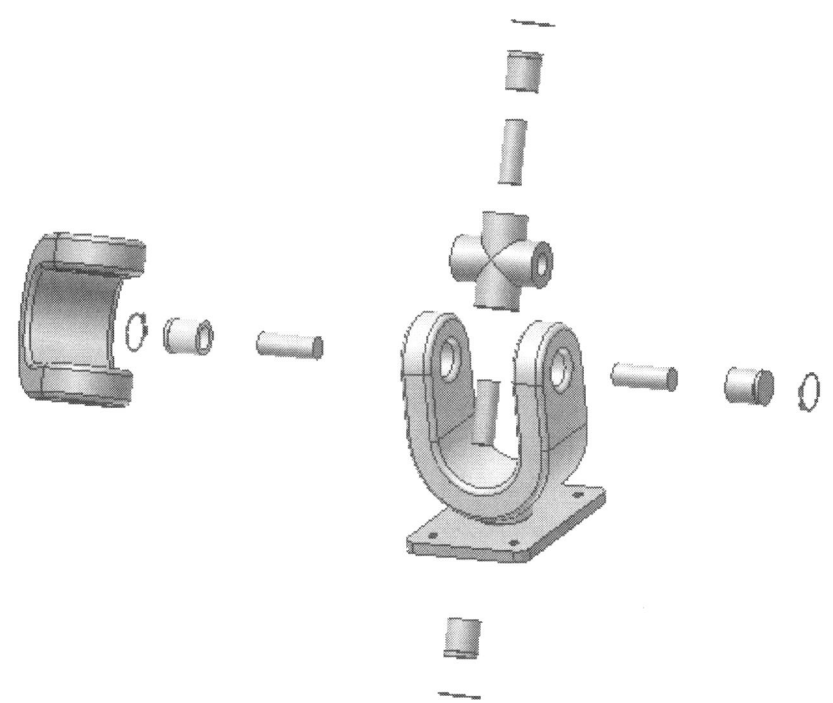

Learning Objective

In this lesson, we will create a yoke assembly consisting of six parts in a bottom up assembly. A bottom up assembly is created by building each part separately in its own file. In many real world situations, project managers may divide components between several designers or drafters. Each person would get one or two parts to create and then the parts would be assembled into an assembly drawing. A bottom up assembly method may also be used by companies that promote modular designs; i.e. the reuse of parts in more than one assembly. Designers should be familiar with both bottom up and top down assembly methods.

Autodesk Inventor Fundamentals

Exercise 17-1
Part 1 – U-Joint

File: New using custom-inches.ipt template (inches)
(Template was created in Lesson 15)

Estimated Time: 30 minutes

Review Skills: Sweep
Revolve
Mirror Feature
Offset Workplane
Extrude
Fillet

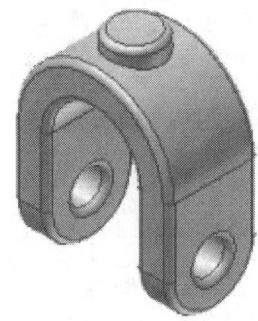

The first part to be created is the U-Joint. We begin by starting a new part file called U-Joint.ipt.

1. Go to **File→New**.

2. Double left click on the ***custom-inches.ipt*** template to open.

3.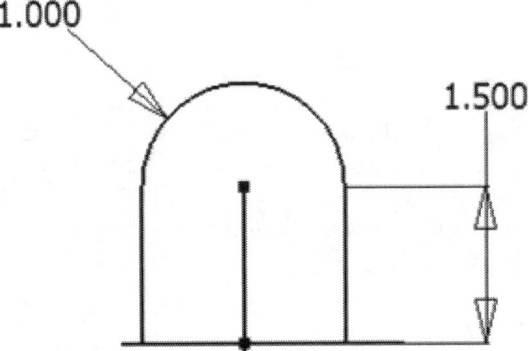

Create this sketch on the XY plane.
We have projected the X and Y axis in order to constrain the sketch.

There should be NO horizontal line between the two vertical lines other than the X axis.

Name this sketch *path*.

4.  Start a new sketch on the Top plane.

17-2

5.

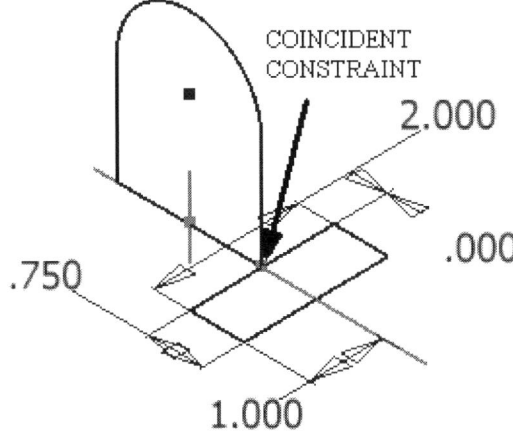

Draw a **2.0 x 0.75** rectangle.
Project the end point of the vertical line and constrain one of the rectangle sides to it using a Coincident constraint.

Project the X-axis and use it to center the rectangle. Align the inner edge of the rectangle with the end point of the path as indicated by the arrow in the image.

The first sketch will be a path.

The second sketch will be a profile.

6. Rename the rectangle sketch to **profile**.

7. Select the **Sweep** tool.

8.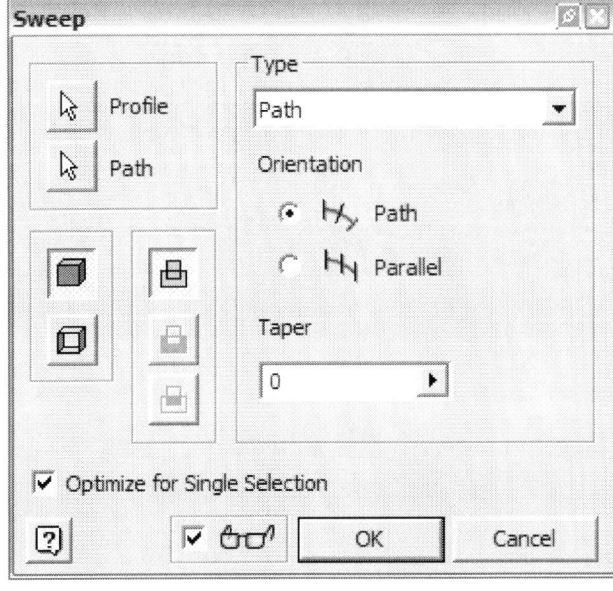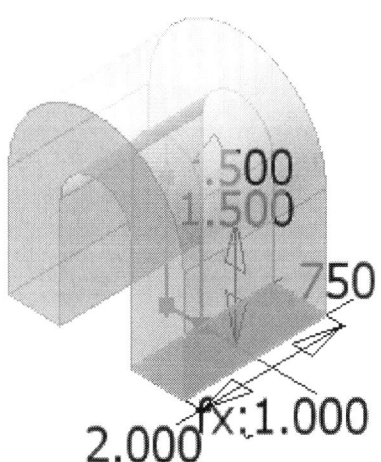

Sweep the rectangle along the path.

17-3

Autodesk Inventor Fundamentals

9. 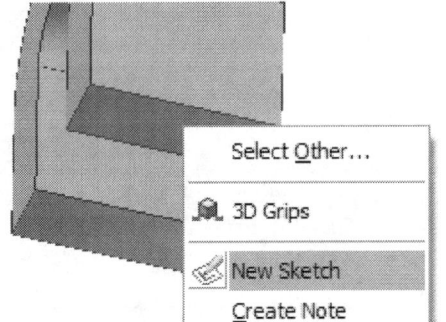 Select the bottom of the sweep feature.
Right click and select **'New Sketch'**.

10. 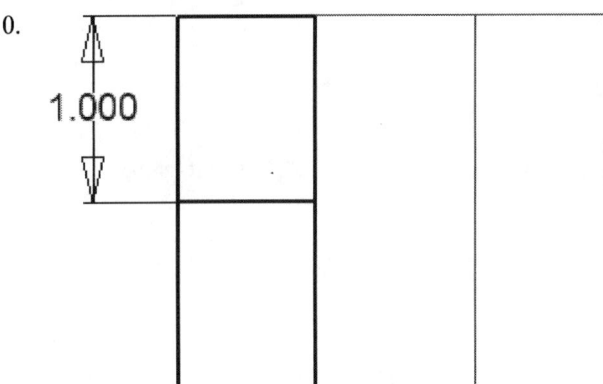 Draw a rectangle on one of the legs.

Use Collinear constraints to control the rectangle.

Set the rectangle to one half the width of the leg.

11. Select the **Revolve** tool.

12.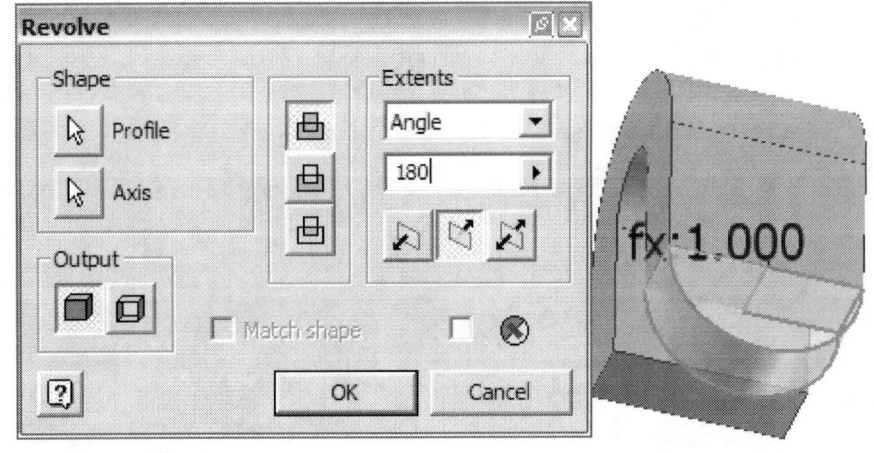

Select the **Centerline** as the **Axis**.
Set the **Angle** to **180** deg.
Set as a **Join**.
Press **OK**.

13. Select the **Mirror** tool.

17-4

Bottom Up Assembly

14. 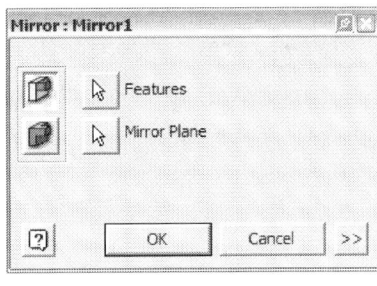 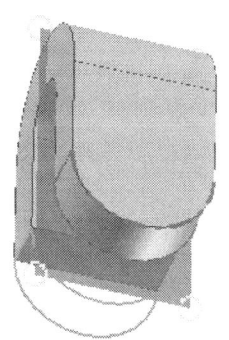 Mirror the revolve about the Right plane.

15. Select the **Hole** tool.

16. Set the Hole as **Drill Thru** with a **Diameter** of **0.75**.

17.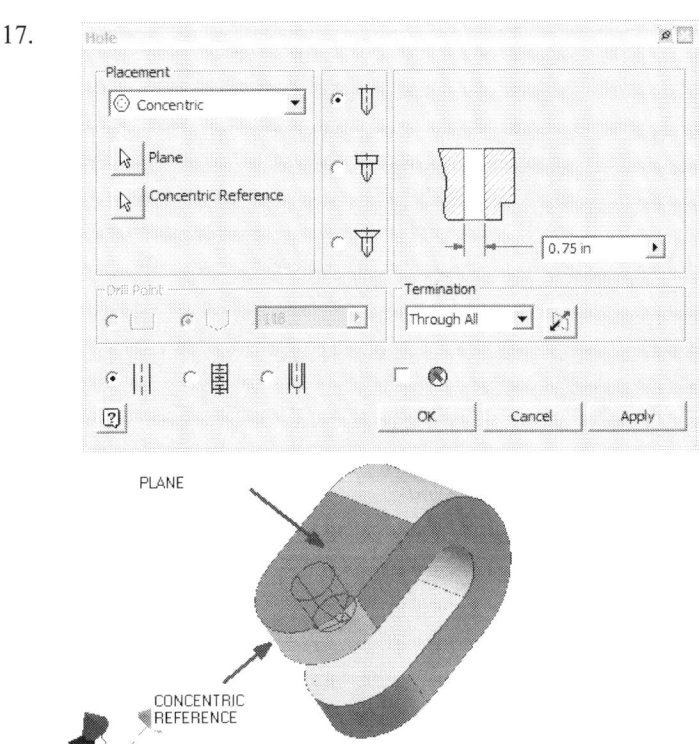

Set the **Placement** to **Concentric**.
Select the face indicated for the plane.
Select the revolve as the concentric reference.
Press **Apply** and **Done**.

18. Create an offset work plane located **3.0 units** above the top plane.

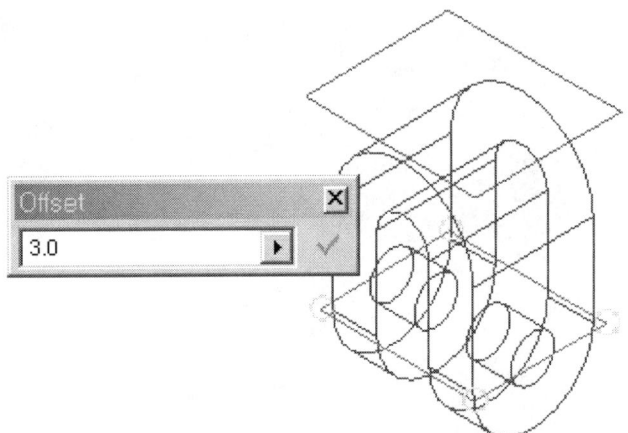

19. Select the offset work plane for a new sketch.

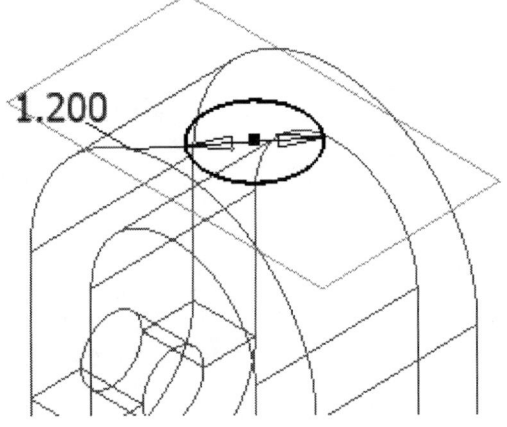

20. Draw a 1.20 diameter circle centered on the U-joint.

    To center, place a coincident constraint between the center of the circle and the center point of the part.

21.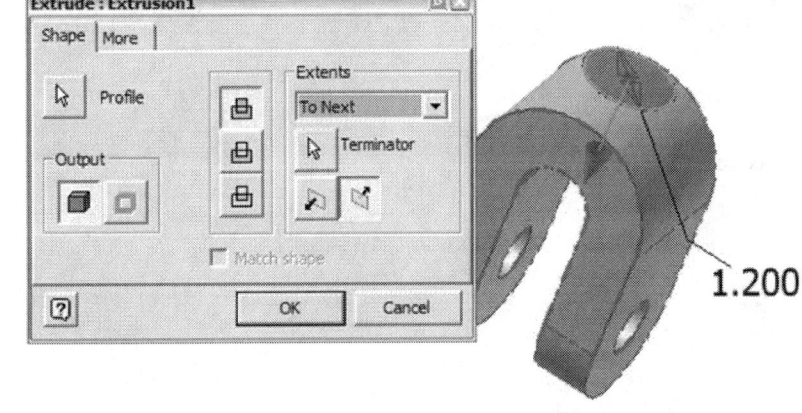

Extrude the circle **To Next**.
Select the outer surface as the terminator.

Bottom Up Assembly

22. Select the **Fillet** tool.

23.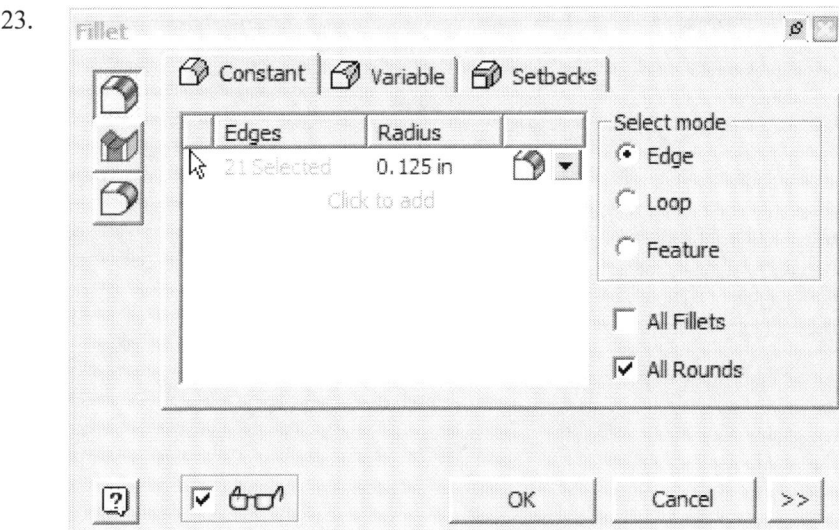

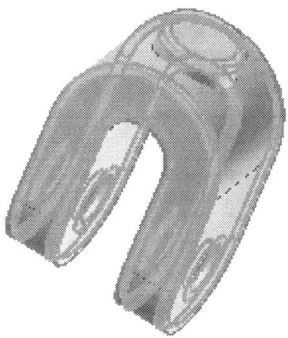

Set the **Radius** to **0.125**.
Place a check mark on the **All Rounds Select Mode**.
Press **OK**.

24. Go to **File→iProperties**.

17-7

25.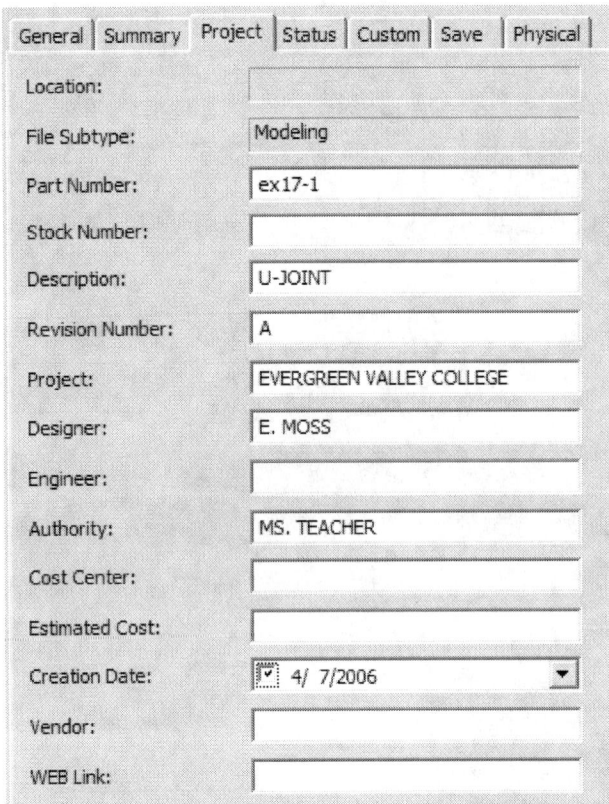

Fill in the **Project** tab with **Part Number, Description, Revision Number,** your school/company name, your name, your teacher/supervisor's name.

Press **Apply**.

This information will be used in any parts list you create.

26. 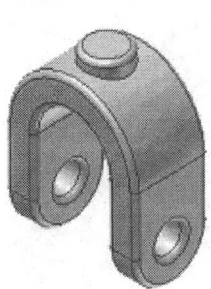   Save the file as *ex17-1.ipt*.

Exercise 17-2
Part 2 – Swivel

File: New using custom-inches.ipt template (inches)
(This template was created in Lesson 15.)

Estimated Time: 15 minutes

Review Skills: Extude
Hole
Circular Pattern
iProperties

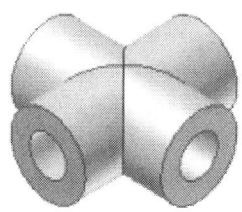

1. Start a new part file using the *custom-inches.ipt* template.

2. 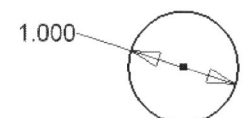 Draw a 1.0 diameter circle constrained on the center point.

3.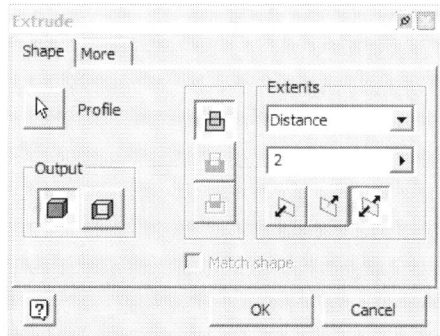

   Extrude 2 units in both directions.

   NOTE: When you extrude using the both directions option, the distance value is the total distance.

   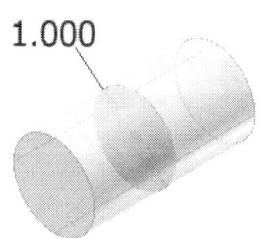

4. 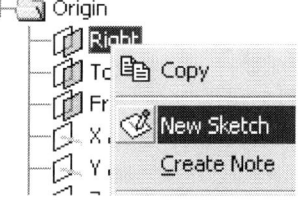 Start a New Sketch on the **Right** plane.

17-9

Autodesk Inventor Fundamentals

5. 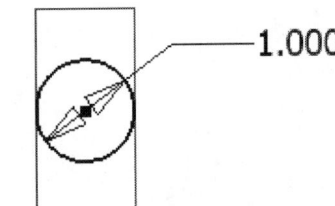 Draw a 1.0 diameter circle constrained on the center point.

6.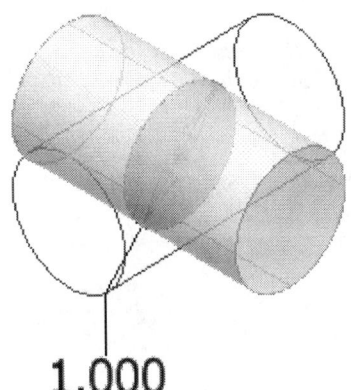

Extrude 2 units in both directions.

NOTE: *When you extrude using the both directions option, the distance value is the total distance.*

7.

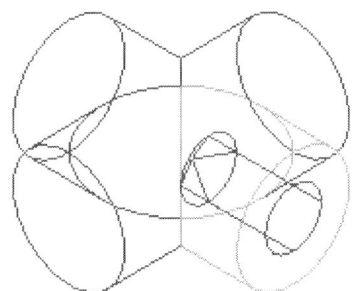

Place a **.50 diameter through hole** through the cylinder.

Enable **iMate** to automatically add an **INSERT** iMate to the hole.

8.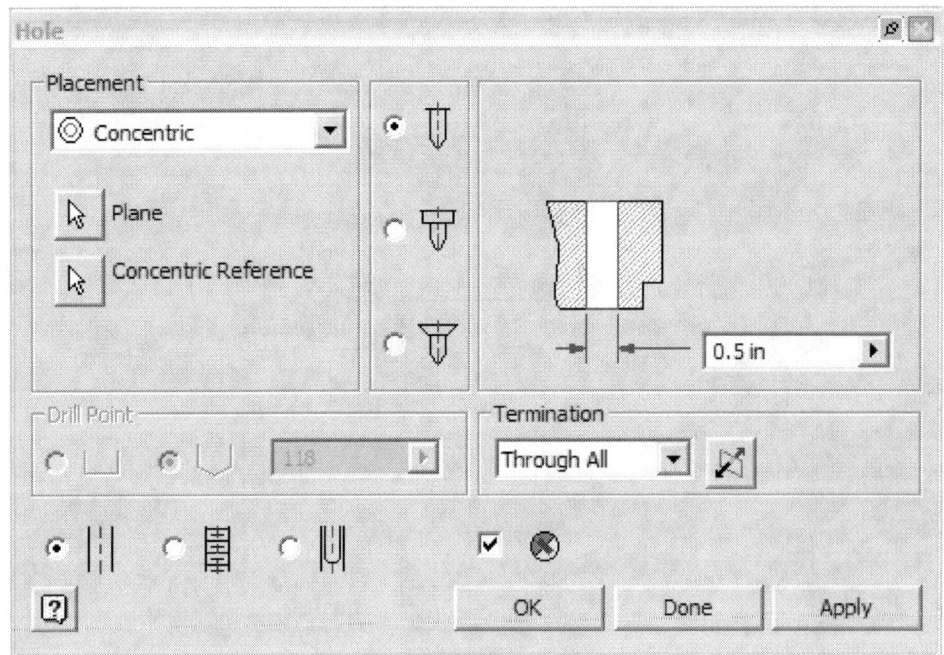

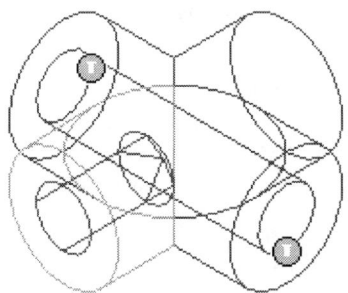

Place a **.50 diameter hole** through the second cylinder using concentric placement.

Enable **iMate** to automatically add an **INSERT** iMate to the hole.

9. Go to **File→iProperties**.

10. Fill in the **Project** tab of the **Properties** dialog.

Location:	
File Subtype:	Modeling
Part Number:	ex17-2
Stock Number:	
Description:	SWIVEL
Revision Number:	A
Project:	EVERGREEN VALLEY COLLEGE
Designer:	E. MOSS
Engineer:	
Authority:	MS. TEACHER
Cost Center:	

11. 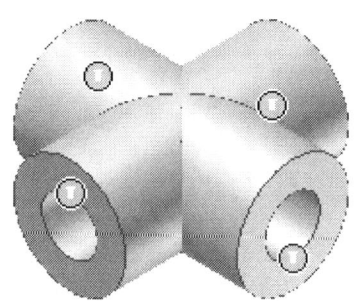 Save as *ex17-2.ipt*.

Exercise 17-3
Part 3 – Plate

File: New using custom-inches.ipt template (inches)
(Template was created in Lesson 15)

Estimated Time: 10 minutes

Review Skills: Extrude
Hole

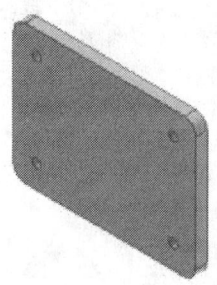

1. Start a new part file using the *custom-inches.ipt* template.

2.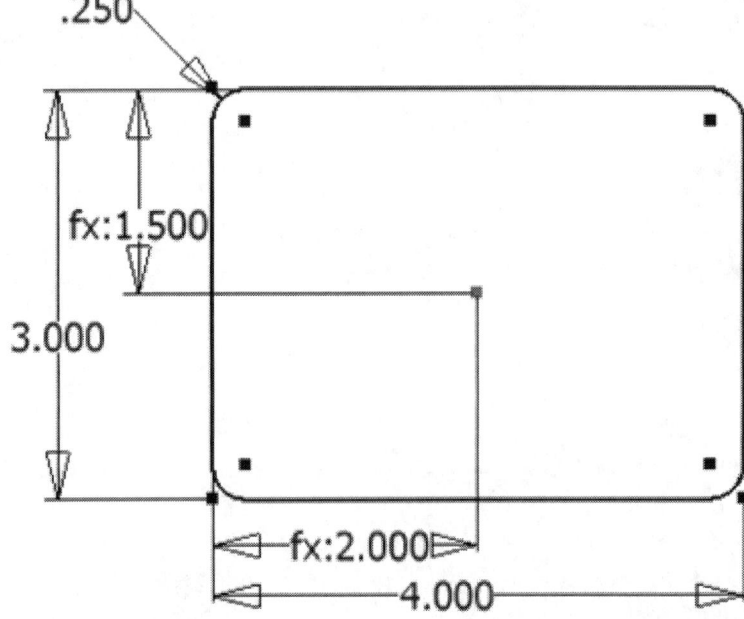

Create the sketch shown.

Bottom Up Assembly

3.

Extrude **.25**.

4. 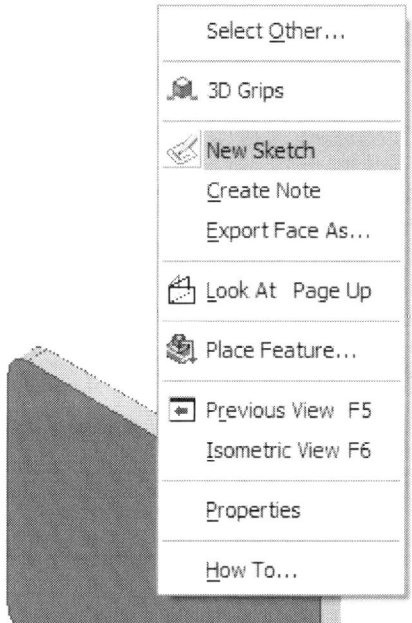 Select the front face for a New Sketch.

5. Draw a rectangle and center it on the plate as shown.

17-15

6.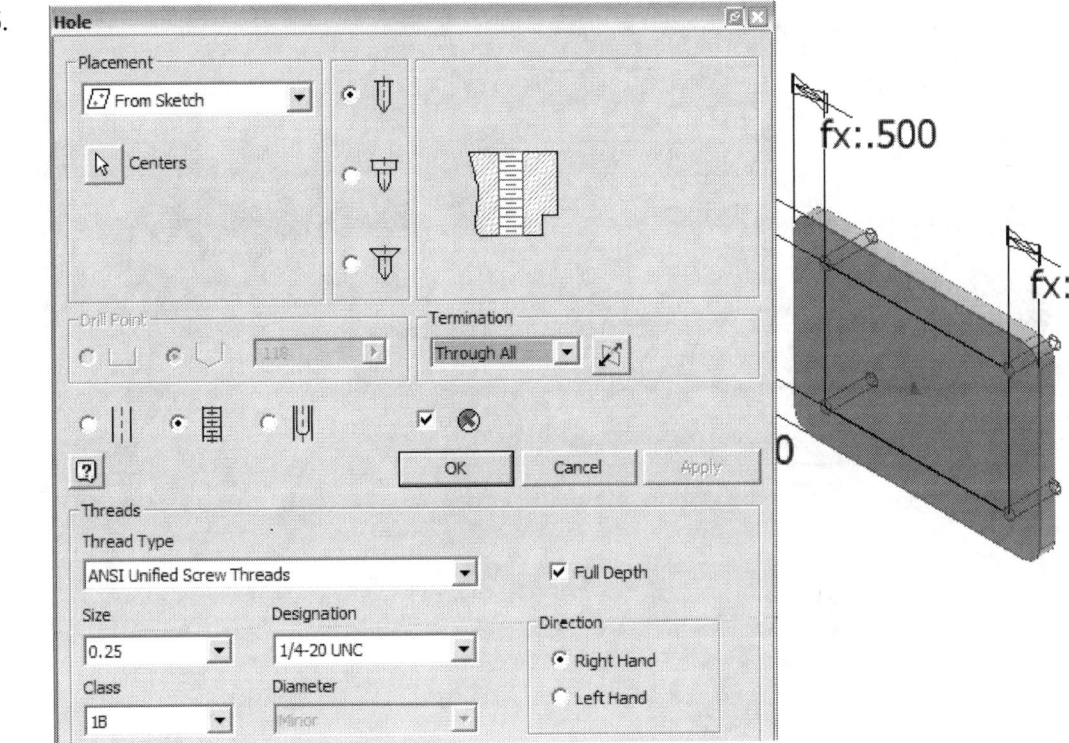

Select the vertices of the rectangle as the center points for the rectangle.
Define the holes as **¼-20 tapped thru**.
Enable the **iMate** option.
Press **OK**.

7. Go to **File→iProperties**.

8. | | |
|---|---|
| Part Number: | EX17-3 |
| Stock Number: | |
| Description: | PLATE |
| Revision Number: | A |
| Project: | EVERGREEN VALLEY COLLEGE |
| Designer: | E. MOSS |
| Engineer: | |
| Authority: | MS. TEACHER |

Fill in the **Project** tab for the file properties.

9. 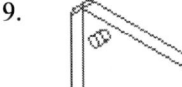 Save the file as *ex17-3.ipt*.

Exercise 17-4
Part 4 – Bushing

File: New using custom-inches.ipt template (inches)
(Template was created in Lesson 15)

Estimated Time: 10 minutes

Review Skills: Extrude
Revolve

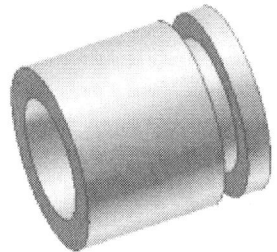

1. Start a new part file using the *custom-inches.ipt* template.

2. 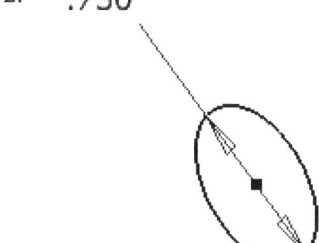 Create a .75 diameter circle constrained to the center point.

3.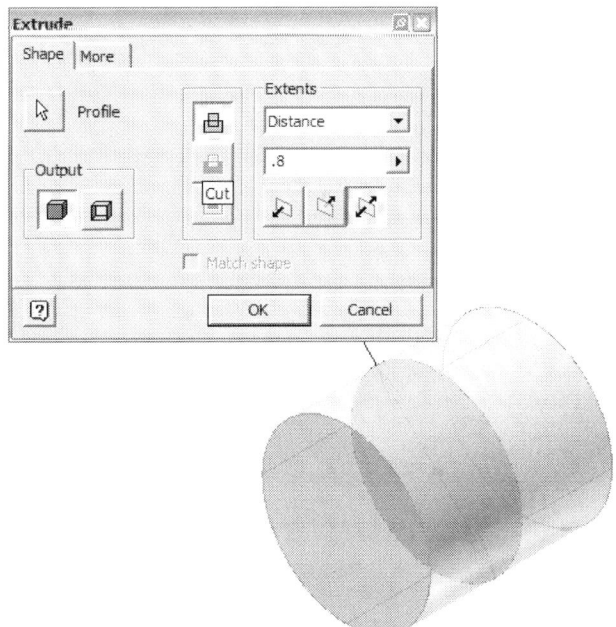

Extrude **.80** as mid-plane.

17-17

Autodesk Inventor Fundamentals

4. 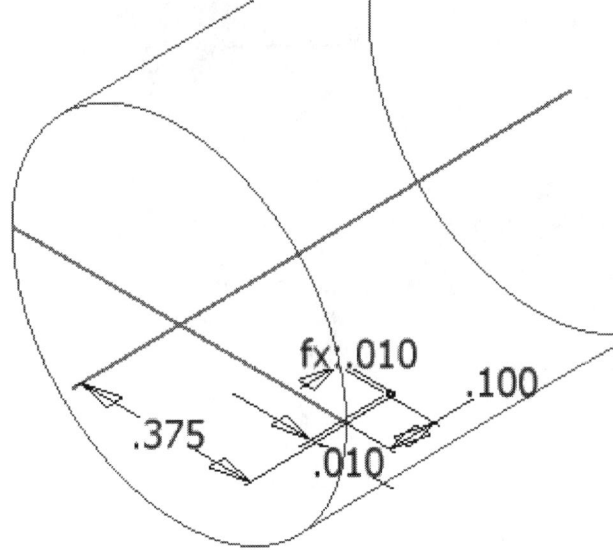 Highlight the **Top** plane.

   Right click and select **New Sketch**.

5.

   Draw a .01 x .01 square and locate it as shown: .375 from the center axis and .10 from the top of the bushing.

6.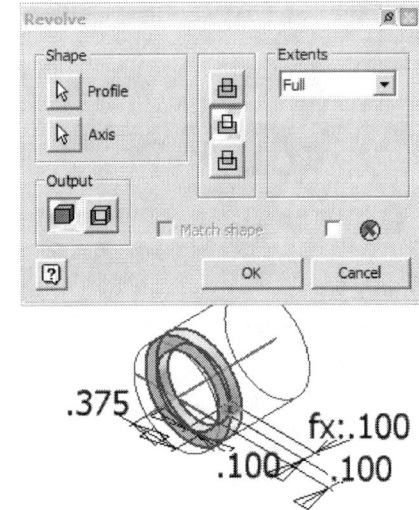

   Revolve as a cut using the central axis.

17-18

Bottom Up Assembly

7.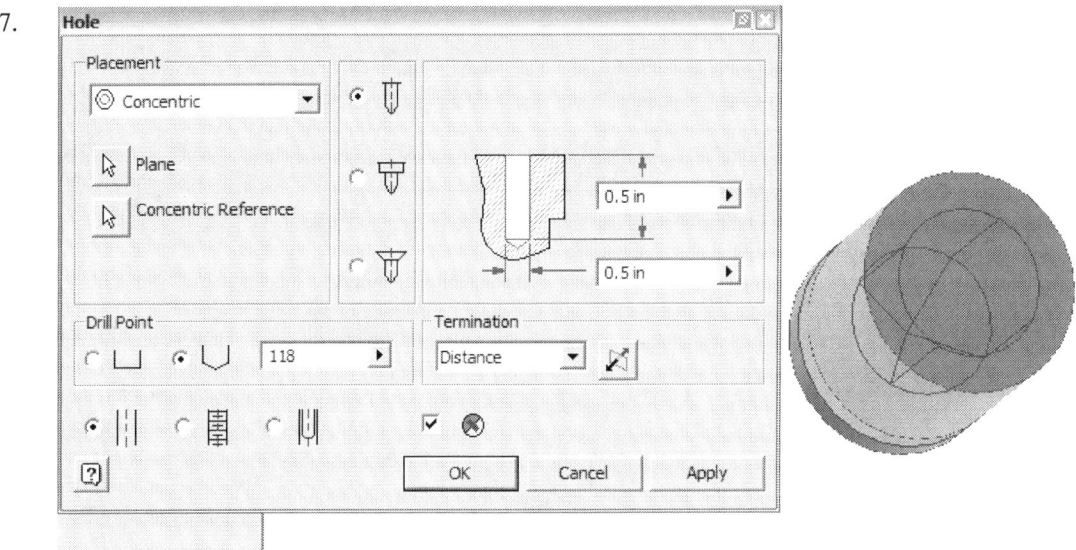

   Create a **.5** drill hole **.5 deep** at the center of the cylinder.
   Enable **iMate**.

8. Go to **File→iProperties**.

9. 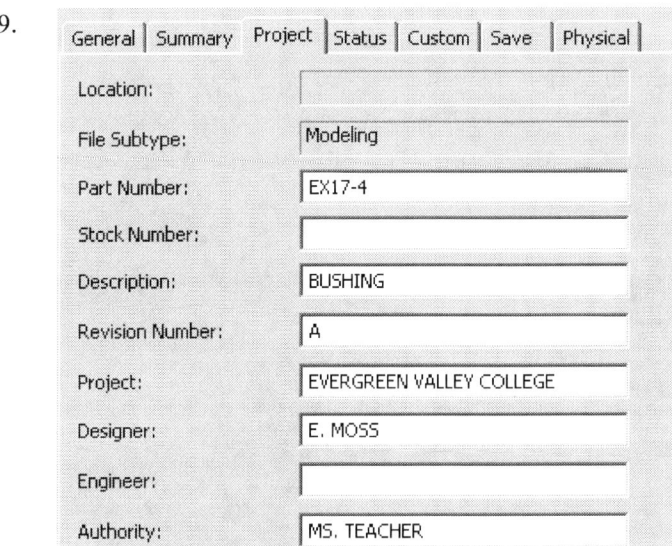 Fill in the properties for the part.

10.
    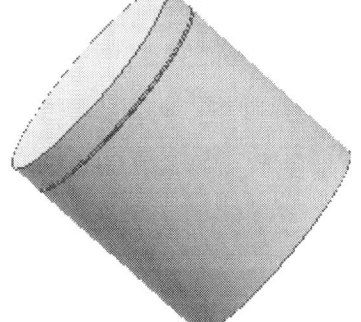

    Save as *ex17-4.ipt*.

17-19

Exercise 17-5
Part 5 – Pin

File: New using custom-inches.ipt template (inches)
(Template was created in Lesson 15)

Estimated Time: 5 minutes

Review Skills: Extrude

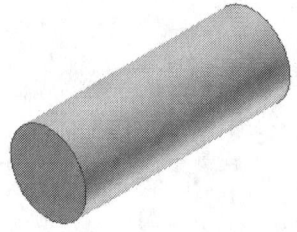

1. Start a new part file using the *custom-inches.ipt* template.

2. 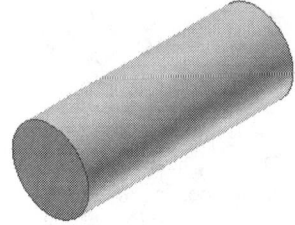 Create a .50 diameter by 1.35L pin.

3. Select the **iMate** tool.

4.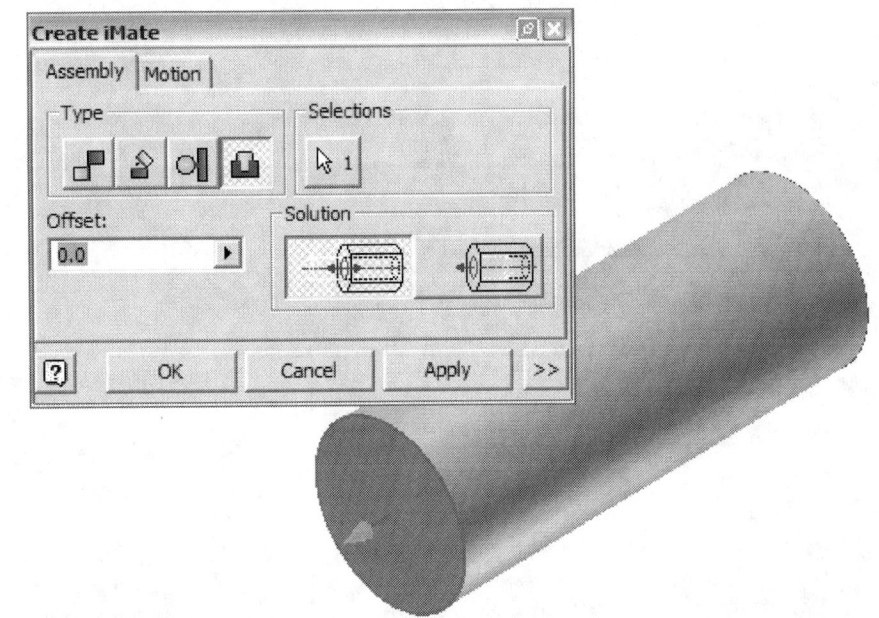

Select the **Insert** option.
Select the end face of the cylinder.
Set the **Offset** to **0.0**.
Press **Apply**.
Repeat for the other side.

Bottom Up Assembly

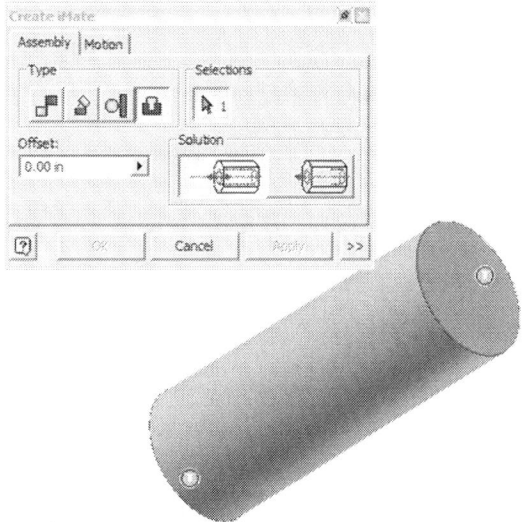

The iMate glyphs will be visible.

5. Go to **File→iProperties**.

6. Fill in the **Properties** for the pin part.

7. Save as *ex17-5.ipt*.

Exercise 17-6
Part 6 – Retaining Clip

File: New using custom-inches.ipt template (inches)
(Template was created in Lesson 15)

Estimated Time: 30 minutes

Review Skills: Extrude
iProperties

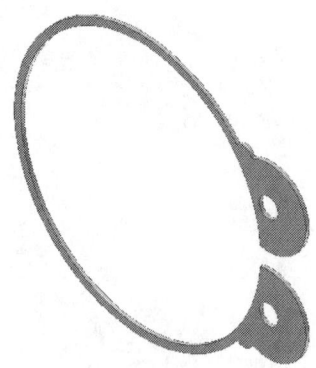

1. Start a new part file using the *custom-inches.ipt* template.

2.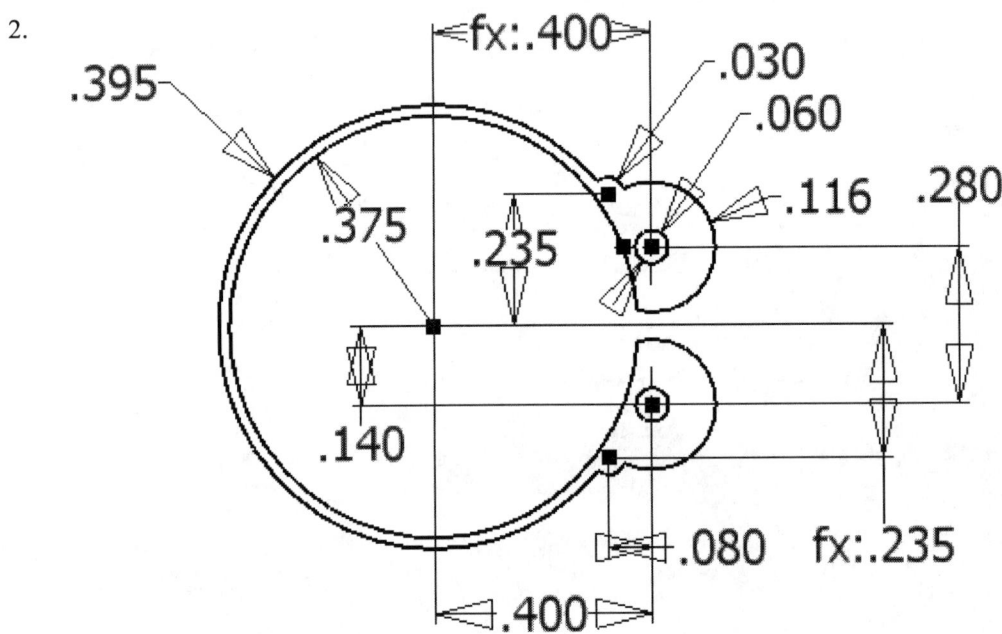

Create the sketch.

3.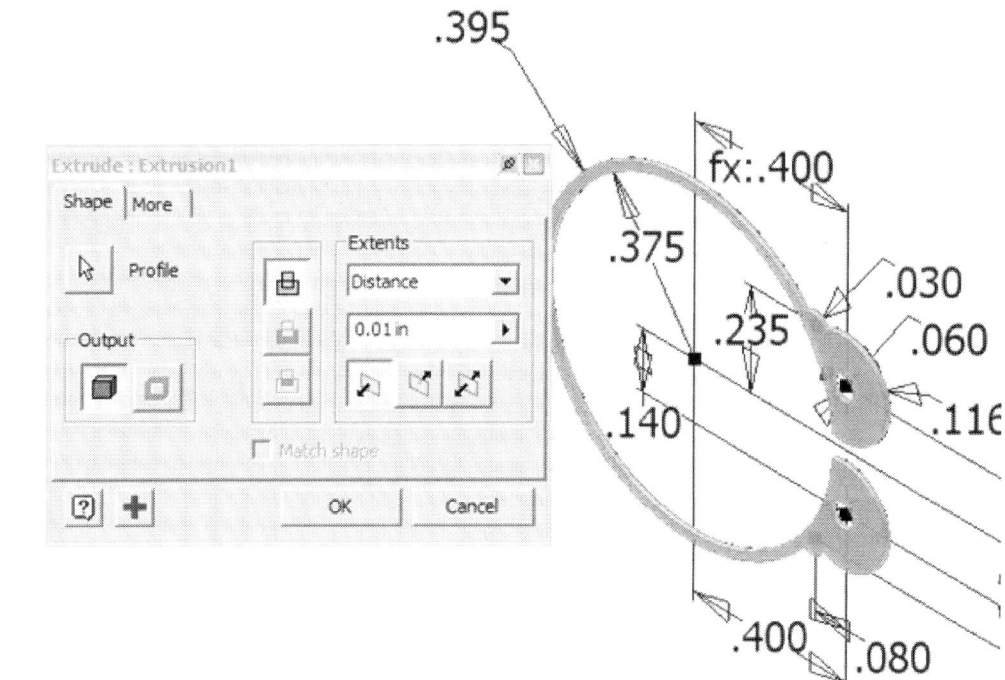

Extrude .01 in.

4. 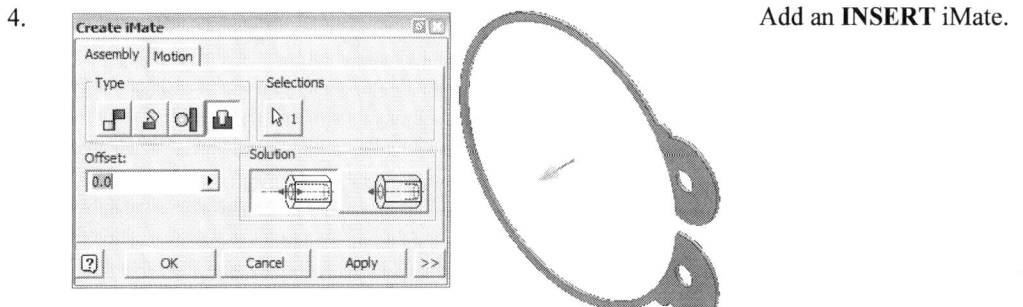   Add an **INSERT** iMate.

5. Go to **File→iProperties**.

6. 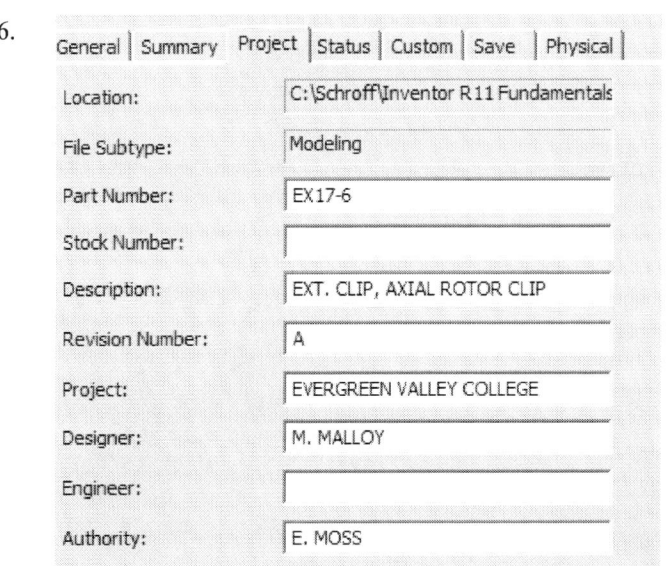   Fill in the **Properties** for the file.

7. 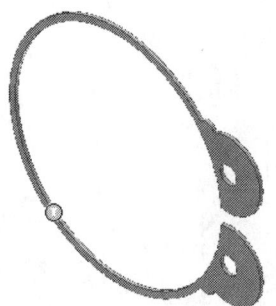 Save as *ex 17-6.ipt*.

We've created all the parts we need to build our model.

Bottom Up Assembly

Exercise 17-7
Assembling the Yoke

File: New using *.iam

Estimated Time: 30 minutes

Review Skills: Insert Constraint
                 Weld Assembly

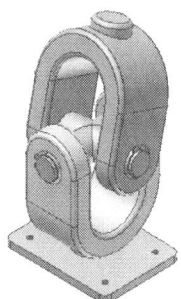

1. 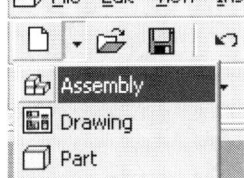 Start a new assembly file.

2. Using Windows Explorer, drag in all the components we just made.

We want to constrain the base plate so it is centered in the assembly and then ground it.

17-25

Autodesk Inventor Fundamentals

3.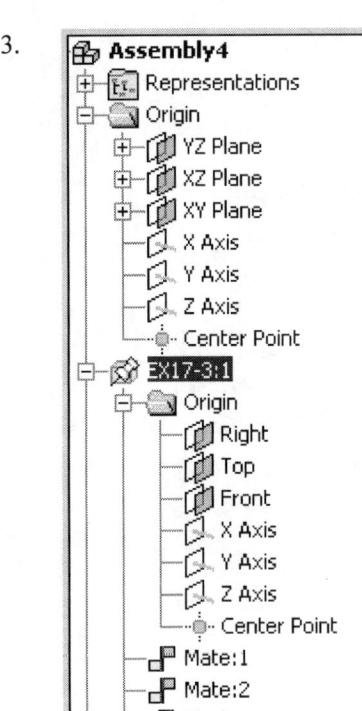

You can expand the reference planes under *ex17-3.ipt* and use them to mate with the assembly work planes.

Then ground *ex17-3.ipt*.

NOTE: *If the ex17-3.ipt is grounded at insertion, you will have to remove the ground constraint before you can constrain it to the assembly work planes.*

We are going to attach one U-Joint to the plate. To do this, we need to modify the plate.

4.

Highlight *ex17-3.ipt*.
Right click and select **Edit**.

All the other components will become opaque.

5.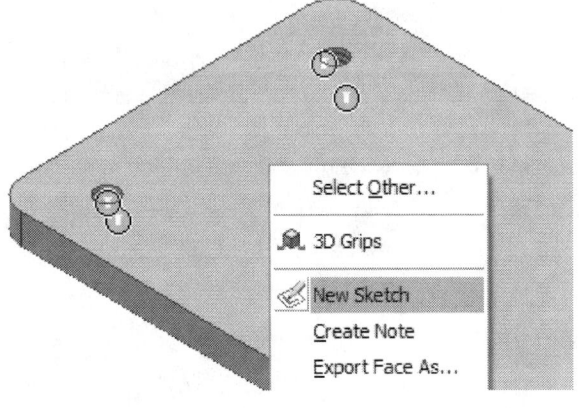

Select the top face of the plate.
Right click and select **New Sketch**.

6.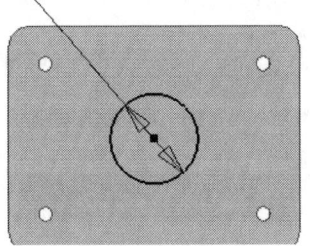

Draw a **1.20** diameter circle.

Constrain to the assembly's center point.

Locate on the top of the plate.

Right click and select **Finish Sketch**.

17-26

7.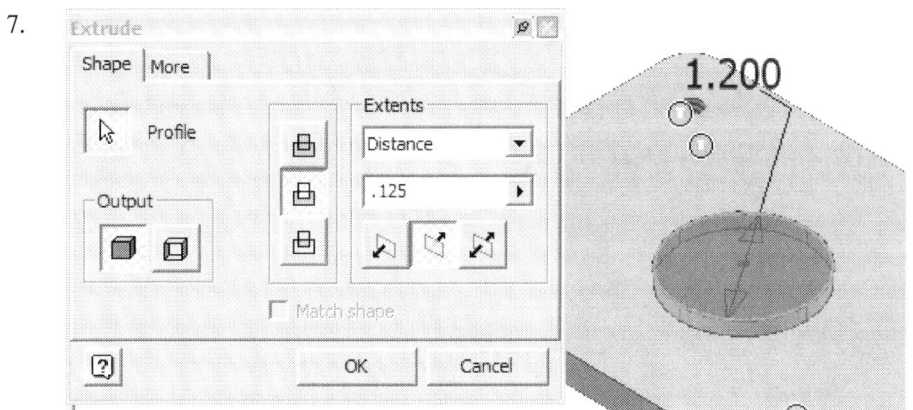

Extrude the sketch as a cut a depth of 0.125.

8.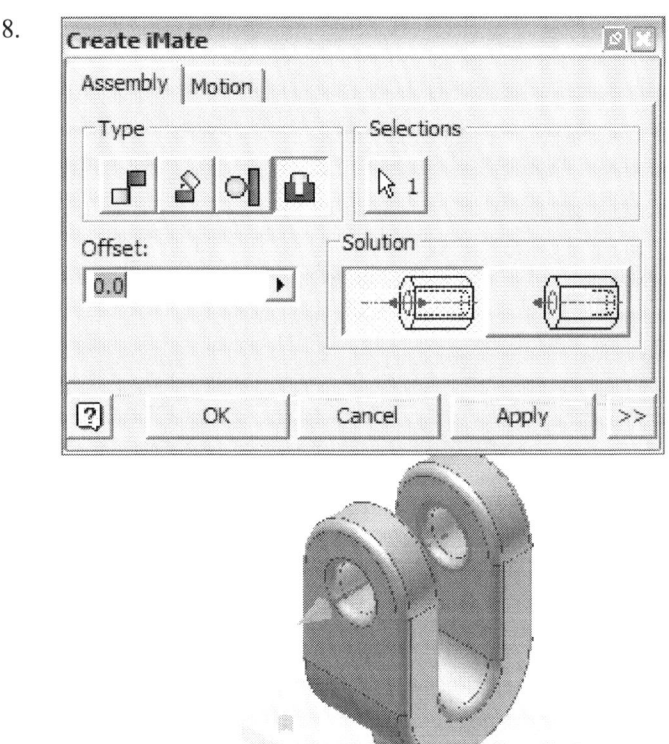

Add an **INSERT** iMate to the holes – both inside and outside.

9. ⇐ Return   Select **Return**.

10.

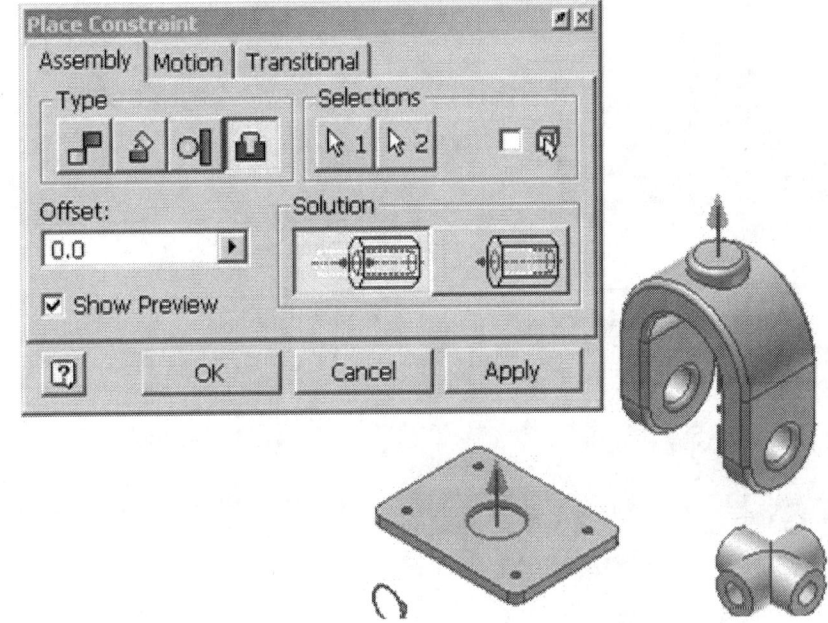

Insert the U-Joint into the preparation cut.

11.  Use a **MATE** constraint between the work plane on the U-Joint and the work plane in the assembly/plate to orient the U-joint properly in the plate.

TIP: When you use Copy and Paste to create additional instances of a component, the pasted component will appear at the position of your cursor.

Bottom Up Assembly

12. 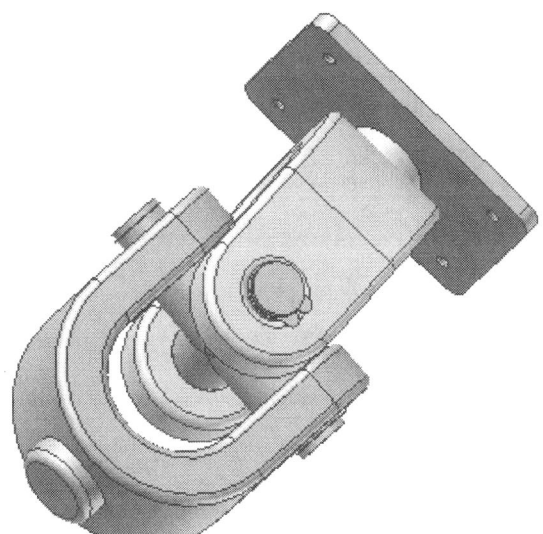 Create a second copy of the U-Joint. You will need four bushings, four retaining rings, and four pins.

    The pins are inserted into the U-Joints. The bushings are placed on top of the pins and then held in place with the retaining rings.

13. **Copy** — Select the component to be copied.
    Right click and select **Copy**.

14. **Paste** — Position your cursor to where you want the copy placed.
    Right click and select **Paste**.

15. Use **Insert** constraints to create the Yoke Assembly.

16. Place your mouse over the top U-Joint and see how the U-Joint moves in the yoke assembly.

17. 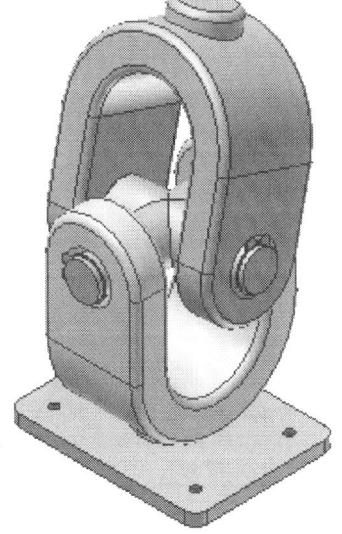 Save the assembly as *ex17-7.iam*.

    When prompted to save the edits to *ex17-3.ipt*, press **OK**.

17-29

Exercise 17-8
Using the Contact Solver

File: ex17-7.iam

Estimated Time: 10 minutes

Review Skills: Contact Solver
Contact Set

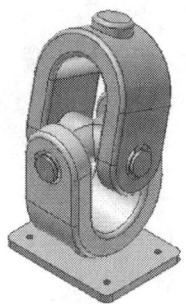

1. Open *ex17-7.iam*.

2. 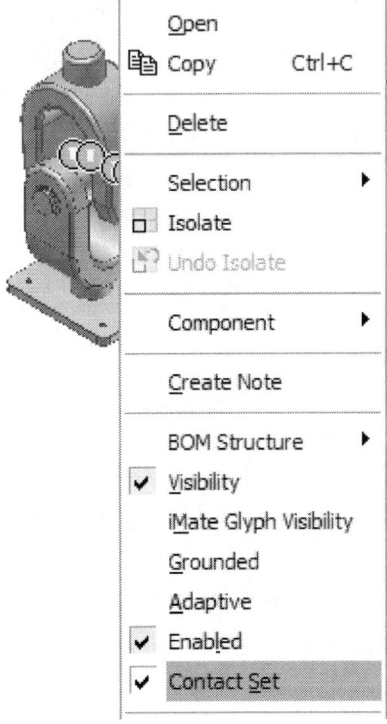   Select the top U-Joint in the model window. Right click and select **Contact Set**.

   This adds it to the contact set to be used by the **Contact Solver**.

3. ex17-1:3   The icon changes in the Browser to indicate a contact set.

Bottom Up Assembly

4. 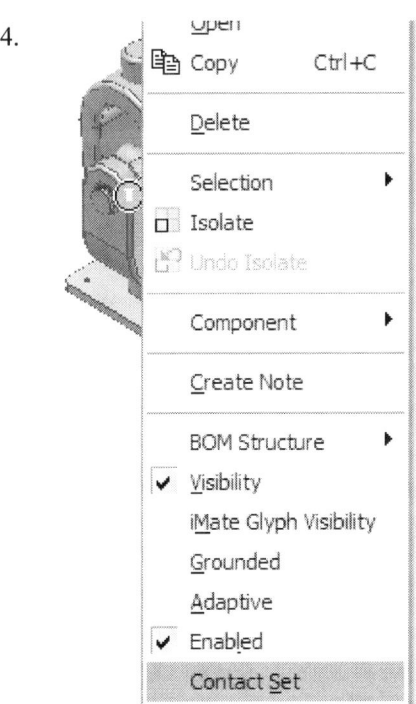 Locate the other U-Joint in the Browser.

   Right click and select **Contact Set**.

5. Go to **Tools→Activate Contact Solver**.

6. Move the top U-Joint around using the mouse.
   Note how the motion is now constrained as you move around whenever the joints collide.

7. Save as *ex17-8.iam*.

Positional Representations are used to create overlay views. The different positions indicate the path moving components and subassemblies can take.

Exercise 17-9
Creating Positional Representations

File: ex17-8.iam
Estimated Time: 30 minutes

1. Open the assembly.

2.

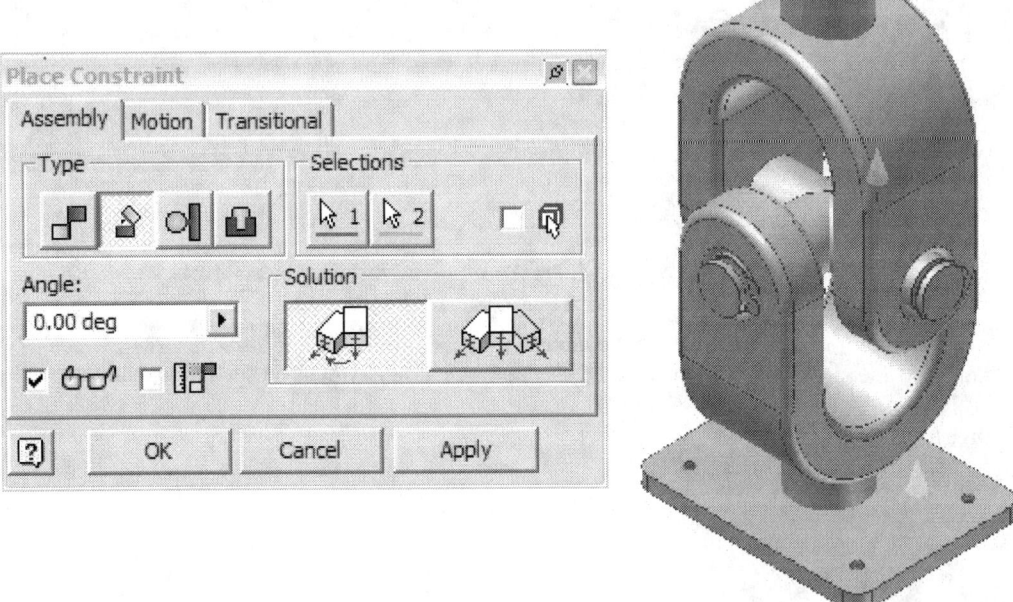

Place an angle constraint between the pin axis and the top of the plate to align the top U-joint.

3. 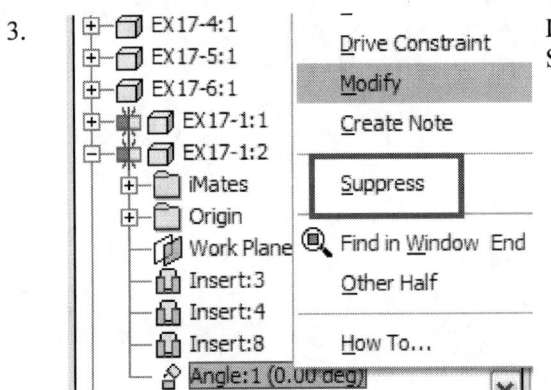 Locate the angle constraint in the Browser. Suppress the angle constraint you just placed.

Bottom Up Assembly

4.

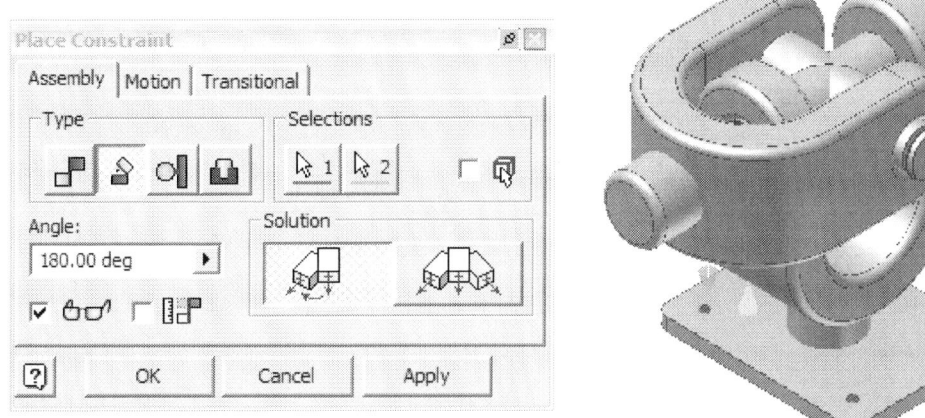

Place an angle constraint between the front face of the top U-joint and the top of the plate. The angle constraint should be a directed angle.

5.

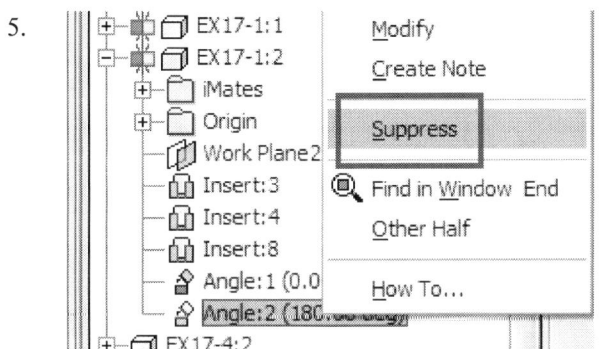

Locate the second angle constraint in the Browser.

Right click and select **Suppress**.

6.

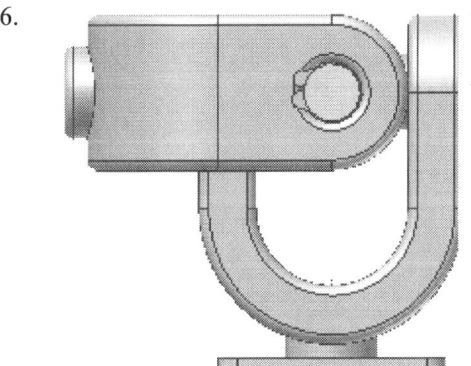

Orient the model so it is facing as shown.

Hint: Use Common View to orient the assembly.

7. 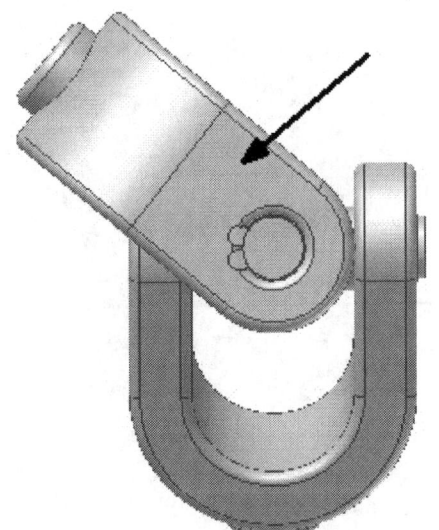 Place your mouse over the top link.
Hold down the left mouse button to move the link from left to right.

8. 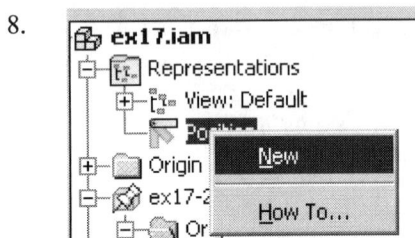 Expand the Representations folder.
Highlight the **Position** category.
Right click and select **New**.

9. The master position is automatically created. This is the default position.
The PositionalRep1 is a copy of the master.

10. Rename the position **Leftside**.
To rename, simply left click on the name to activate the text edit mode.

11. Highlight the **Position** category.
Right click and select **New**.

12. Rename the Position **Middle**.

13. 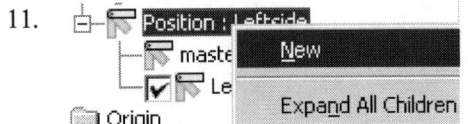 Highlight the **Position** category.
Right click and select **New**.

Bottom Up Assembly

14. Rename the Position **Rightside**.

15. You should have four positions defined:
    Master
    Leftside
    Middle
    Rightside

16. Highlight the **Leftside** position.
    Right click and select **Activate**.

17. A checkmark should appear next to the Leftside position to indicate that it is active.

18. Highlight the part *ex17-1:1* that is the top link.

    Right click on the **suppressed Angle constraint** and select **Override**.

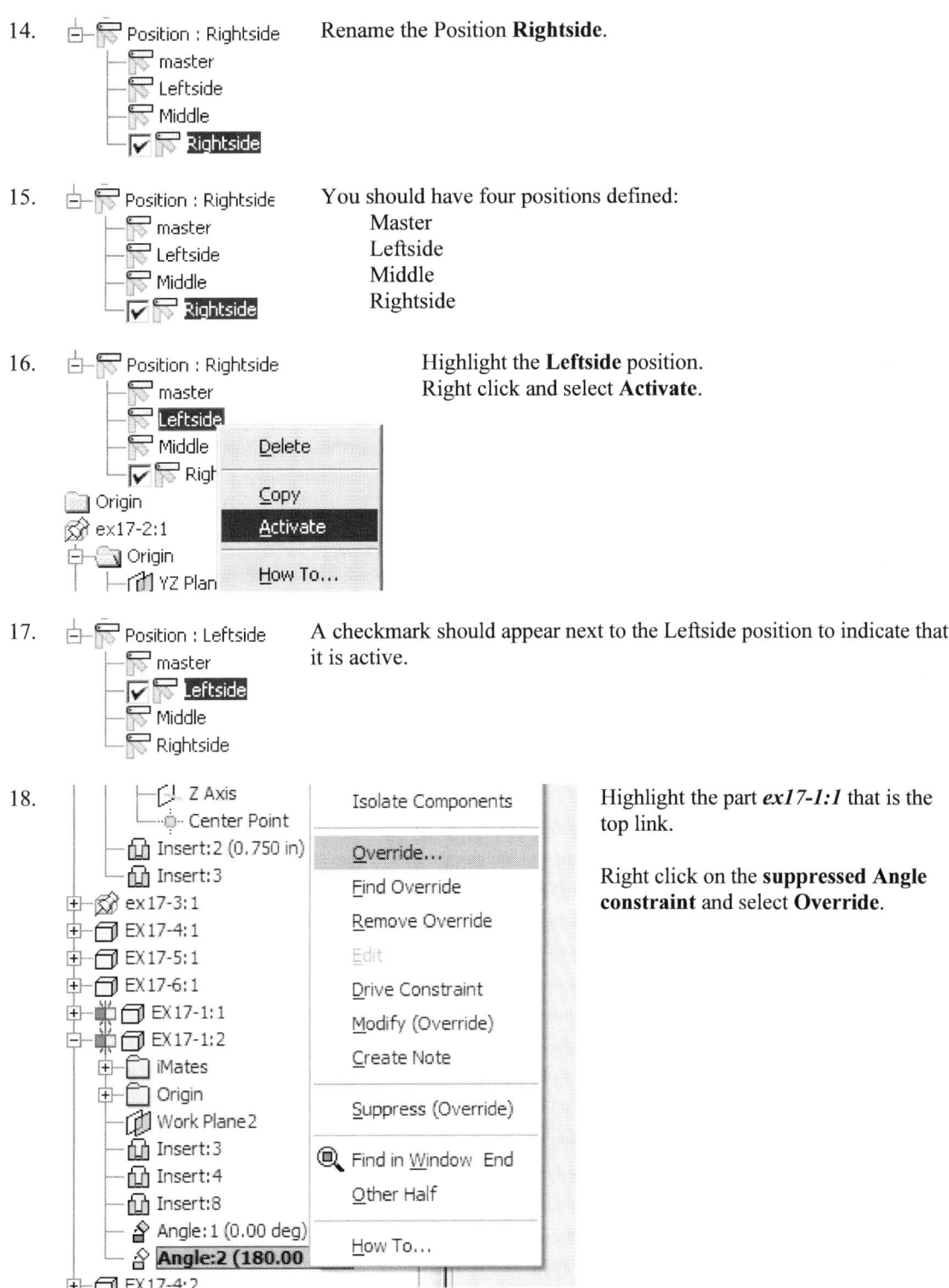

19.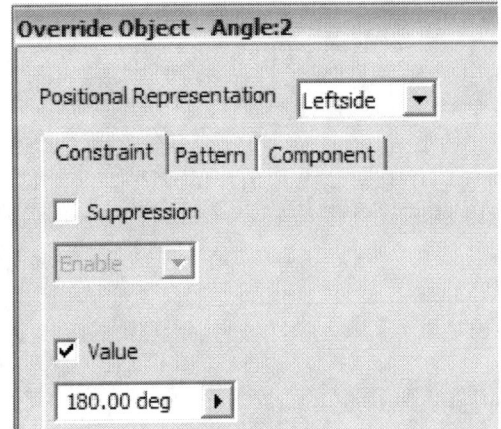

    Select the **Constraint** tab.

    You should see **Leftside** as the active **Positional Representation** at the top of the dialog.

    Make sure **Suppression** is disabled.
    Enable the **Value** box.
    Press **Apply**.

20.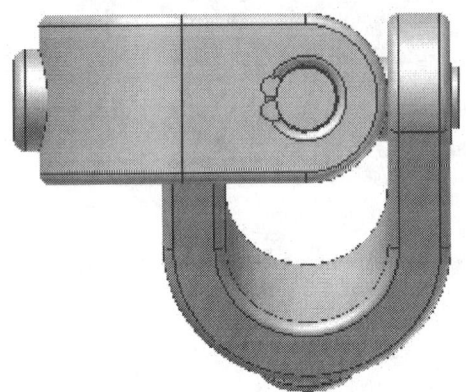

    The link flips to the left side.

21.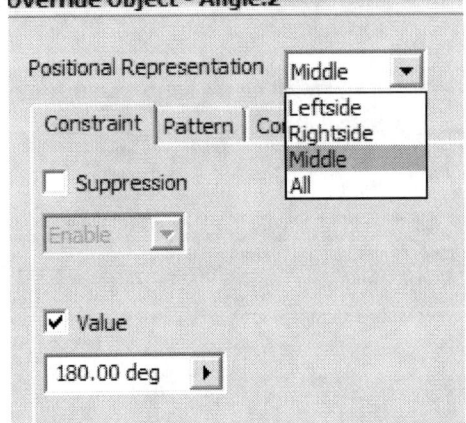

    Under **Positional Representation**, select **Middle** from the drop-down list.

22. 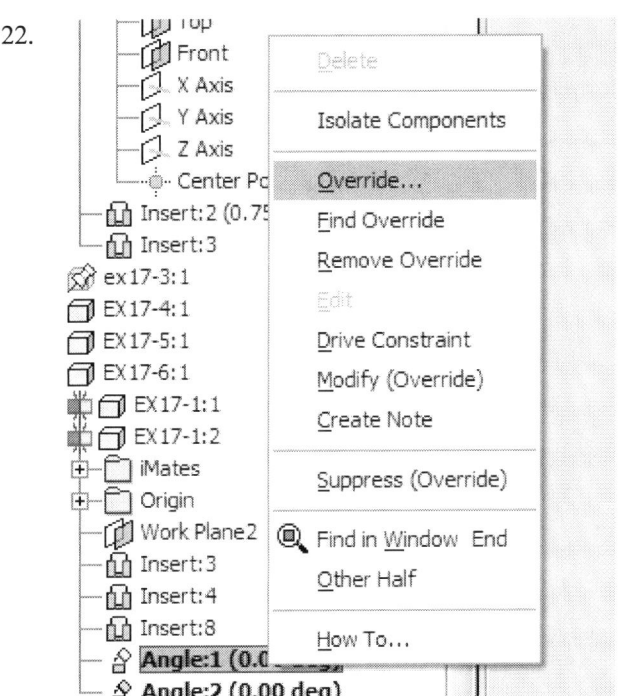 Suppress the second angle constraint.
Enable the first angle constraint.

Highlight the first angle constraint.
Right click and select **Override**.

23. 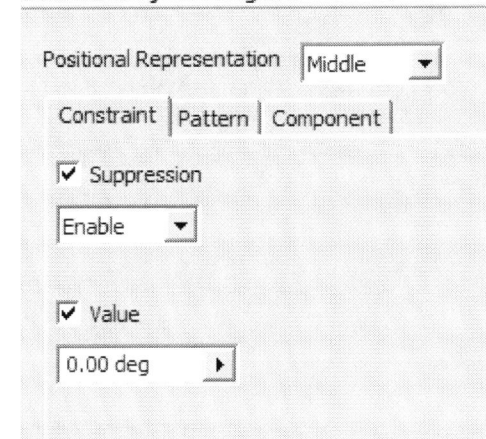 Select the **Constraint** tab.

Select the **Middle Positional Representation**.

Enable **Suppression**.

Select **Enable** from the drop-down. This will enable the Angle constraint.

Enable **Value**.

Set the angle to **0.00**.

Press **Apply**.

NOTE: *The link will not move until the representation is activated.*

Autodesk Inventor Fundamentals

24. 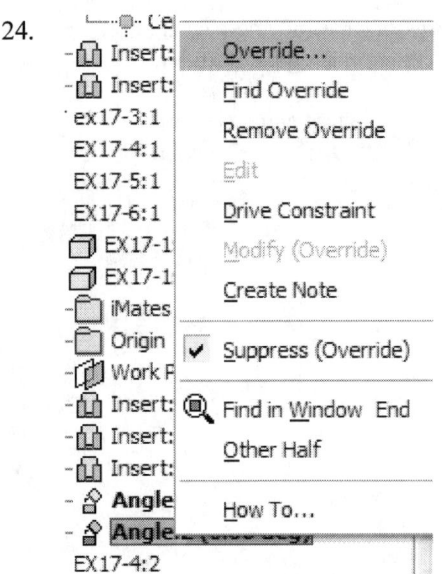 Highlight the second angle constraint.
Right click and select **Override**.

25. Under Positional Representation, select **Rightside** from the drop-down list.

26. Select the **Constraint** tab.

Under **Suppression**:
Enable **Suppression**.
Select **Enable** from the drop-down. This will enable the Angle constraint.

Under **Value**:
Set the angle to **0.00**.

Press **Apply**.

NOTE: The link will not move until the representation is activated.

27. Close the dialog.

28. 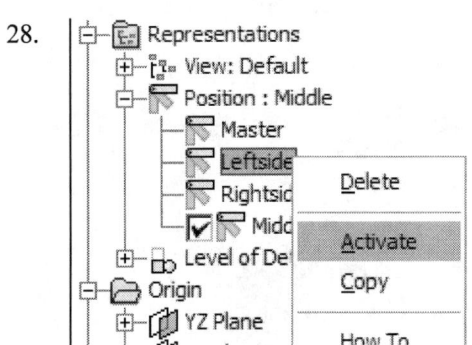 Highlight the **Leftside** position.
Right click and select **Activate**.

The link should shift to the left side.

17-38

Bottom Up Assembly

29. 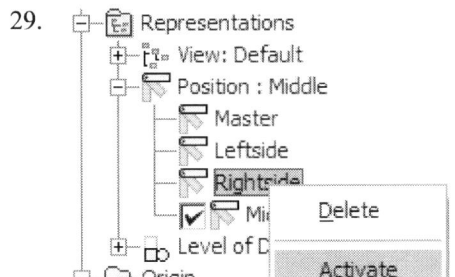 Highlight the **Middle** position.
Right click and select **Activate**.

30. 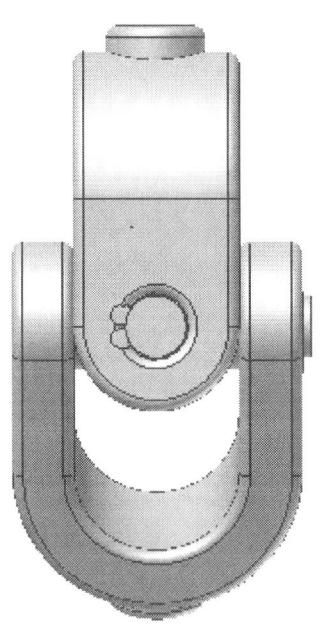 The link should shift to the middle/center position.

31. 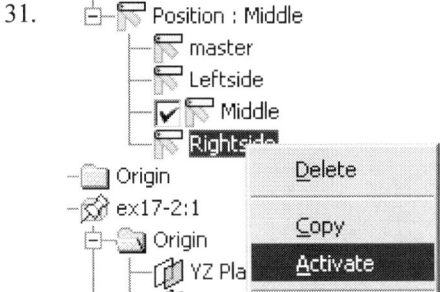 Highlight the **Rightside** position.
Right click and select **Activate**.

17-39

32. 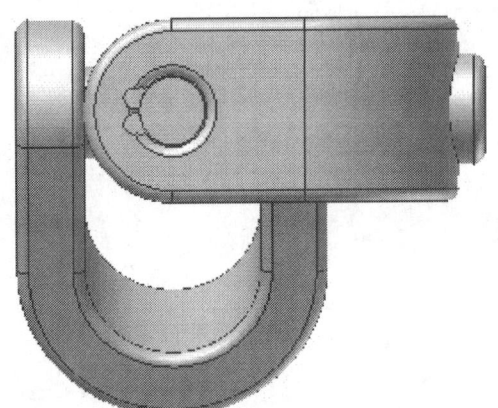   The link should shift to the right position.

You have created three positional representations, which can be used for overlay views.

33. Save as *ex17-9.iam*.

TIP: In order for this exercise to work properly, the angle constraint was applied using a Directed Angle. The angle values you may need to apply could be different depending on the initial angle value applied.

Autodesk Inventor Fundamentals

Lesson 18
Top–Down Assembly

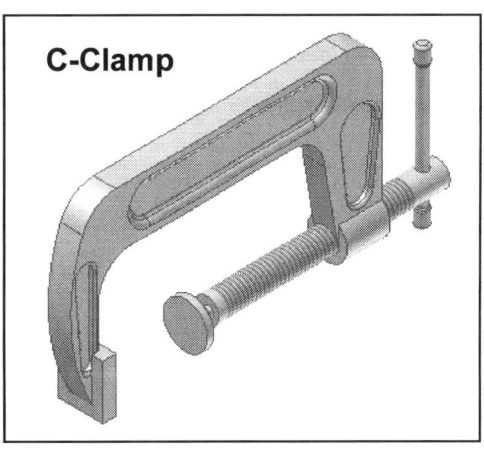

Learning Objective

In this lesson, the user will create a top-down assembly. A top-down assembly means that all the parts are created in the same file. A top-down approach allows the designer to check for interference and make changes quickly on the fly.

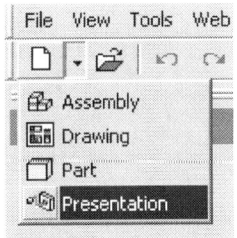

When you first open Inventor, it gives you four choices...to build an assembly, to create a drawing, create a presentation or to build a part.

Inventor provides users the flexibility to create an assembly using the top down approach or the bottom up approach.

Exercise 18-1:
C Bracket

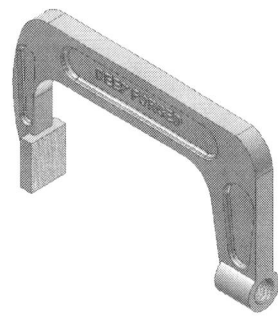

File: New Assembly (Standard using Inches)
Estimated Time: 60 minutes

This part requires use of the following tools:

- Extrude
- Mirror Feature
- Fillet
- iFeatures
- Hole
- Copy Feature
- Project Geometry

Let's create a C-clamp assembly using the top down approach. We start by opening an assembly file.

18-1

1. 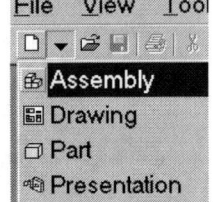 Start a new Assembly file.

We can create a New Component using four methods:

Menu		Insert→Create Component
In the Graphics Window		Right click and select 'Create Component'
In the Panel Bar		Select Create Component
Assembly toolbar		Create Component

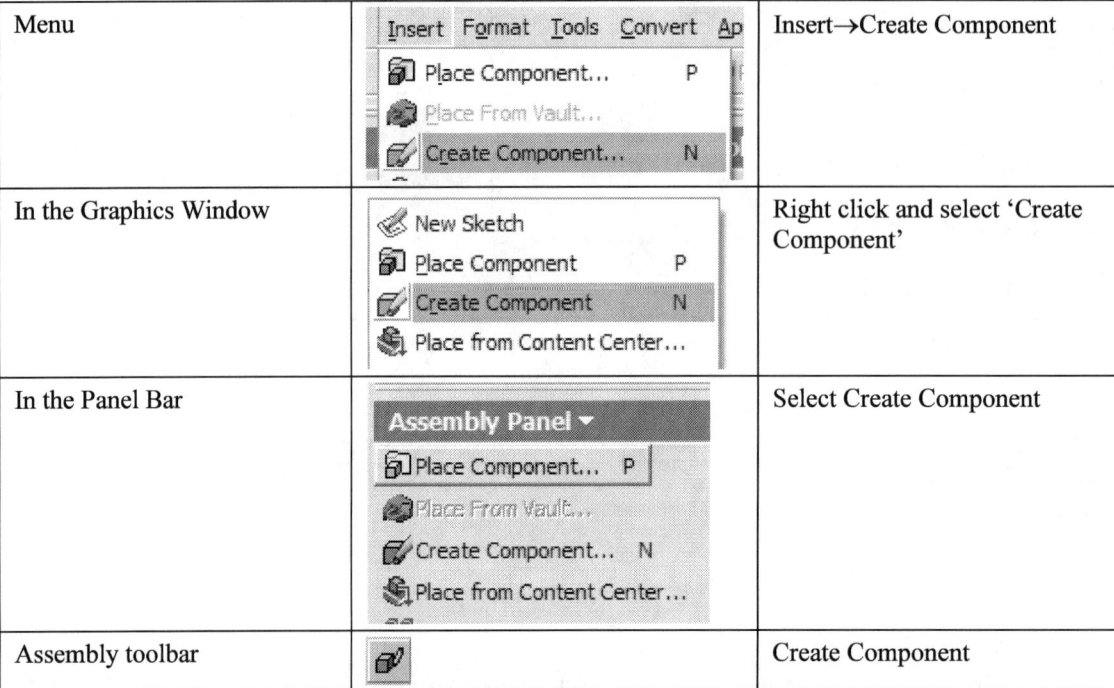

2. Start a new component using one of the above methods.

3. 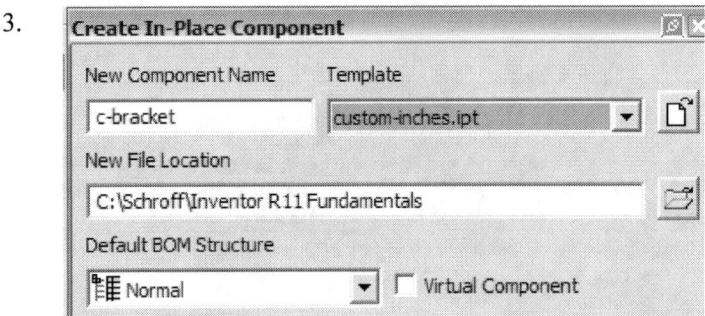 Now a new dialog box comes up.

We name our component *'c-bracket'* and note it will automatically create the external drawing file so the part can be re-used in future assemblies. We can use the Browse button to locate the subdirectory where the file will be saved.

Under the template drop down, we can select the *custom-inches template* we created in Lesson 15.

At this point, we have named our drawing file, defined it as a part, selected a template, and located it in our working directory.

Top Down Assembly

4. 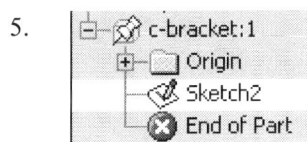 Pick the **XY Plane** in the Browser to constrain your part to the XY Plane.

5. 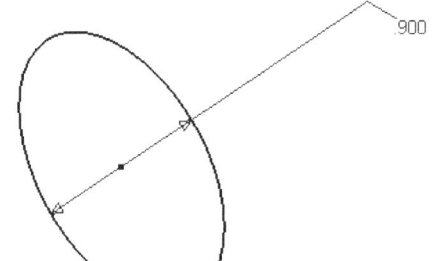 Note that the Assembly has its own origin, axes and planes. Each part is assigned an origin, plane and axes as well.

   Also, note that the first part we create is automatically grounded in the assembly. This is indicated by the push-pin next to the part name.

6. 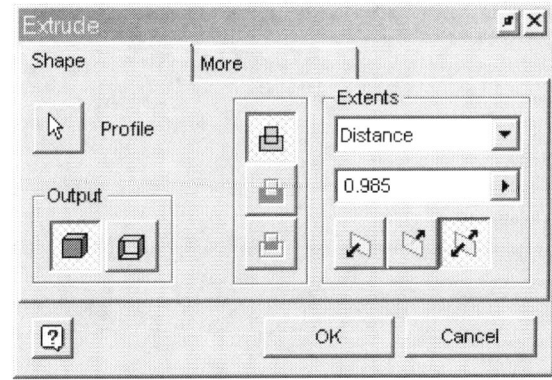 Use the **Project Geometry** tool to project the Center Point into the current sketch.

   Draw a circle with the center constrained at the projected center point.

   Set the circle's diameter to **0.9**.

7. 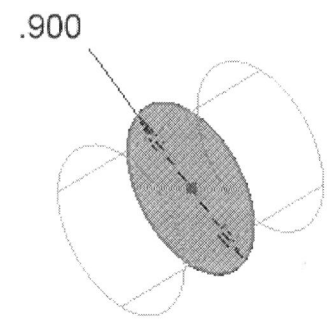 Extrude the circle using mid-plane a distance of **0.985**.

TIP: If you select 'Finish Edit' that means you are finished creating your component. Inventor to returns you to the top assembly. To return to editing the component, select the component in the Browser, right click and select 'Edit'. If you wish to edit the part outside the assembly, right click and select 'Open'.

18-3

Autodesk Inventor Fundamentals

8. 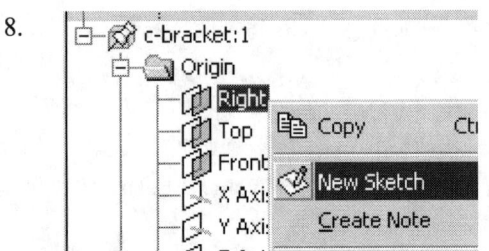 Select the **Right Plane** in the Browser.
Right click and select **New Sketch**.

9. 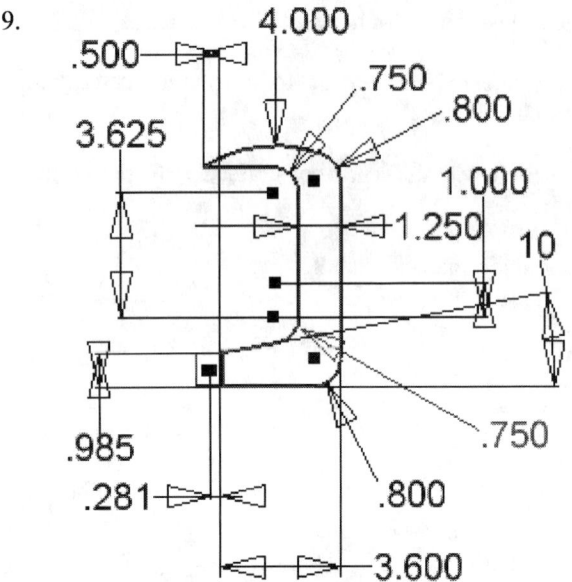 Draw the shape of the clamp as shown.

Use **Sketch Doctor** to help you diagnose any problems with your sketch.

10.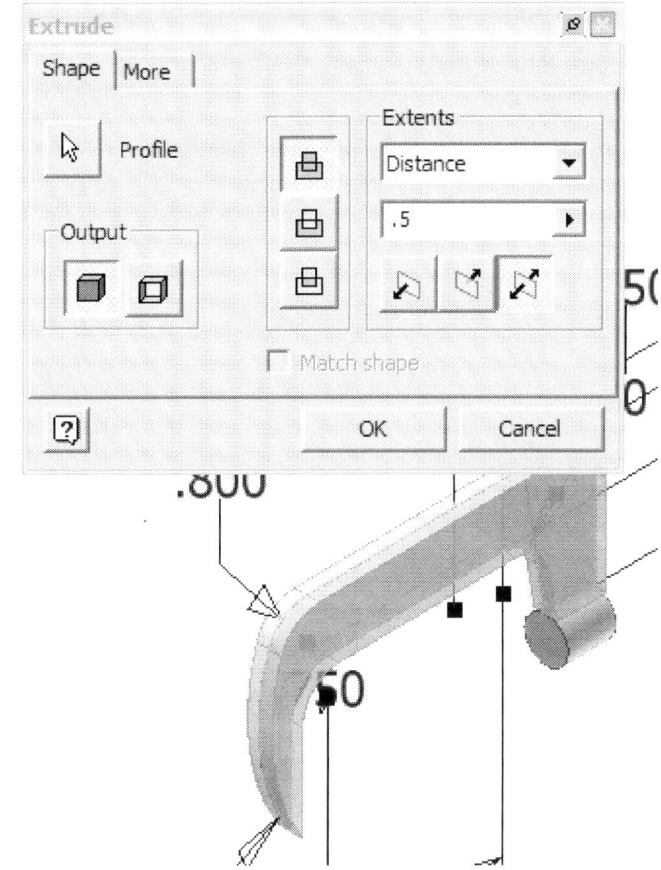

Extrude the sketch in both directions a distance of **0.5**.

11. 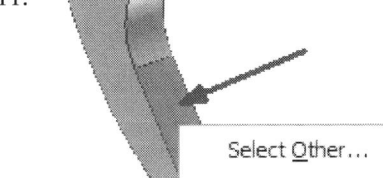 Select the inside plane indicated.

Right click and select **New Sketch**.

12. Draw a rectangle on the end of the selected plane and center it as shown.

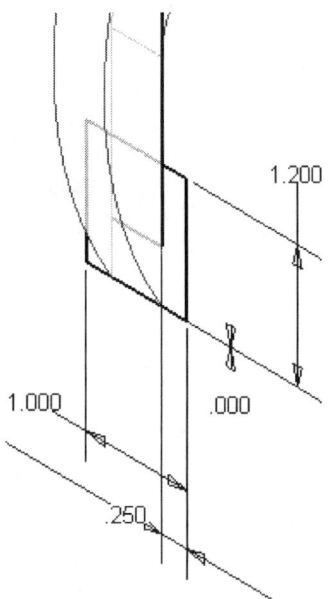

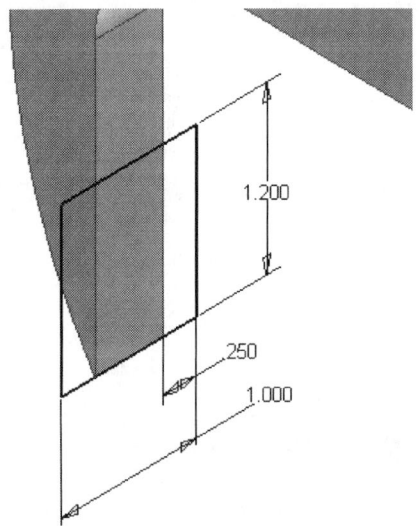

The sketch is shown from a different viewpoint to help you visualize how it is placed.

13. Extrude the rectangle in two directions a distance of **0.3**.

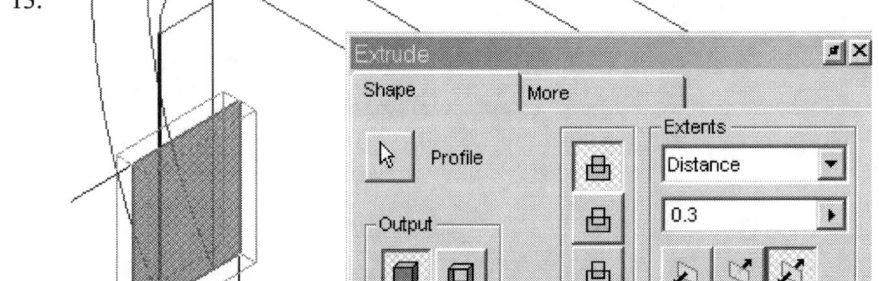

14. Select the **Insert iFeature** tool from the Features toolbar.

Top Down Assembly

15. 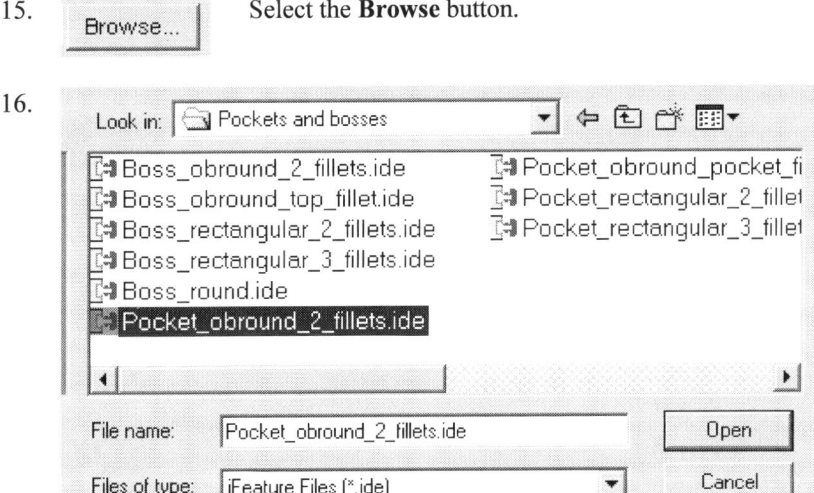 Select the **Browse** button.

16.

Locate the iFeature called ***Pocket_obround_2_fillets.ide*** in the *Pockets and bosses* folder.

Press **Open**.

17. 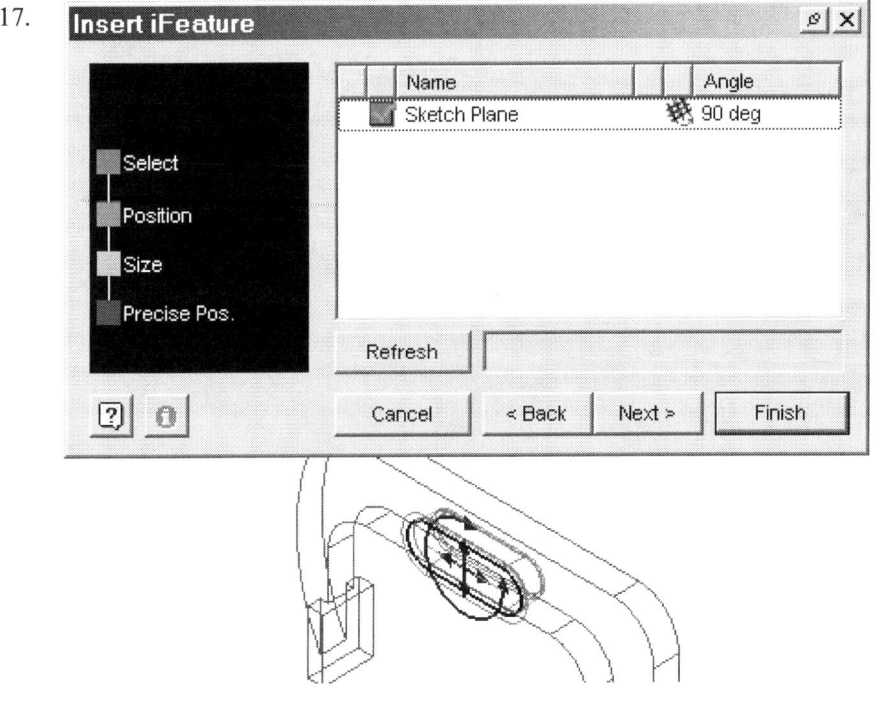   Select the flat surface and rotate the iFeature so it lies horizontally.

Press **Next**.

18-7

18. Set the **Length** to **3.75**.
    Set the **Width** to **0.75**.
    Set the **Depth** to **0.125**.
    Set the **Radius_interface** to **0.03 in**.

    Press **Next**.

Name	Value
Length	3.75 in
Width	0.75 in
Depth	0.125 in
Draft	5.00 deg
Radius_pocket	0.063 in
Radius_interface	0.03 in

19. Enable the **Activate Sketch Edit immediately**.

    Press **Finish**.

20.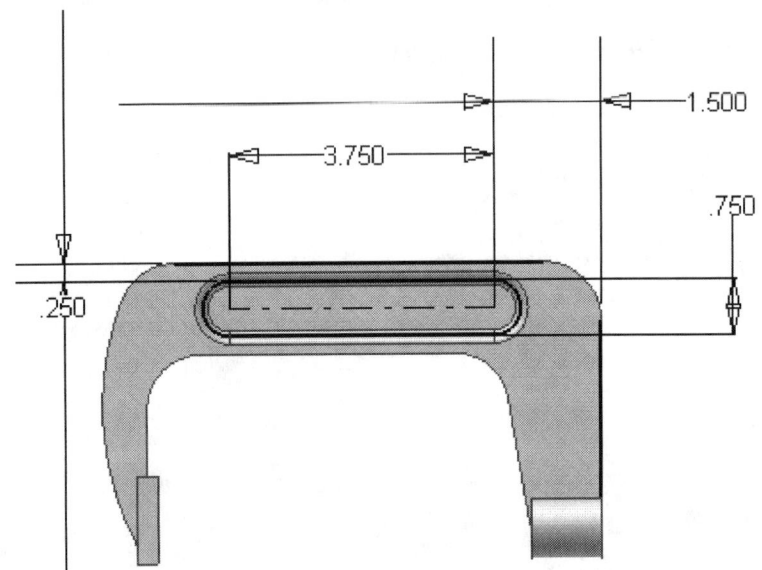

    Add a 0.25 dimension in the vertical direction.

    Add a 1.50 dimension in the horizontal direction.

    Exit Sketch Mode to finish.

21.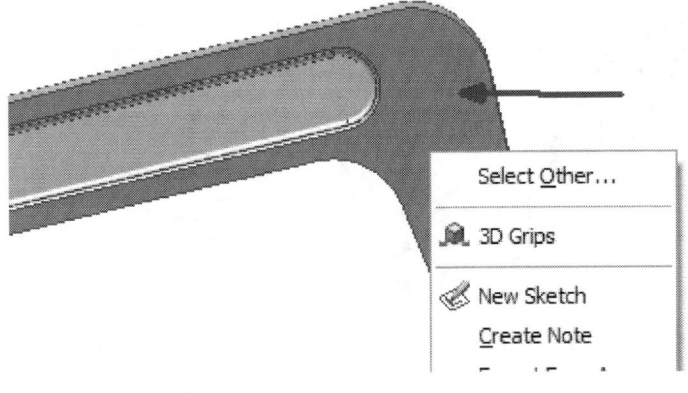

    Select the plane indicated. Right click and select **New Sketch**.

TIP: See if you can use the 'Show Dimensions' option of the Edit Dimension box to add the dimensions.

22. Create the sketch shown.

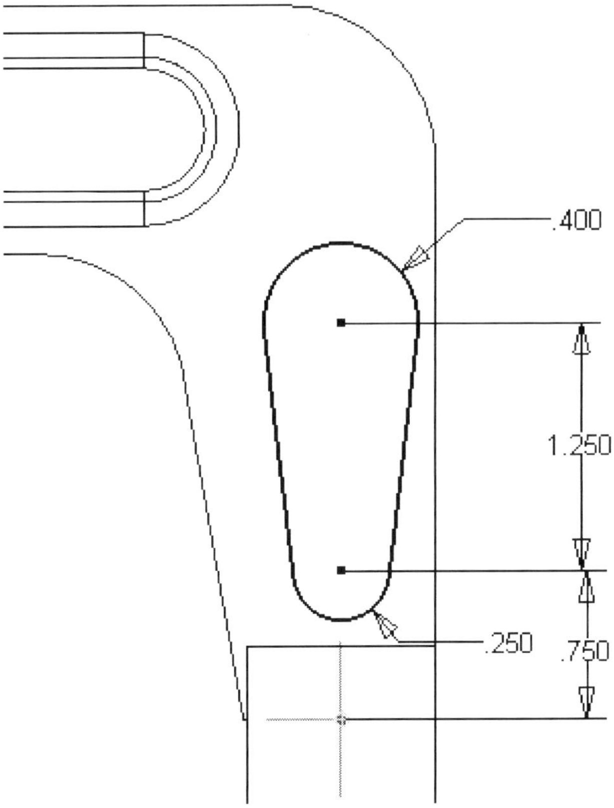

The center point was projected into the sketch.

A vertical constraint was then added to line up the center points of the arc with the sketch center point.

23. Create a cut with a depth of **0.125**.

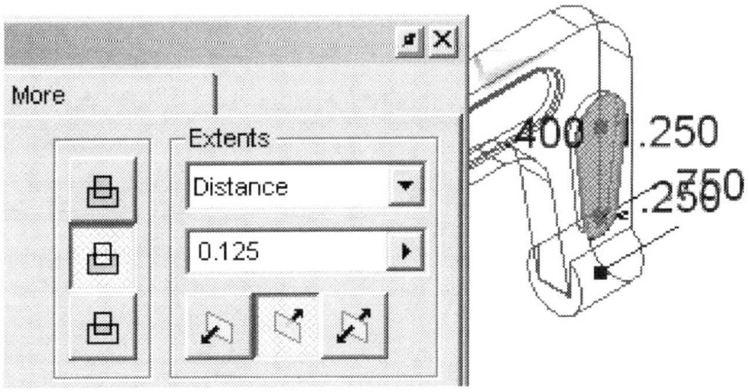

24. Define a taper angle of **-5 degrees**.

Press **OK**.

25. Select the front face again and press **New Sketch**.

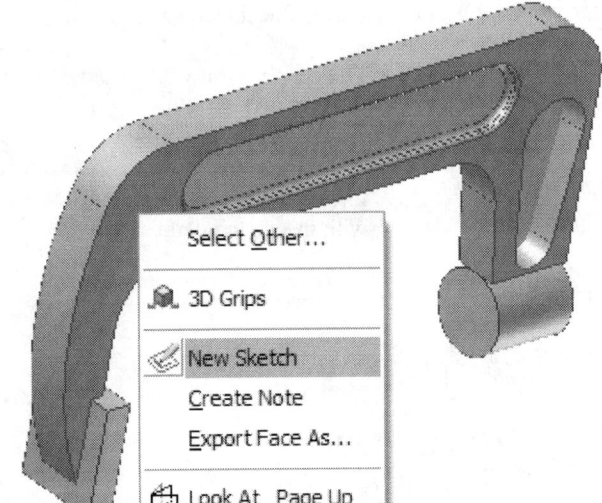

26. Create the sketch shown.

    The vertical line is set to a centerline style.

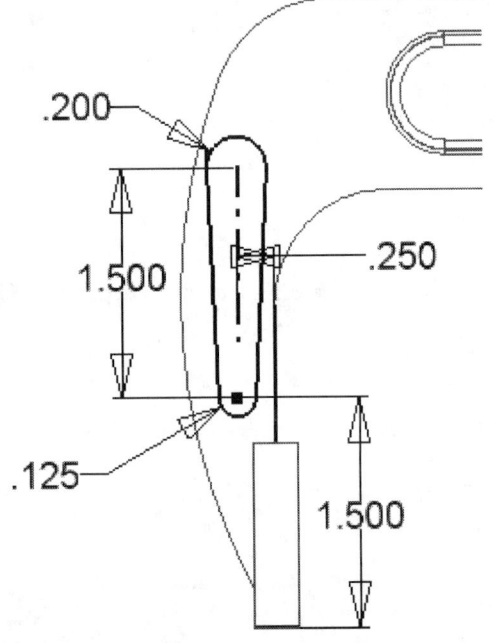

27. Create a cut **0.125 deep**.
    Select the **More** tab.
    Set the taper to **-5 degrees**.

    Press **OK**.

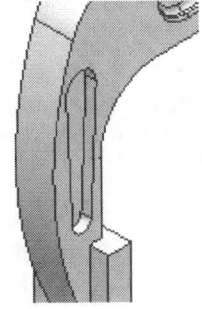

18-10

Top Down Assembly

28. 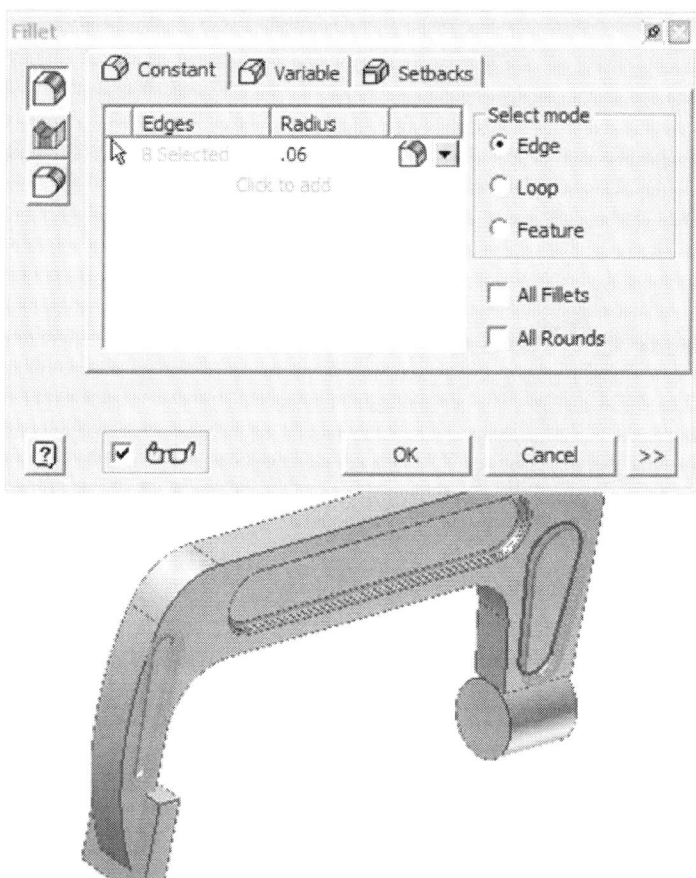 Add a **0.06** fillet to the edges of the two cuts you just made.

29. Turn the **Visibility ON** for the Right/YZ Plane.

30. 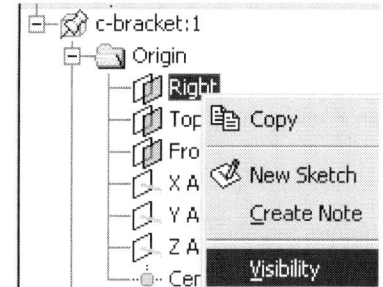 Select the **Mirror Feature** tool.

18-11

31.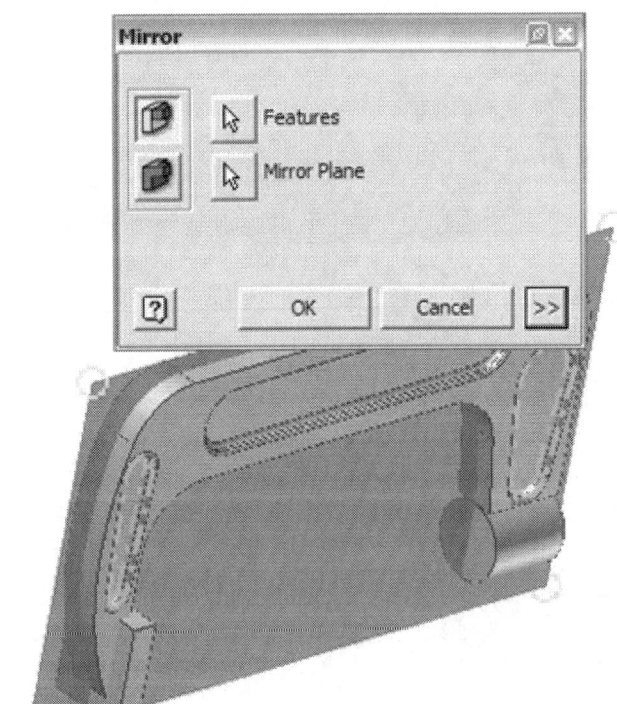

Select the two cut features.

NOTE: Do not select the Fillet.

These can be selected in the Browser or on the model.

Select the **YZ Plane**.
Press **OK**.

32.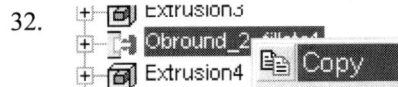

Highlight the iFeature in the Browser.

Right click and select **Copy**.

Top Down Assembly

33.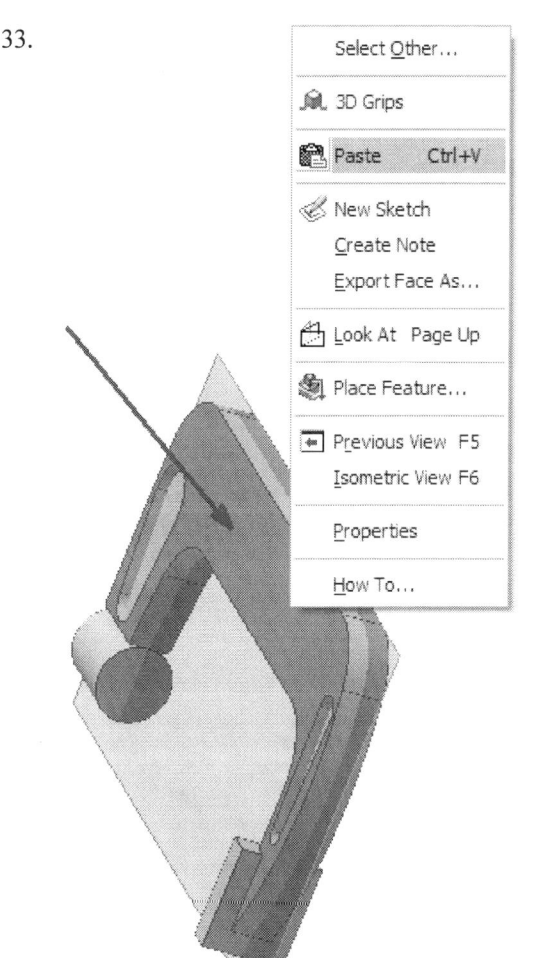

Select the plane where the iFeature is missing.

Right click and select **Paste**.

34.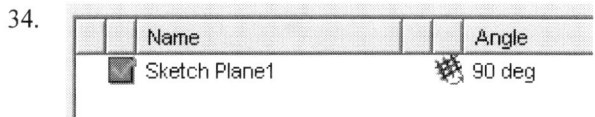

Set the **Angle** to **90 deg**.

Press **Finish**.

35. Select the sketch for the iFeature in the Browser.
Highlight the sketch under the Obround_2_fillets2 features we just pasted.

Right click and select **Edit Sketch**.

18-13

36. 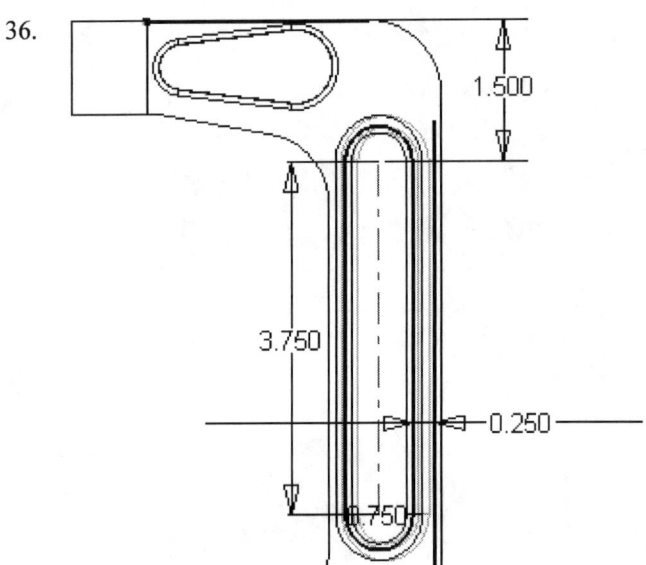 Delete the vertical and horizontal dimensions that constrain the feature.

    Add a 1.50 vertical dimension and a .250 horizontal dimension.

    Exit the sketch mode.

37.  Turn off the **Visibility** of the YZ Plane.

38. 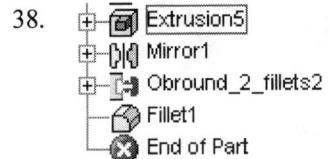 It's a good idea to keep all our fillets defined together.

    Fillets are considered "children" or dependent features because they rely on a "parent" feature being defined first.

    We can move the fillet feature to the bottom of the part so we can add fillets to the mirrored cuts.

    Simply highlight the Fillet and drag it to the spot just above the End of Part icon.

39.  Once the fillet is in the new position, right click and select **Edit Feature**.

Top Down Assembly

40.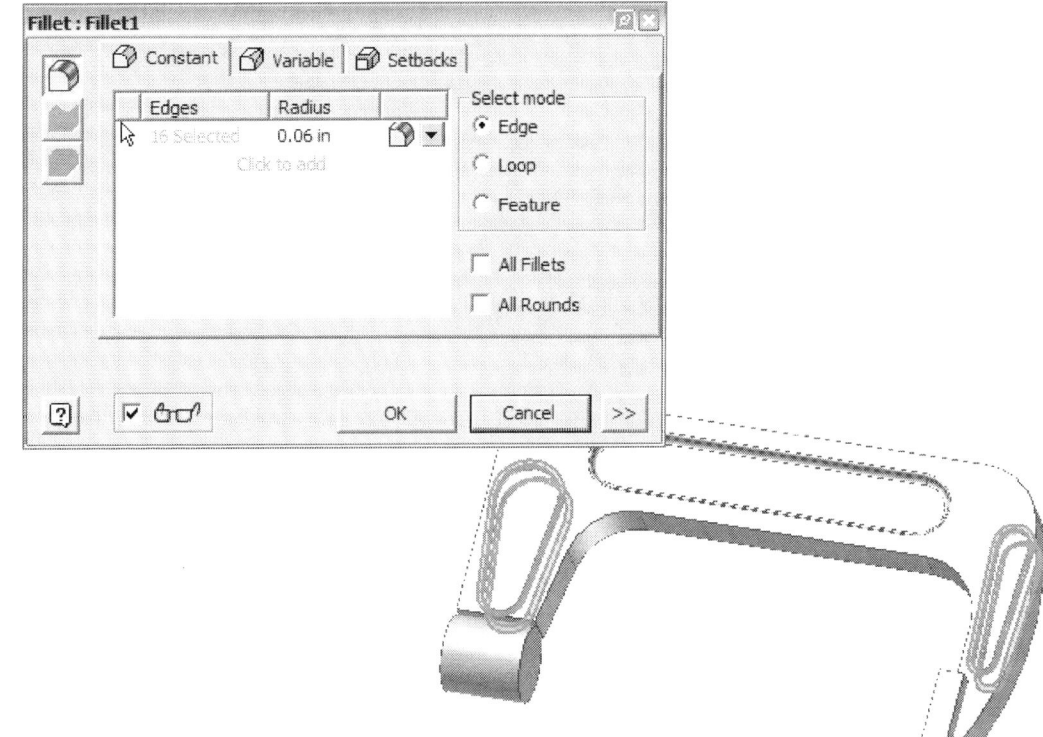

Add **0.06** fillets to the mirrored cuts.

41.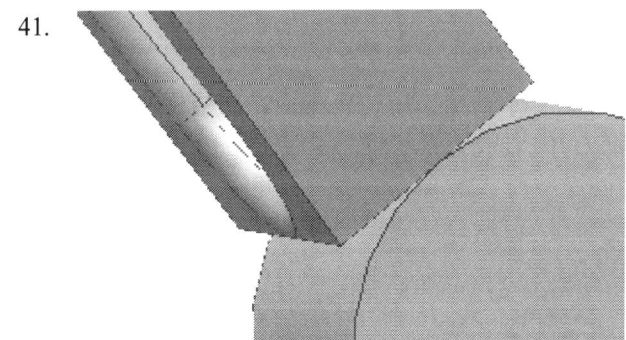

If you have a gap between the cylindrical feature and the arm, you can close it using the following steps:

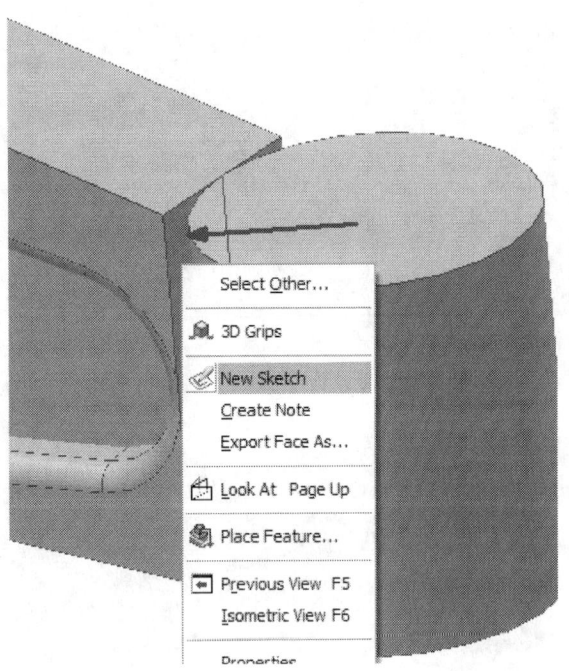

Select the bottom of the flat face for a New Sketch.

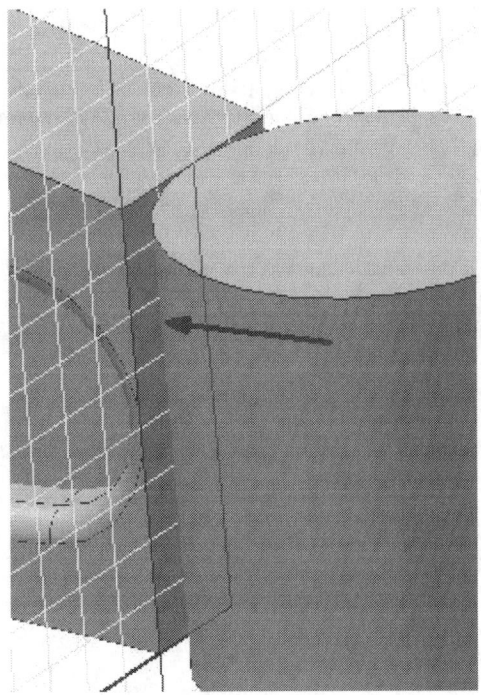

Project the entire face by simply left clicking on it.

Top Down Assembly

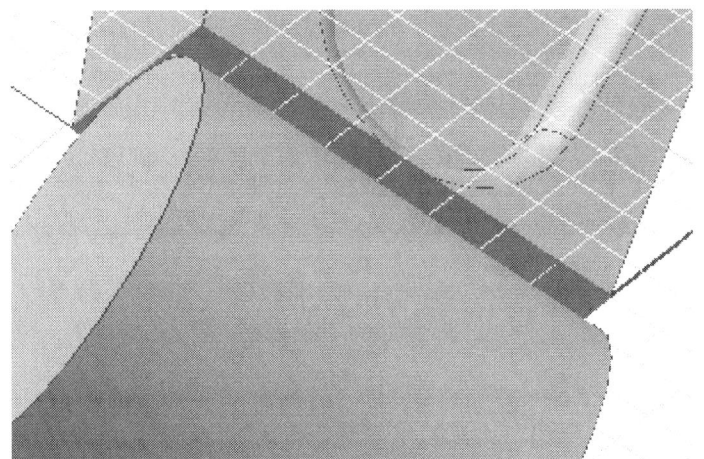

Rotate the view around to confirm that all edges have been projected.

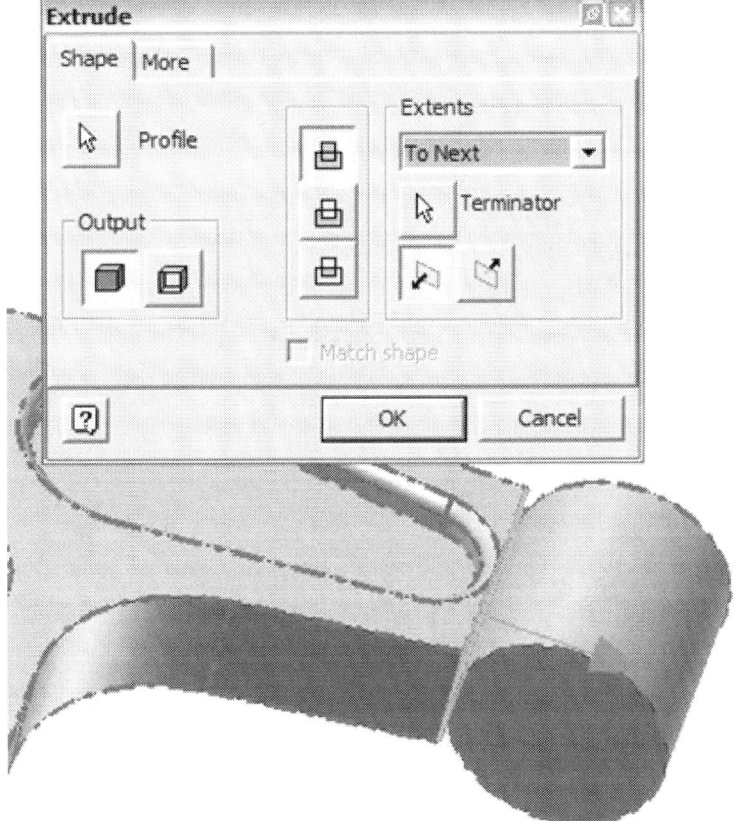

Extrude using **To Next** and select the cylinder as the **Terminator**.

Autodesk Inventor Fundamentals

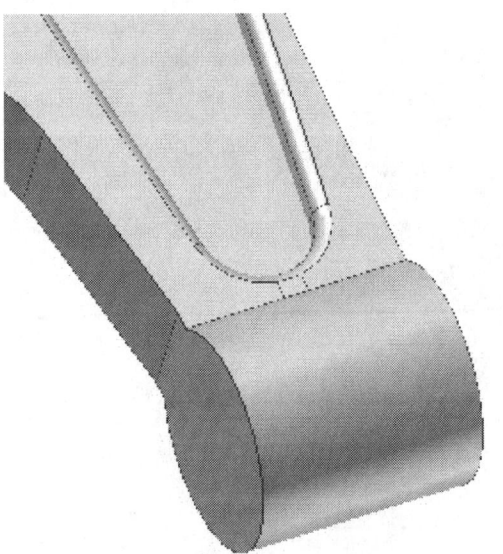

The gap is now filled in.

42. Select the **Hole** tool.

TIP: When you copy a feature you can make it Independent or Dependent. Dependent features get their dimension values from the source feature.

TIP: In an assembly file, any planar surface of any part can be used as a sketch plane.

43.

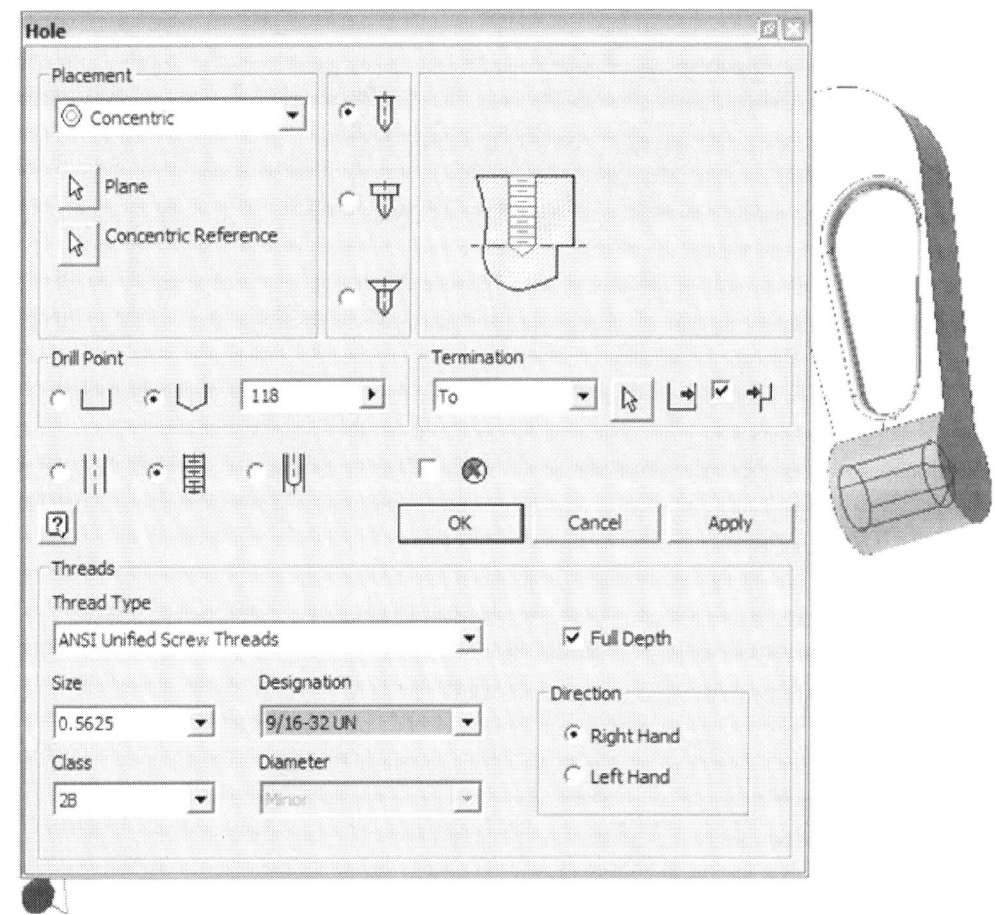

Set the **Placement** to **Concentric**.
Set the flat face of the cylinder as the plane.
Set the round face of the cylinder as the concentric reference.
Set the **Termination** to 'To' and select the other cylinder end.

Enable **Tapped**.
Enable **Full Depth**.
Set the **Thread Type** to **ANSI**.

Set the **Nominal Size** to **0.5625**.
Set the **Pitch** to **9/16-32 UN**.
Set the **Class** to **2B**.
Set the **Diameter** to **Major**.
Press **OK**.

44.

Select the slot face.

Right click and select **New Sketch**.

Autodesk Inventor Fundamentals

45. Select the **Create Text** tool.

46.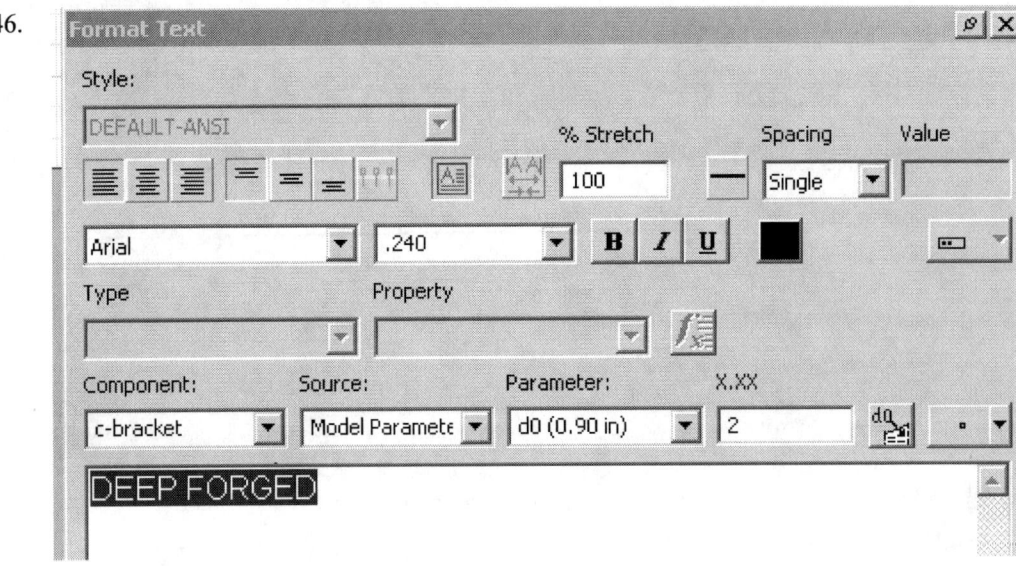

Enter **Deep Forged** for the text value.

47.

Position the text so it is centered in the slot.

48. Select the **Emboss** tool.

18-20

49.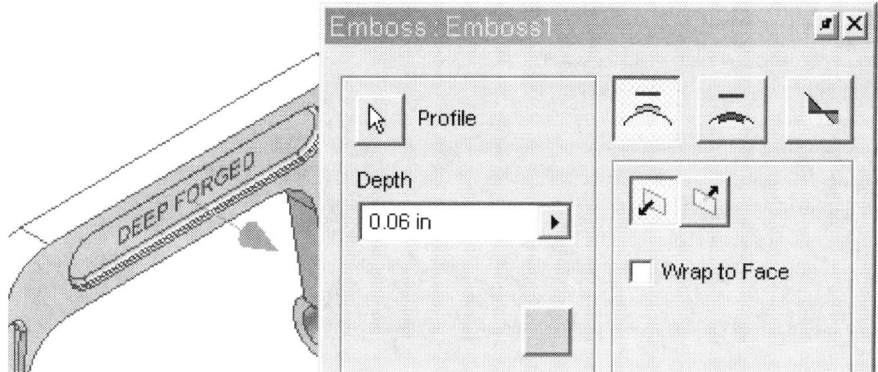

Select the text to use as the profile.

Set the **Depth** to **0.06**.
Set to **Emboss from Face**.
Press **OK**.

If you mirror the embossed text to the other side, you will see that the text does not reverse properly.

50. 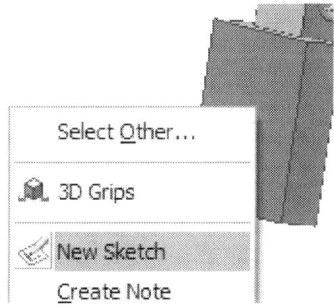 Select the flat surface of the rectangle.

   Right click and select **New Sketch**.

51. Project the edges of the rectangle by left clicking on the face.
    Close the loop.

18-21

Autodesk Inventor Fundamentals

52. 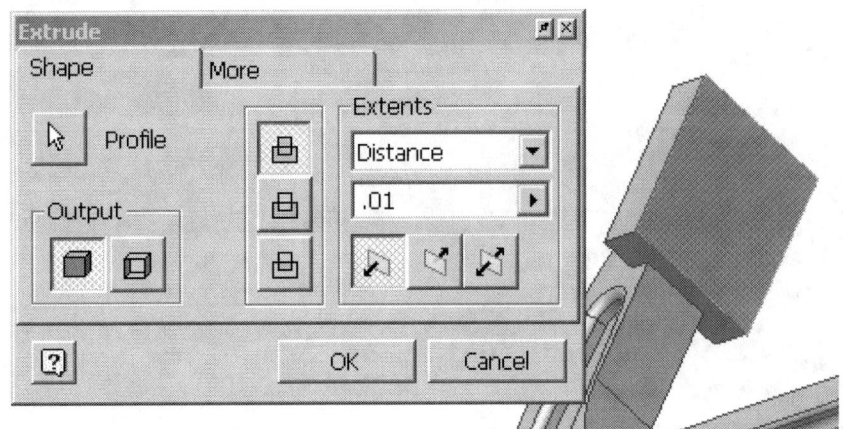   Extrude .01.

53. Rename the extrusion **texture**.

54.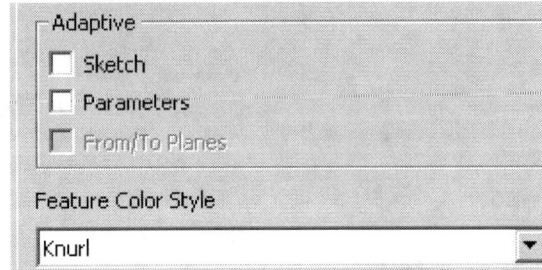

Highlight the texture extrusion.
Right click and select '**Properties**'.
Set the **Feature Color Style** to **Knurl**.

NOTE: Remember the Knurl material was created in Exercise 15-3.

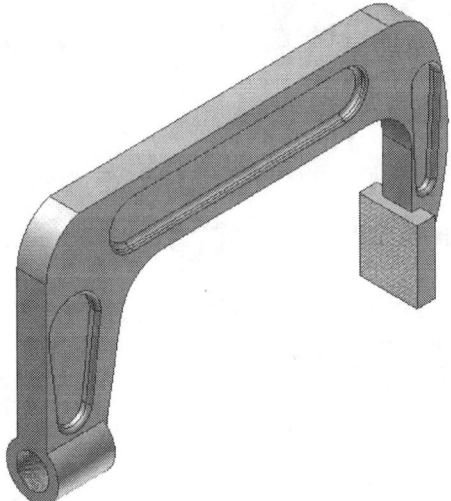

We now have a textured pattern on the rectangular plate.

You may have to rotate the part to see the texture.

55. 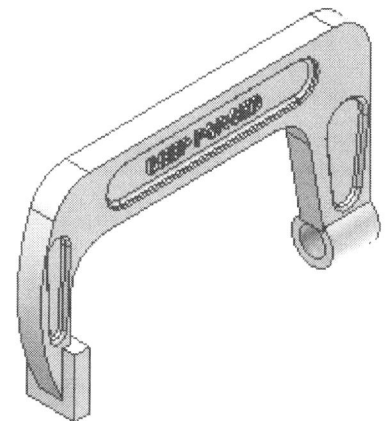 Our completed C bracket.

Use the **Return** button to exit Edit Part mode.

56. Save the assembly file as *ex18-1.iam*.

TIP: We can control visibility of parts by selecting the part in the Browser or by selecting the part in the drawing window.

Exercise 18-2:
Clamp Bolt

Estimated Time: 60 minutes

This part requires use of the following tools:

- Extrude
- Mirror Feature
- Fillet
- Thread
- Hole
- Copy Feature
- Project Geometry
- Revolve

1. Use the **Create Component** tool from the Assembly toolbar to start our next part.

2.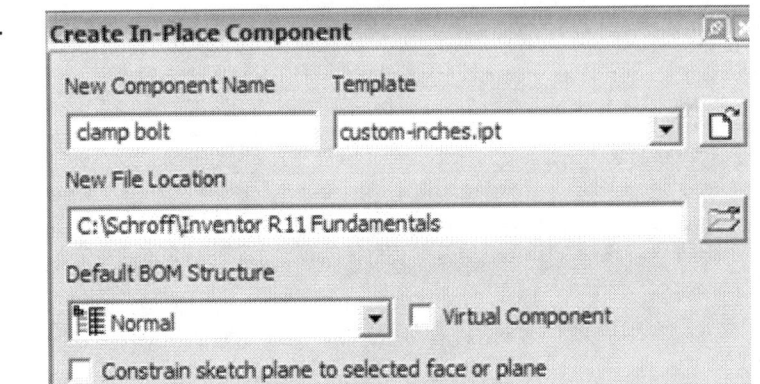

   Use your custom-inches template created in Lesson 15.

   Notice the 'Constrain sketch plane to selected face or plane' is NOT enabled.

3. Name our new part '**clamp bolt**'.
   Select the *custom-inches.ipt* template.
   Press **OK**.

4.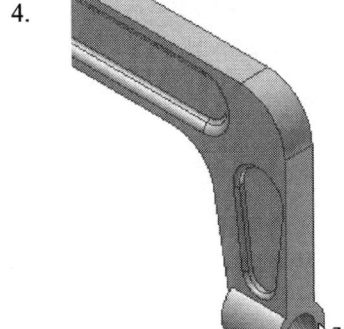

   Select the cylinder end as the plane to use to start sketching our new part.

   This will also automatically select that face for the first sketch.

18-24

Top Down Assembly

5. Project the center point of the clamp bolt.

6. 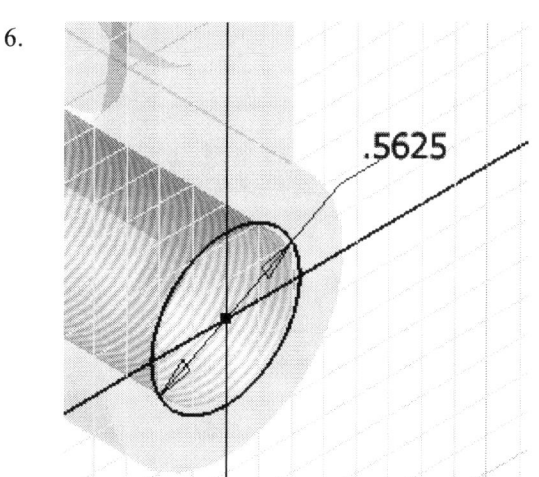 Draw a 0.5625 circle constrained to the center point.

7.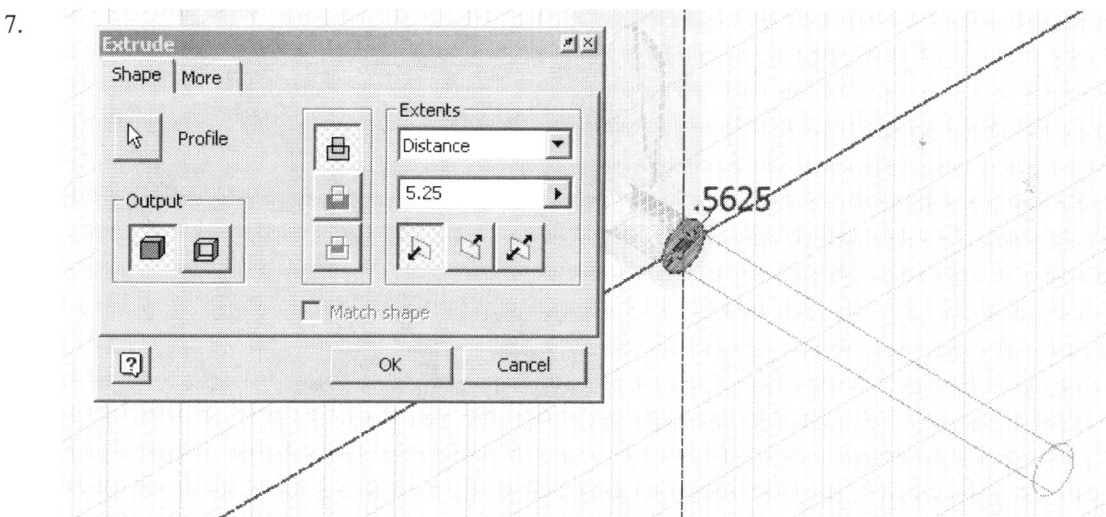

Extrude the circle 5.25 inches.

8. 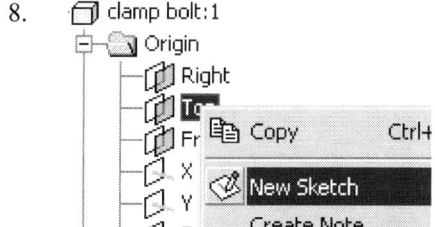 Select the **Top/XZ Plane** for a New Sketch.

18-25

9.

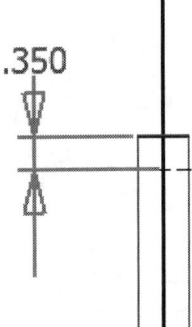

Place a **Point, Hole Center** on the end of the cylinder.

Project the Z-axis.

Add a **Coincident** constraint between the point and the projected axis.

10.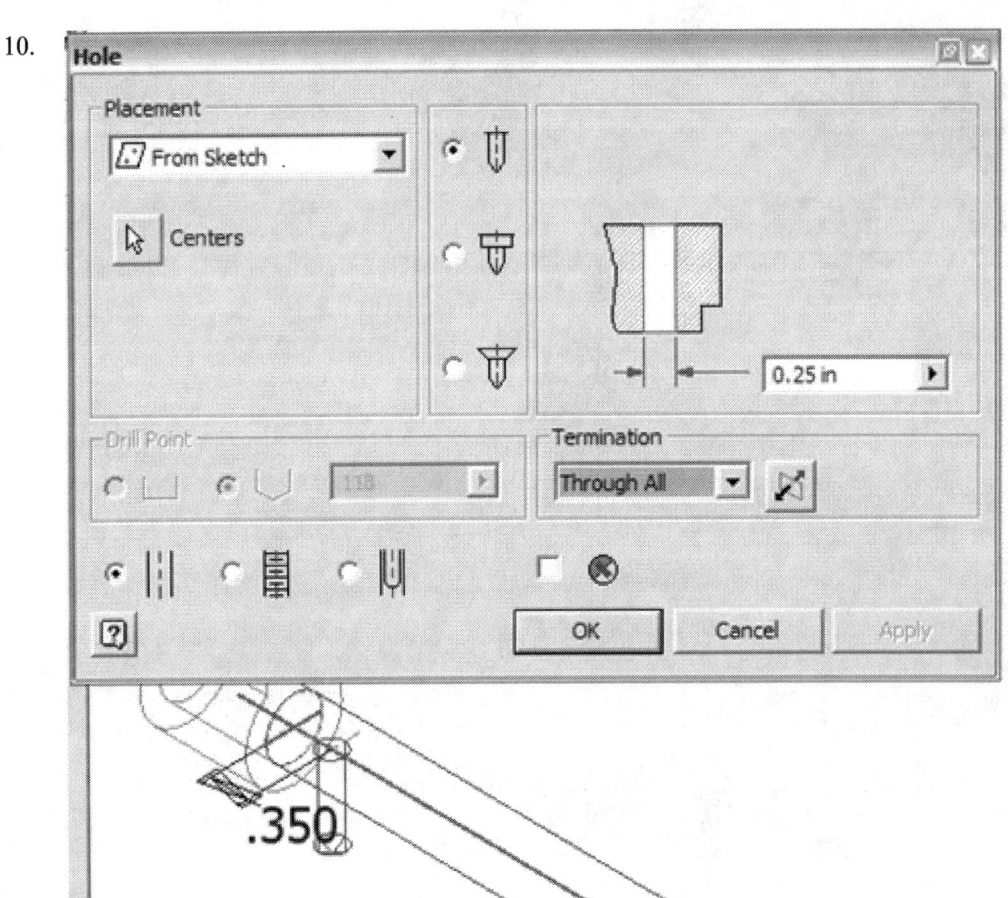

Select the **Hole** tool.
Set the **Placement** to **From Sketch**.
Set the **Type** to **Drilled**.

Set **Termination** to **Through All**.
Set the **Diameter** to **0.25**.
The hole will not be tapped.

Press **OK**.

11.  We need to create a hole for the other side. We can share the sketch between the two holes.

    Highlight the Sketch under the Hole in the Browser. Right click and select **'Share Sketch'**.

12.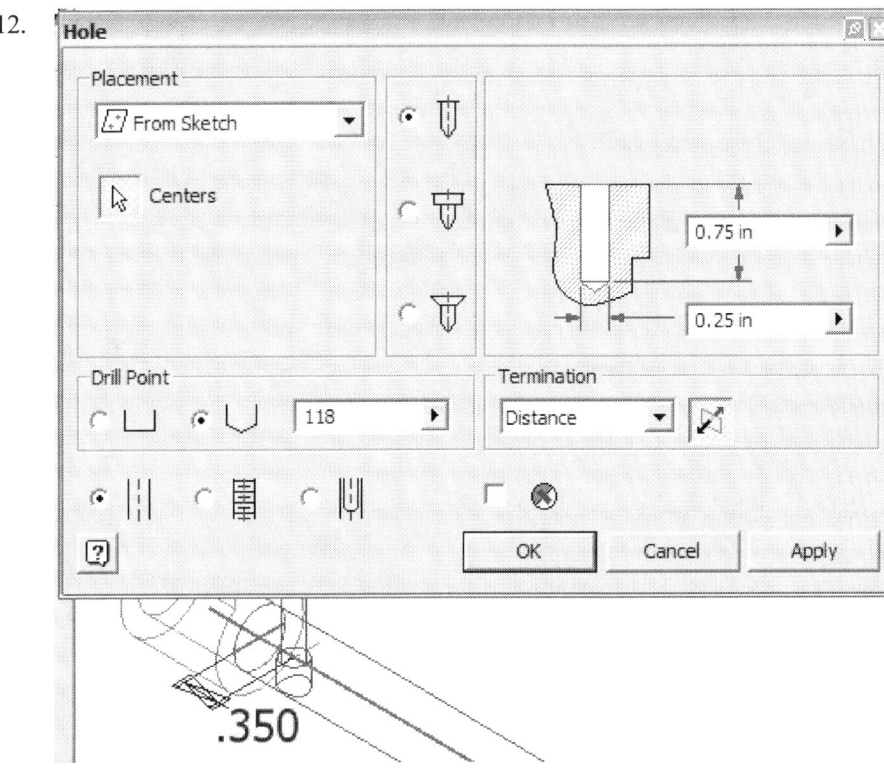

    Start the **Hole** tool.

    Select the **Point, Hole Center**.

    Change the direction for the hole.

    Press **OK**.

13. 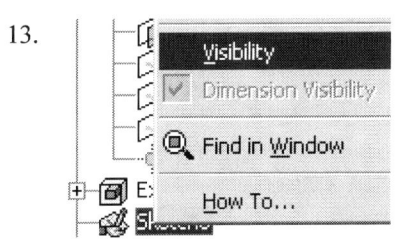 When we enabled Sketch Sharing, Inventor automatically turned on sketch visibility.

    To turn it off so you don't see the sketch, highlight the sketch, right click and disable **'Visibility'**.

18-27

Our model so far.

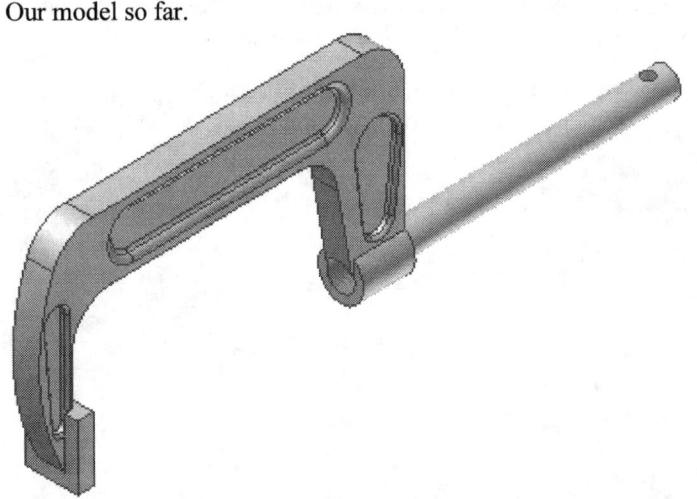

14. Select the **Thread** tool on the Features toolbar.

15.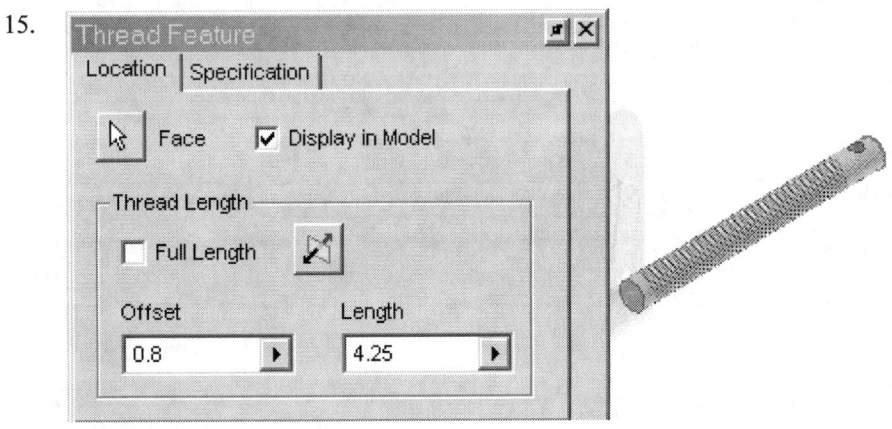

Disable **Full Length**.
Set the **Length** to **4.25 in**.
Set the **Offset** to **0.8**.
Select the Face of the cylinder.
The offset will depend on which end you are closest to when you press the left mouse button, so select the end closest the small 0.25 hole we added.

16.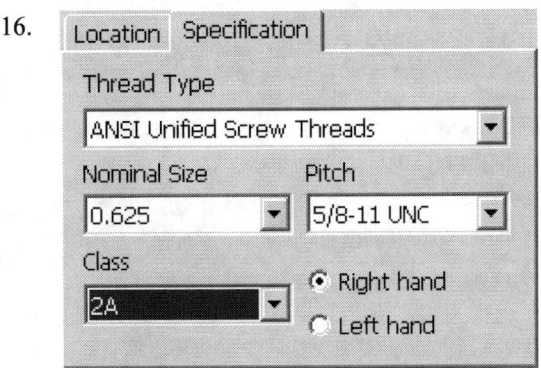

Select the **Specification** tab.
Set the **Pitch** to **5/8-11 UNC**.
Set the **Class** to **2A**.
Press **OK**.

17. Use the Return button to exit out of the Edit Part mode.

18-28

Top Down Assembly

18. Use the **Move Component** tool from the Assembly toolbar to move the clamp bolt away from the C bracket.

TIP: Double clicking on the clamp bolt will automatically switch you to Part Edit mode on the selected part.

Our two parts.

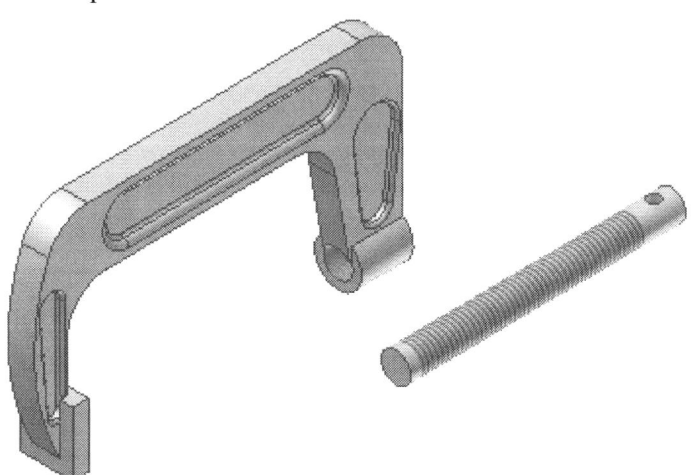

19. 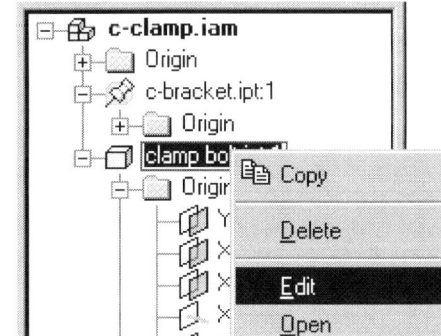 Highlight the clamp bolt in the Browser and right click to **Edit**.

20. 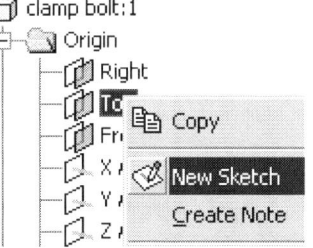 Highlight the Top/XZ Plane. Right click and select **New Sketch**.

21. 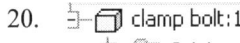 Use the **Project Geometry** tool to project the Z Axis and X axis onto the Sketch.

18-29

Autodesk Inventor Fundamentals

22. Create the sketch shown.

Constrain the sketch to the Z axis and X axis.

When creating this sketch, sketch it big and away from the projected axes. Use the dimensions to control the size and shape the sketch.

Once you constrain the sketch, you can test it by using your mouse to try and drag it out of shape.

23. Use the **Revolve** tool.
Select the **Z Axis** to revolve about.
Set the **Extents** to **Full**.
Press '**OK**'.

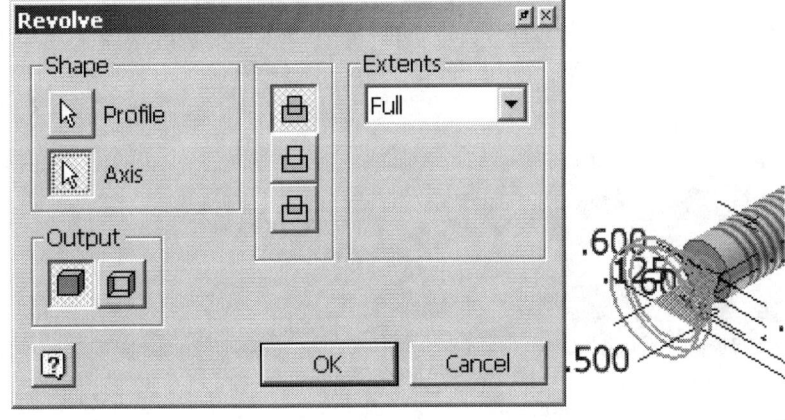

24.  Add a work axis on the small hole so you can add an insert constraint.

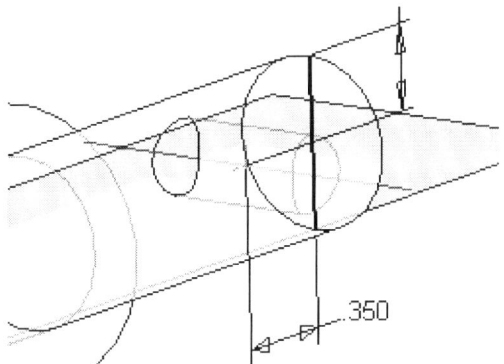

To create the work axis:

Start the **Work Axis** command, then select the hole. The axis will place perfectly.

Our model so far.

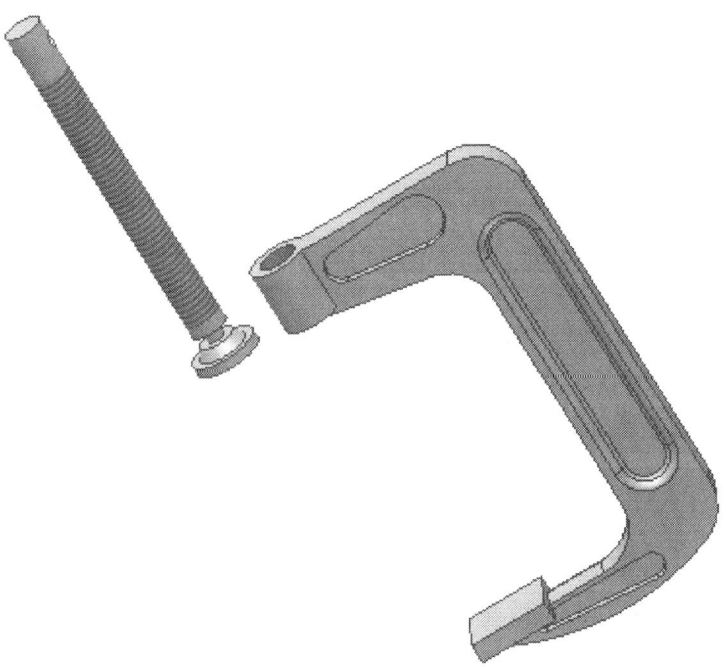

Use the **Return** key to exit out of Part Edit mode.

25. Save your assembly as *clamp assembly*.

Autodesk Inventor Fundamentals

Exercise 18-3:
Handle

Estimated Time: 30 minutes

This part requires use of the following tools:

- Revolve
- Fillet

1. Select the **Create Component** tool from the Assembly toolbar.

2.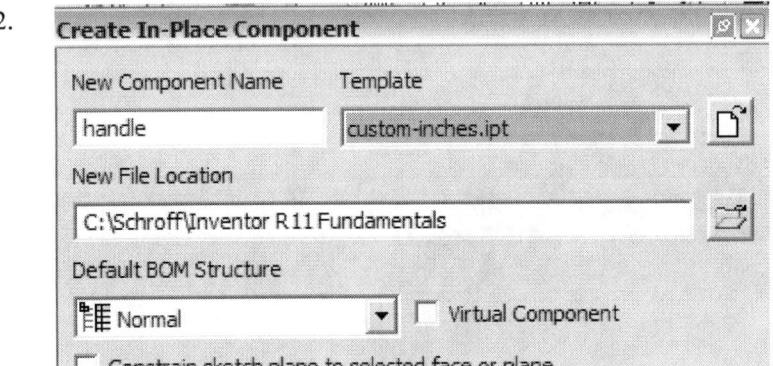

   Name the new part **Handle**.

   Use the *custom-inches* template.

   Disable **Constrain sketch plane to selected face or plane**.

   Press **OK**.

3.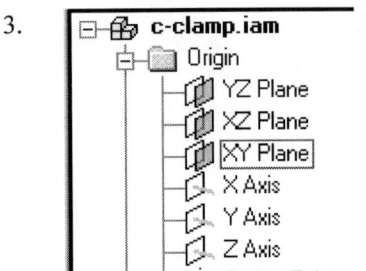

   Select the XY Plane for the sketch plane under the assembly in the Browser.

4.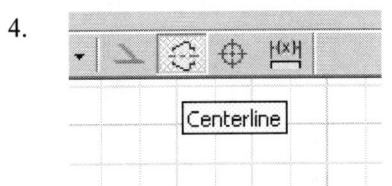

   Draw a vertical line.

   Select the line and define as a **Centerline** under the Style toggle.

5. Verify that the **Style** is set to **Normal** before you start sketching again or all your geometry will be defined as Centerline.

18-32

Top Down Assembly

6. 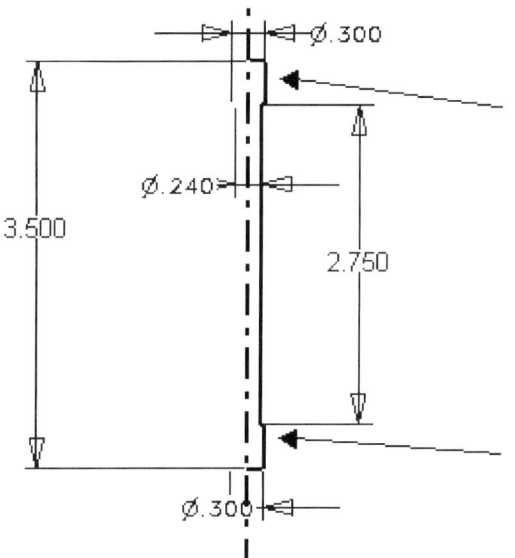 Create the sketch shown.

Set the two vertical lines shown as equal.

7. Use the **Revolve** tool to revolve the sketch around the centerline.

8. 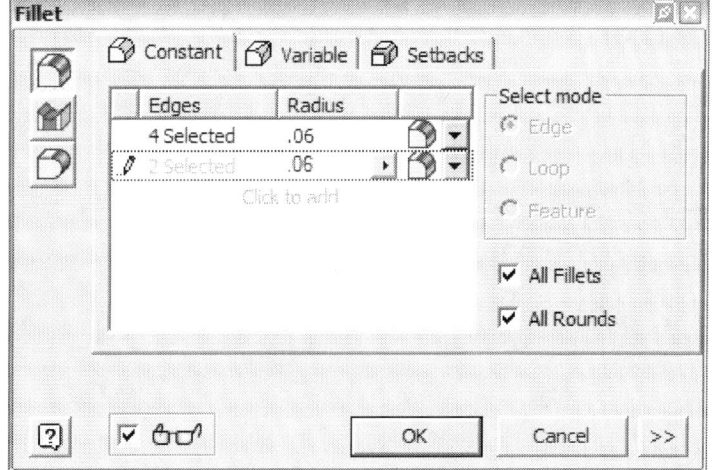 Add a **0.06** fillet to all the edges.

Remember the fastest way to do that is to enable All Fillets and All Rounds.

Press '**OK**.'

 Our completed handle.

9. Use the **Return** button to exit Edit Part mode.

18-33

Exercise 18-4:
Clamp Assembly

Estimated Time: 10minutes

1. Add assembly constraints.

2.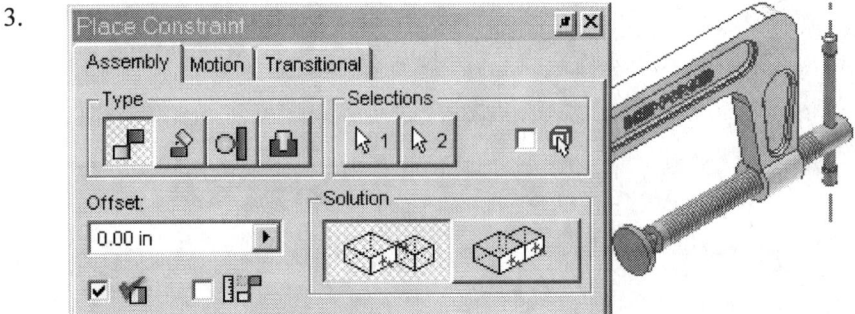

   Add an **Insert Constraint** between the C-clamp and the clamp bolt.

   Set the **Offset** to **-2.5**.

   Press **Apply**.

3. Use a **Mate** constraint between the axis through the handle and the axis through the small hole in the bolt.

4.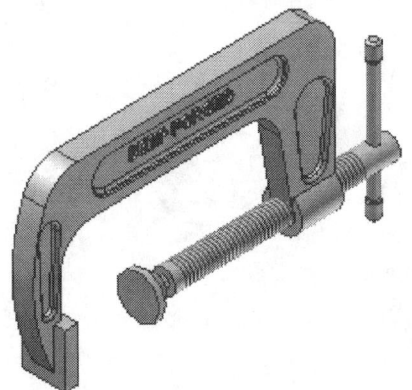

   Our completed assembly.

   Save the file as *c-clamp.iam*.

Exercise 18-5:
Using a Drive Constraint

File: c-clamp.iam
Estimated Time: 10 Minutes

1. Open *c-clamp.iam*.

2. 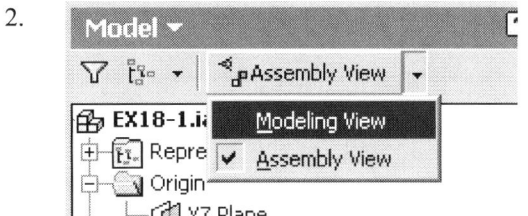 Enable **Modeling View** in your Browser.

3.  Now all your constraints are located under a single folder called *Constraints*.

   Locate the **Insert** constraint for the clamp bolt.

4. Right click and select **Drive Constraint**.

5. 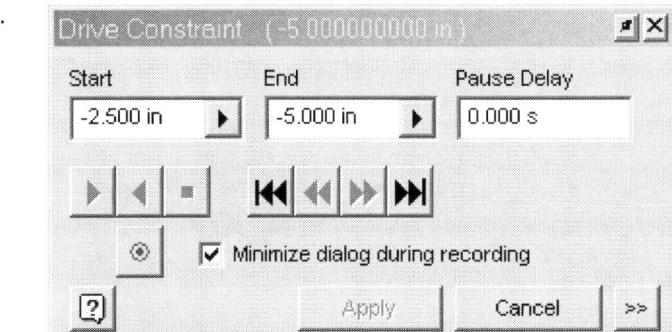 Set the **Start** to **-2.50 in**.
   Set the **End** to **-5.0 in**.

   Press the **>> More** button.

6. 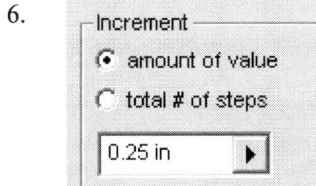 Set the **Increment** to **amount of value**.
   Set the value to **0.25 in**.

   (This sets how much the component will move in each step.)

18-35

7. Press the **Maximum** and **Minimum** buttons to see how the bolt appears at the maximum and minimum positions.

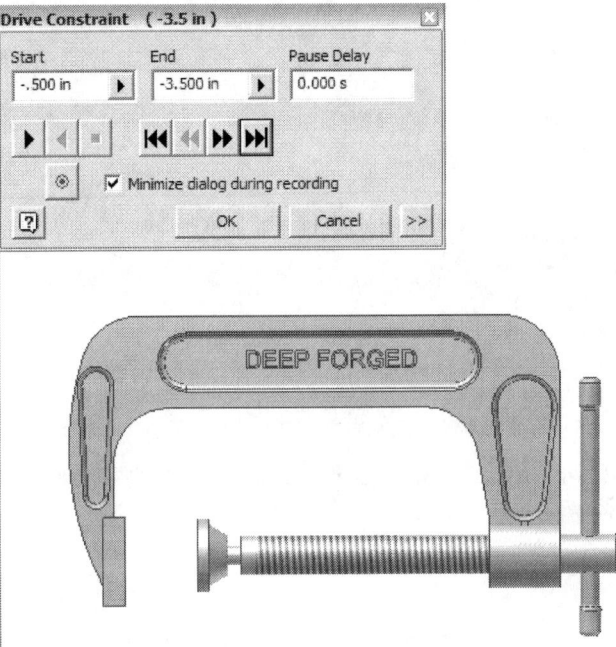

Adjust the **Maximum** and **Minimum** values so that the bolt moves in and out properly.

8. Select **Forward** and **Backward** to see how the bolt moves using the step.

9. Save the file.

Lesson 19
Presentations

Learning Objectives

The user will learn how to:

- Create a Presentation File
- Create an Exploded View
- Create an Animation

Presentation views are used to:

- Create exploded views that can be used for assembly instructions
- Create an animation to show how components interact
- Turn off the visibility of components in a large assembly to make it easier to identify components

Each presentation file can contain as many animations, exploded views, stylized views, section views, etc. that you need for an assembly. Once you set up the views, you can then insert them into your drawing layout.

NOTE: Presentation files can only link to an assembly file.

You can create exploded views either by manually moving components around or automatically expanding the distance between the parts.

TIP: Presentation views do not recognize assembly constraints for any purpose other than creating the first automatic explosion. You can manually tweak (move) a component along any axis or rotate it.

Autodesk Inventor Fundamentals

Exercise 19-1:
Creating an Exploded View

File: New Presentation using Standard.ipn
Estimated Time: 15 minutes

1.  We start by selecting **Presentation** from the file drop down.

    The Presentation Panel has four tools:
   - Create View
   - Tweak Components
   - Precise View Rotation
   - Animate

The Create View tool prompts the user to select an assembly file. The assembly file is then linked to the presentation file. If you update the assembly, any changes will automatically be reflected in the presentation file.

2. Select the **Create View** tool.

TIP: The first view that you add associates the presentation file to an assembly. You can add as many presentation views as needed. All measurements in the presentation file assume the same units as the selected assembly.

Presentations

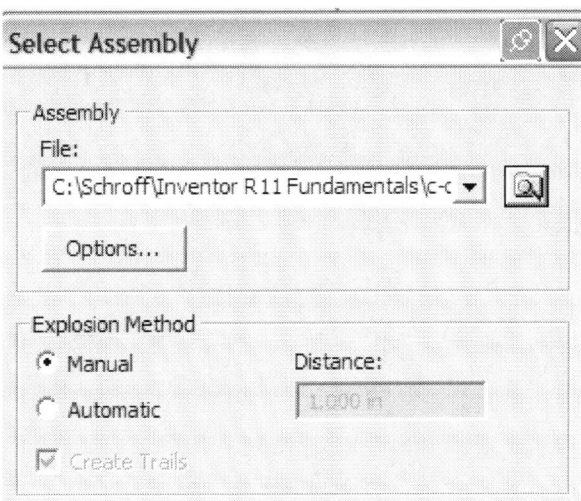

File	Specify the assembly file to use in the presentation. The drop-down list shows any open assembly files. You can also browse for an assembly file using the Browse button or type the path and file name in the edit box.
Design View	This field lists all design views available in the selected assembly file. Design views are saved views stored in the assembly file.
Explosion Method	Manual – Places the selected design view of the assembly in the graphics window. You can then add manual tweaks to move and rotate the components. Automatic – Sets the tweak distance between the components based on the assembly constraints applied.
Distance	Sets the value to be used when Automatic is enabled.
Create Trails	This displays a dotted line to indicate the travel path between components. This option is only available when Automatic is selected.

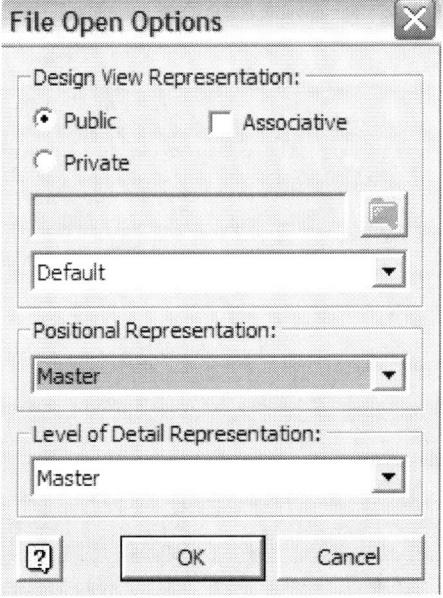

File Open Options:

Select the Design View Representation.

If the assembly has positional representations defined, select a positional representation.

Autodesk Inventor Fundamentals

TIP: To create several presentation views from the same design view, add the first view, then use Copy and Paste to create additional copies. You can then modify the copies independently.

3. Locate the *c-clamp.iam* file.

   Select the **Manual** method.

   Press **OK**.

4. Your assembly appears in the graphics window.

5. Highlight the explosion in the Browser.
   Right click and select '**Copy**'.

6.  Highlight the top assembly.
   Right click and select '**Paste**'.

19-4

Presentations

7. 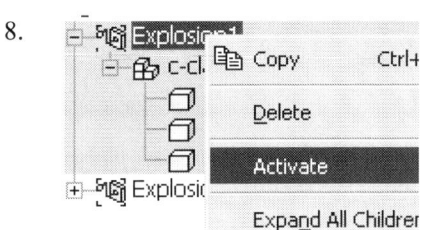 We have two exploded views available in the presentation file.

8. 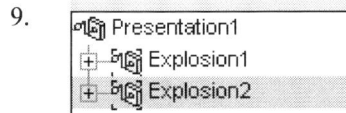 Select the first exploded view.
Right click and select '**Activate**'.

9. The first explosion will be used to create an exploded view. The second explosion will be used to create an animation showing how the c-clamp works.

To change the location of a component in the graphics window, we add a "tweak".
To add a tweak:

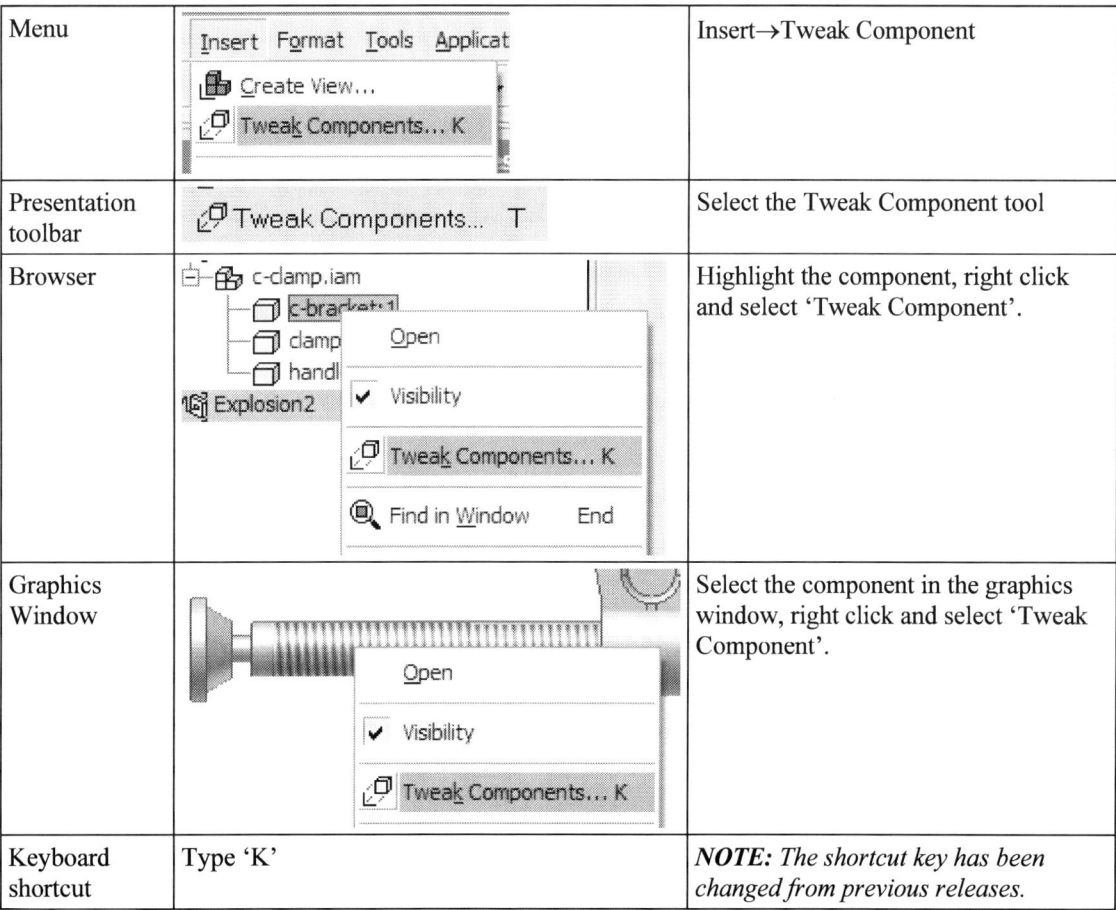

19-5

TIP: You can tweak more than one component at a time. Hold down the Control key to select multiple components.

10. Tweak Components... T   Select the **Tweak Components** tool.

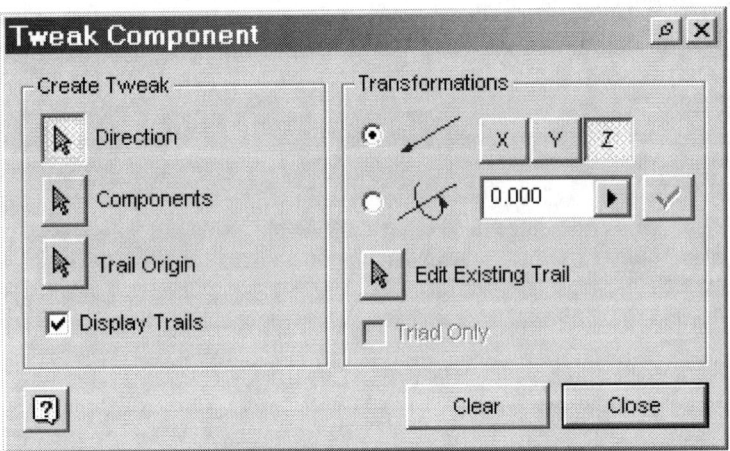

Direction	Select the edge or center point on the assembly to establish the direction to move/rotate – this selection places an XYZ axis at the selection point.
Transformation	Select which axis to move along for a transform Or enable rotation and select number of degrees to rotate
Components	Select the component(s) to be moved/rotated
Trail Origin	Set the starting point for the trail
Display Trails	Sets trails visible to indicate the path taken by a tweaked component
Edit Existing Trail	Allows the user to modify an existing trail
Triad Only	Sets the rotation of selected components when the axis is rotated
Clear	Resets the selections so you can add another tweak.

TIP: You can also clear the current selection by clicking anywhere (but on a part) in the graphics window or Browser

11. 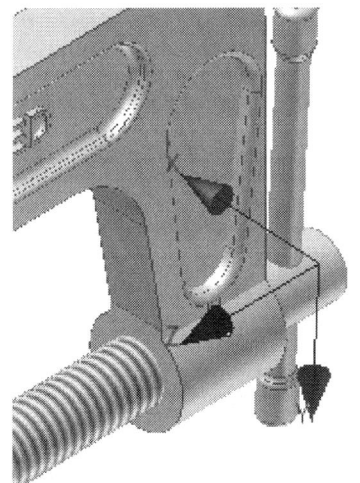 When you select the clamp bolt, you can see how the UCS is orientated. You can determine which axis you should select.

    Select the **Y axis**.
    Select the clamp bolt.
    Set the **Distance** to **-1.2**.
    Select the **Green Check** button to apply the tweak.

    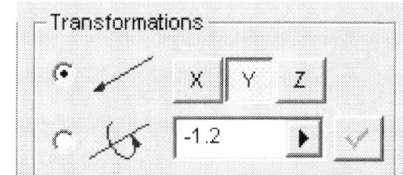

12. Press **Clear** to reset the Tweak dialog.

13. Select the handle.
    Right click and select **Tweak Components**.

    Move the bolt down and below the bolt.
    To do this, pick an axis and simply drag the bolt into the desired position.

    Press the check mark to apply the tweak.

    Press **Close**.

    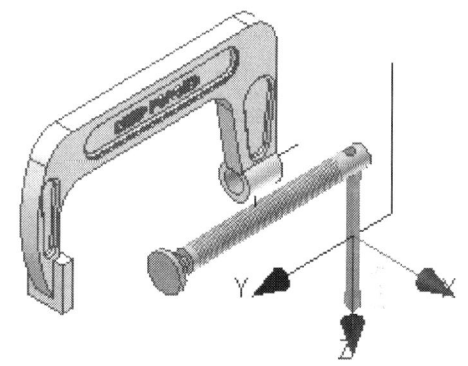

Autodesk Inventor Fundamentals

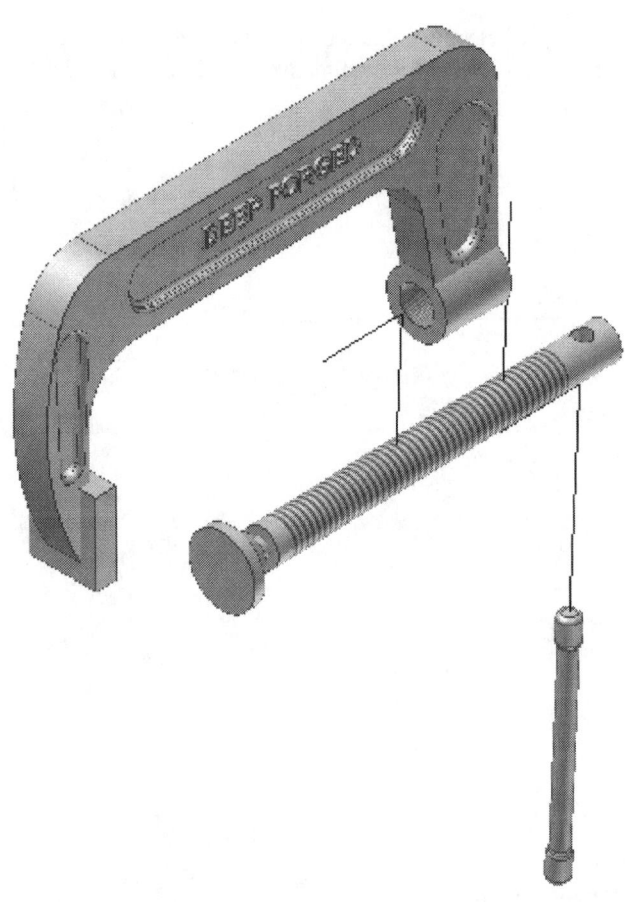

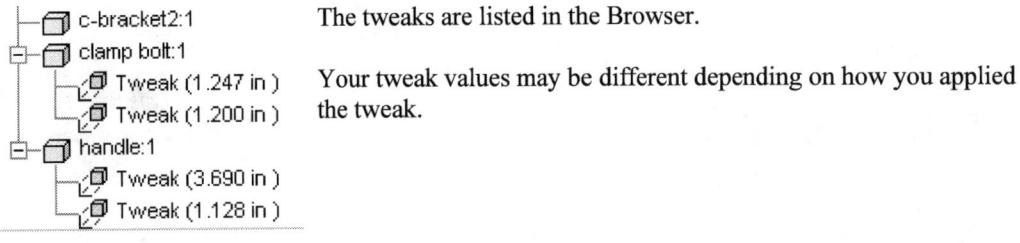

The tweaks are listed in the Browser.

Your tweak values may be different depending on how you applied the tweak.

14. If you highlight a tweak in the Browser, the tweak value will appear in an edit field. You can then modify the value if needed to adjust the tweak.

15. This view can be used as an exploded view in our assembly drawing.

Save the file as *ex19-1.ipn*.

TIP: You can select components and then select the Tweak tool and the highlighted components are automatically selected to apply the tweak.

19-8

Exercise 19-2:
Creating an Animation

File: Ex19-1.ipn
Estimated Time: 15 minutes

1. Open *ex19-1.ipn*.

2. 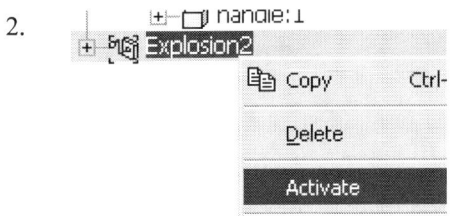 Activate **Explosion2**.

3.  Highlight the clamp bolt and handle in the Browser.

   Right click and select **Tweak Components**.

4. 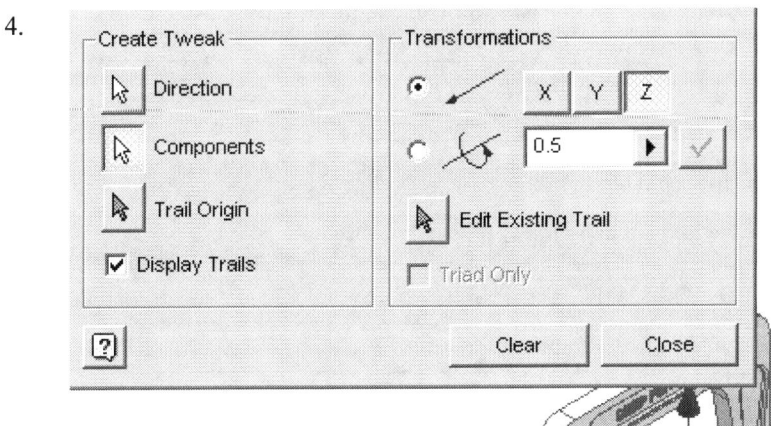 Pick the center of the circular face on the bolt to locate the triad.

   Select the Z-axis. Press and drag using the mouse to move the bolt and handle into the C-clamp.

   When the value displayed is **.5**, press the green check to apply the tweak.

5. 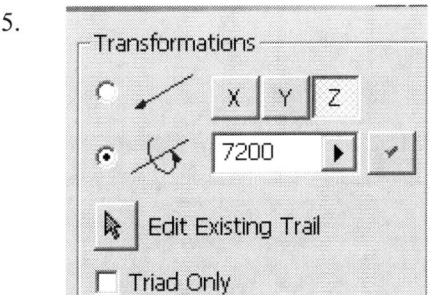 Enable the **Rotation** option and enter a value of **7200**. Press the green check to apply the rotation value.

   Close the tweak dialog.

19-9

6. 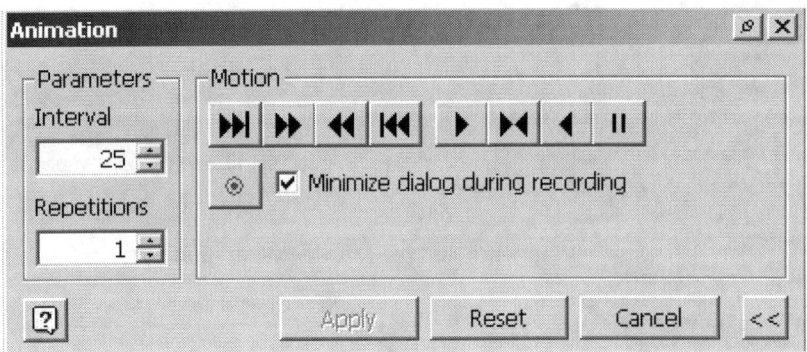 We see the tweaks we applied in the Browser.

7. ![Animate] Select the **Animate** tool.

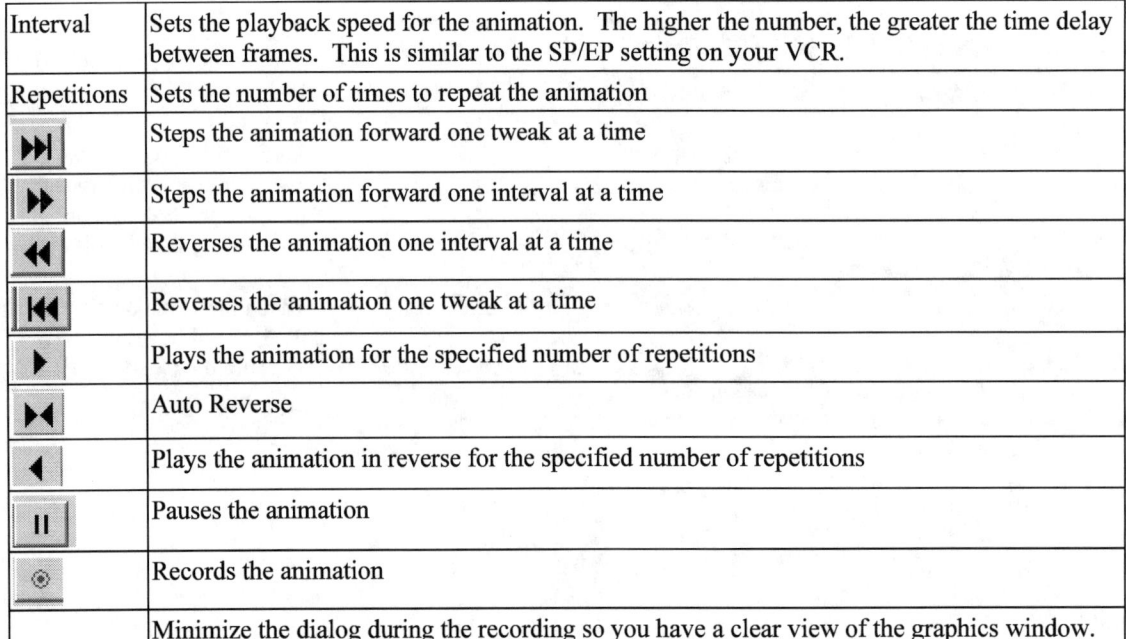

Interval	Sets the playback speed for the animation. The higher the number, the greater the time delay between frames. This is similar to the SP/EP setting on your VCR.
Repetitions	Sets the number of times to repeat the animation
▶▶|	Steps the animation forward one tweak at a time
▶▶	Steps the animation forward one interval at a time
◀◀	Reverses the animation one interval at a time
|◀◀	Reverses the animation one tweak at a time
▶	Plays the animation for the specified number of repetitions
▶|	Auto Reverse
◀	Plays the animation in reverse for the specified number of repetitions
‖	Pauses the animation
●	Records the animation
	Minimize the dialog during the recording so you have a clear view of the graphics window.

If you select the More option of the dialog, you can re-order the tweaks, group them so parts move together and ungroup them.

Presentations

8. Select the **More** button.

9. Highlight the components.

1	handle:1	Tweak (7200.00 deg)
1	clamp bolt:1	Tweak (7200.00 deg)
2	handle:1	Tweak (0.500 in)
2	clamp bolt:1	Tweak (0.500 in)

Note that some components are moved in step 1 and some are moved in step 2.

10. Select the **Group** button so the components move together.

11.
Sequence	Component
1	handle:1
1	clamp bolt:1
1	handle:1
1	clamp bolt:1

Note that the sequence for the components is now the same number.

11. Press **Apply**.

12. Press the **Play** button to see how the c-clamp moves.

13. Use the **Reset** button to re-initialize your animation.

14. Save the file as *ex19-2.ipn*.

TIP: Turn off the visibility of the 3D indicator before you create your animation to make your animation look more professional. This is controlled under Tools→Application Options in General Tab.

19-11

Exercise 19-3:
Recording an Animation

File: Ex19-2.ipn
Estimated Time: 30 minutes

1. Open *ex19-2.ipn*.

2. Select the **Animate** tool.

3. Set the '**Reset**' button to make sure the assembly is set at interval 0.

4. Select the **Record** button.

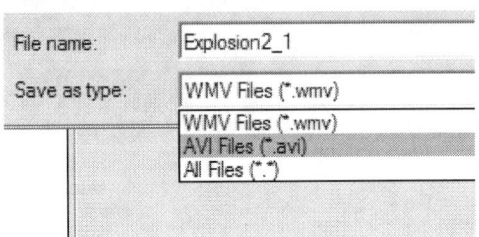

Animation files can be saved as *avi* or *wmv* file formats.

5.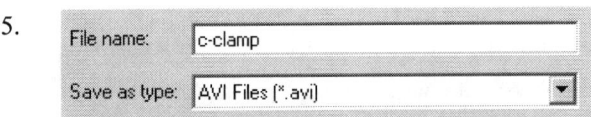

Name your animation **c-clamp**.
Press **Save**.

6.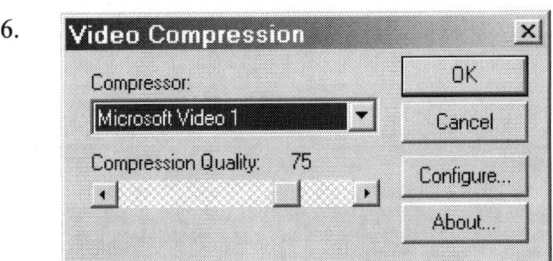

Set the **Compressor** to **Microsoft Video 1**.

This ensures that the avi you create will be able to played using Microsoft software.

Press **OK**.

7. Press the **Play** button.

8. When the bolt reaches the end of the travel, select the **Record** button again to stop the recording.

9. Locate the *c-clamp.avi* file using Windows Explorer.

10. Double click on the file to play the animation.

11. Save as *ex19-3.ipn*.

Presentations

Exercise 19-4:
Changing Views in an Animation

File: ex19-3.ipn
Estimated Time: 30 minutes

1. Open *ex19-3.ipn*.

2. 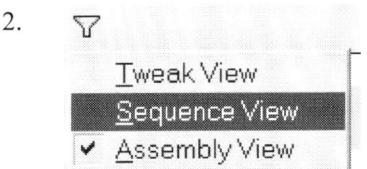 Under the **Filter** icon drop-down, we can select three different view styles for our Browser.

   Select **Sequence View**.

3. 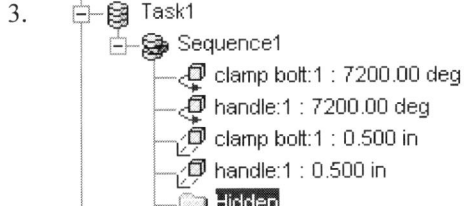 Under the second explosion, you should see one Task in the Browser and one Sequence. This is the motion of the bolt going into the clamp.

4. 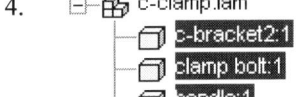 Highlight all the components.

5. 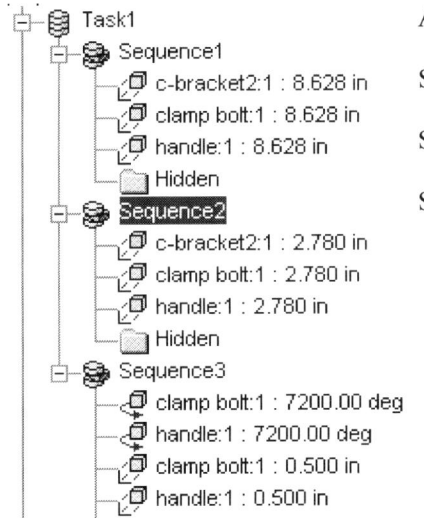 Add two more tweaks.

   Sequence 1 should move the entire assembly down a distance.

   Sequence 2 moves the assembly forward a distance.

   Sequence 3 turns the bolt into the c-clamp.

6. 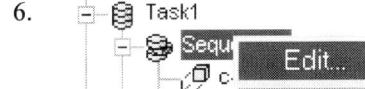 Highlight Sequence 1 in the Browser.
   Right click and select **Edit**.

19-13

7. Rotate the view so you see the back view of the clamp assembly.
    Press **Set Camera**.
    Press **Apply**.

8. Select **Sequence2** from the Sequence drop-down.

9. Turn the view back around and press **Set Camera**.

    Press **Apply**.

10. Select **Sequence3** from the Sequence drop-down.

11. Zoom into the view for a close-up.

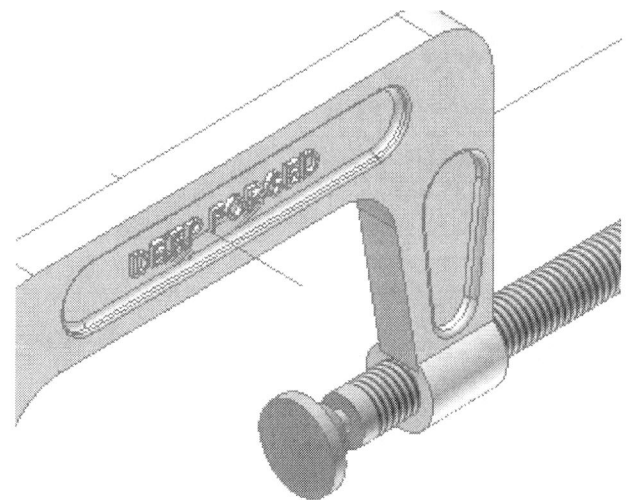

12. Press **Set Camera**.

    Press **Apply**.

13. Select the **Animate** tool.

14. Press **Play Forward**.

15. You see that your camera views now change with each sequence.

16. Save as *ex19-4.ipn*.

Autodesk Inventor Fundamentals

Exercise 19-5:
Precise View Rotation

File: Ex19-4.ipn
Estimated Time: 10 minutes

1. Open *ex19-4.ipn*.

2. 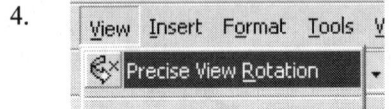 Select the **Precise View Rotation** tool.

3. 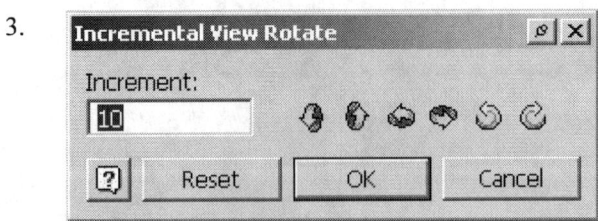 This tool allows you to rotate your assembly by a specified increment in the direction selected.

    Select each button in turn to see how your model rotates.

4. The Precise View Rotation dialog can also be launched from the View menu when you are in an active Presentation file.

    You can set up a view using the Precise View Rotation tool to be used in an assembly drawing.

5. Close without saving.

 TIP: New trails are generated using the tweak values. If the original trail has been moved, the trails may not match. Changing a trail does not alter the path of a tweak.

Exercise 19-6:
Managing Trails

File: ex19-3.ipn
Estimated Time: 15 minutes

1. Open *ex19-3.ipn*.

You can control the visibility of trails in an exploded view three ways:

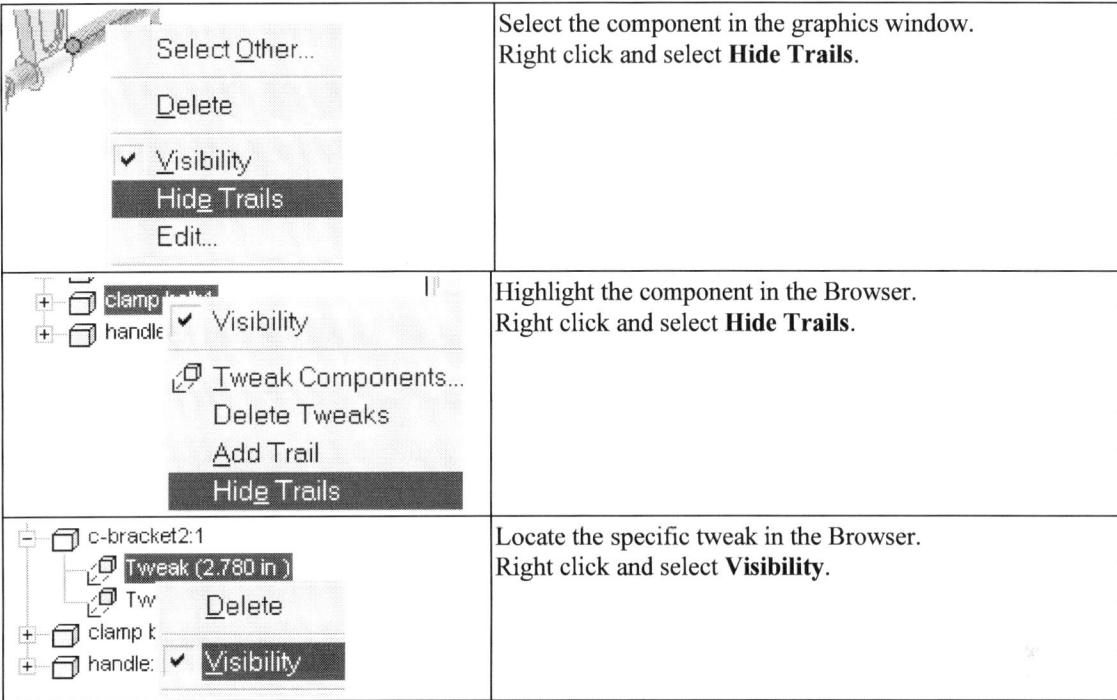

	Select the component in the graphics window. Right click and select **Hide Trails**.
	Highlight the component in the Browser. Right click and select **Hide Trails**.
	Locate the specific tweak in the Browser. Right click and select **Visibility**.

You can add a trail to an existing tweak to clarify relationships between components.

2. 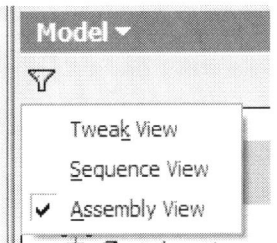   Switch to an **Assembly View** under the Filter.

19-17

3. 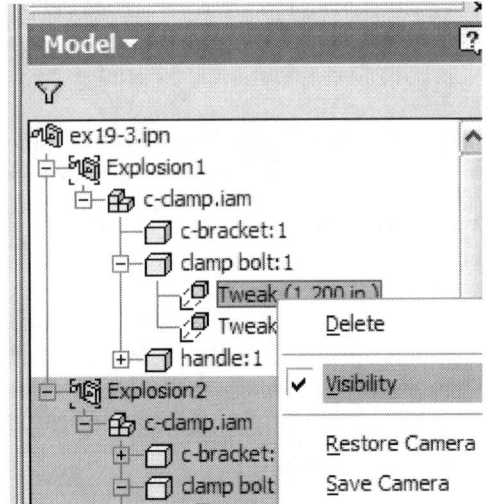 Activate the first explosion.

4. Locate a transitional tweak under the clamp bolt. A transitional tweak will show inches not degrees; a tweak showing degrees is a rotational tweak.

   Right click and enable **Visibility**.

5. 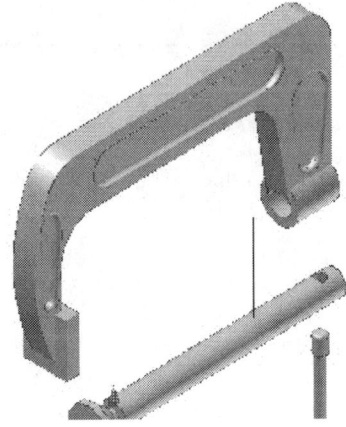 A line appears to indicate the path the component traveled.

Presentations

6. 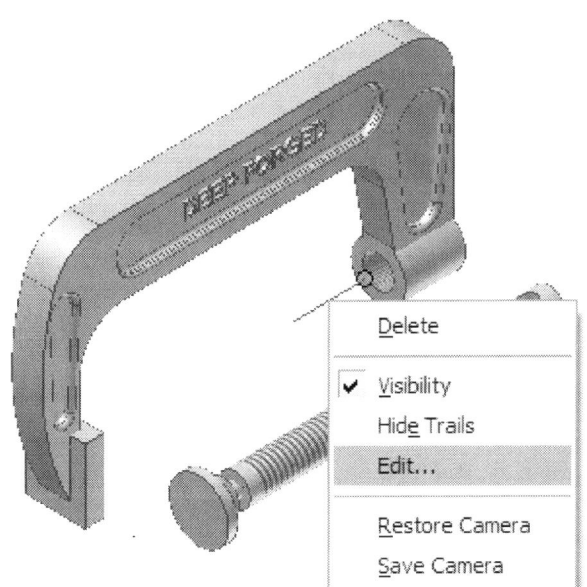 Select the trail in the display window.

   Right click and select **Edit**.

7. 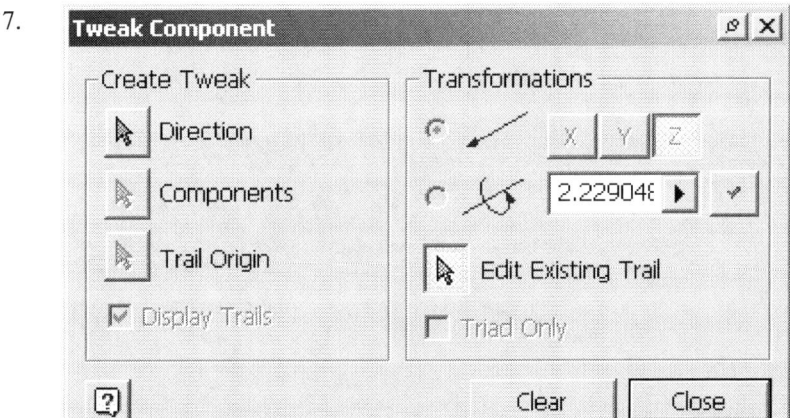 The **Tweak Component** dialog appears.

8. 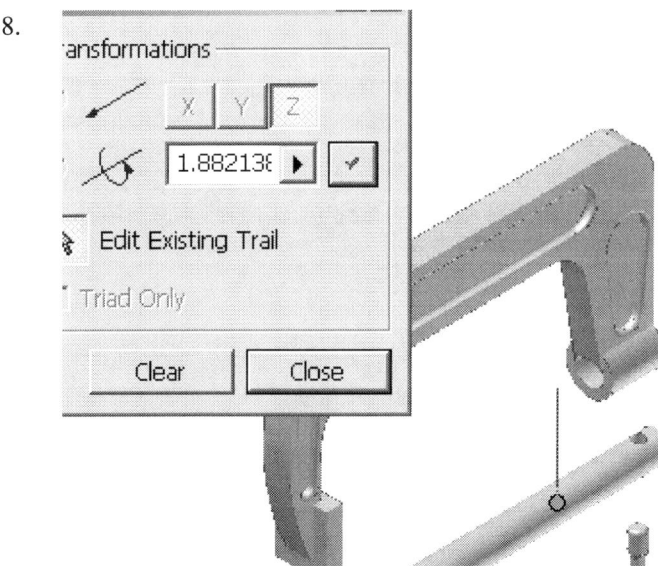 If you place your left mouse over the green grip on the trail and drag the mouse up and down, note that the value of the tweak changes in the dialog.

   Adjust the location of the component.

   Close the dialog.

9. 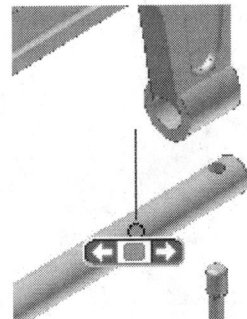 You can also move the components simply by mousing over the tweak grip to activate it, then grabbing and dragging with the mouse.

10.  Highlight **Explosion2** in the Browser. Right click and select **Copy**.

11. Highlight the presentation file name. Right click and select **Paste**.

At this point we have three different explosion views in our presentations. This can be confusing to know which is which, so we can rename our views to help us keep track of what is going on.

12. Simply click on the view name to change the field to an edit box (similar to Windows Explorer).

Name your views: **Exploded View**, **Animation**, and **Assembled**.

13. 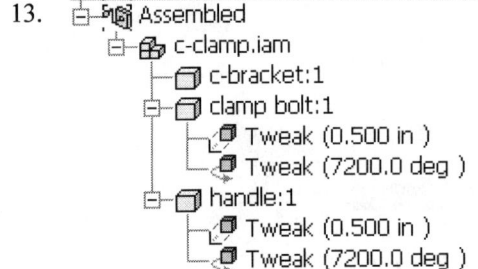 Activate the **Assembled View**.
Expand it so you can see the contents.
We need to delete the tweaks so the parts stay assembled in the view.

Presentations

14. Select the **Tweak View** from the Filter drop-down.

15. You can expand each tweak to see the components it affects.

16. Switch back to an **Assembly View**.

17. Select the handle.
    Right click.

    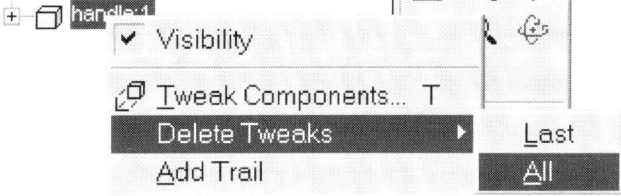

    Note that **Delete Tweaks** contains a submenu that allows you to delete the last tweak applied or all the tweaks applied to that component.

    Select **Delete Tweaks→All**.

18. Delete all the tweaks for the clamp bolt.

    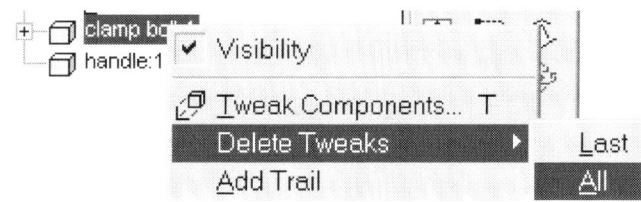

19. Delete all the tweaks for the c-bracket.

    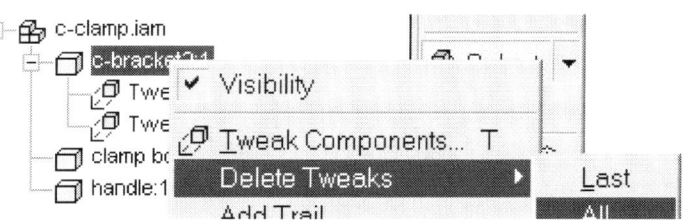

19-21

20.
```
c-clamp.iam
 ├── c-bracket2:1
 ├── clamp bolt:1
 └── handle:1
```
You notice the +/- symbol is no longer listed next to components since you deleted the tweaks.

21. Save as *ex19-6.ipn*.

Presentation Toolbar

Button	Name	Description
Create View...	Create View	Links a view of a selected assembly file to a Presentation file.
Tweak Components... +T	Tweak Components	Move or rotate components to position them to make it easy to see how parts are assembled.
Precise View Rotation	Precise View Rotation	Allows the user to precisely rotate the view to get the desired orientation.
Animate...	Animate	Create avi files to demonstrate how to build assemblies or how assemblies operate.

Review Questions

1. Identify the tool shown.

    A. Move Component
    B. Rotate Component
    C. Add a Tweak
    D. Animate

2. Select here to speed up or slow down an animation.

3. Select here to Record an Animation

4. Select here to see all the tweaks added.

5. Identify the tool shown.
    A. Rotate component
    B. Animate Presentation
    C. Precise View Rotation
    D. Move component

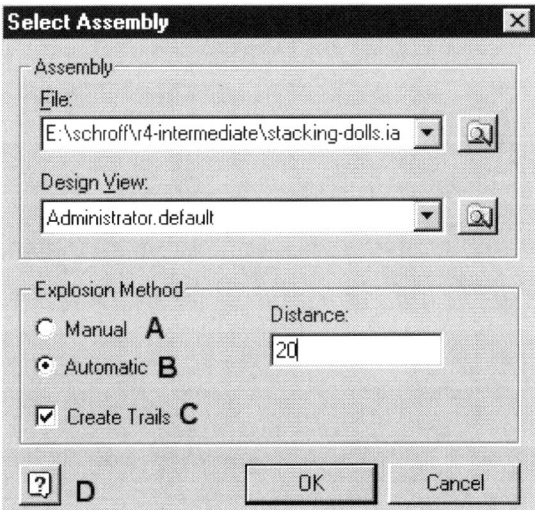

6. Select here to bring up HELP on Presentations.

7. Select here to automatically place Trails.

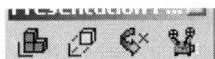

8. Name the toolbar shown:

    A. Animation
    B. Scene
    C. Assembly
    D. Presentation

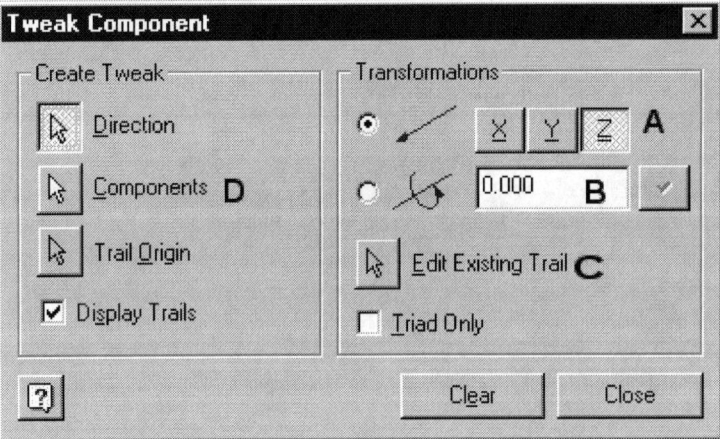

9. To add a rotating motion to a component, select here.

10. To select the axis the component will move along for a tweak, select here.

11. Presentation files are linked to Part files.

    A. True
    B. False

12. To speed up an animation, increase the number of Intervals.

    A. True
    B. False

13. When creating an Animation, there is no need to Reset between Plays.

    A. True
    B. False

14. The tool shown is:

    A. Animate
    B. Mickey Mouse
    C. Movie
    D. Camera

ANSWERS: C; 2) A; 3) C; 4) D; 5) C; 6) D; 7) C; 8) D; 9) B; 10) A; 11) B; 12) B; 13) B; 14) A

Lesson 20
Rendering

Learning Objectives

The user will learn how to:

- Create a Rendering
- Create an Exploded View
- Create an Animation

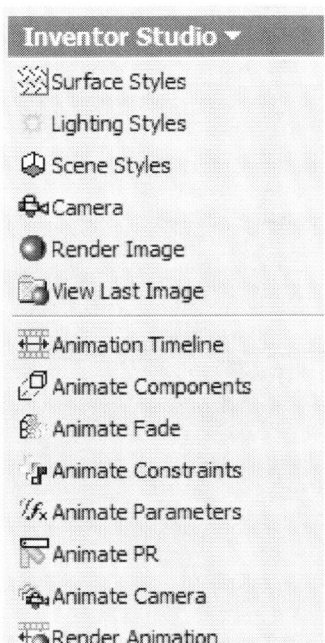

Inventor R10 introduced a new rendering module called Inventor Studio.

This module is available in either part or assembly files.

To access the Studio module, go to Applications→Inventor Studio.

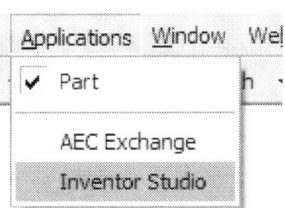

A new set of tools appears.

TIP: You can switch between the Part/Assembly environment and the Studio environment in order to make changes to the part/assembly; add, edit, or remove constraints, or define new views.

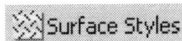

Surface Styles

Surface Styles are stored in the Styles Library. Once you create a new surface style and save it to the library, it is available to any part or assembly in your project.

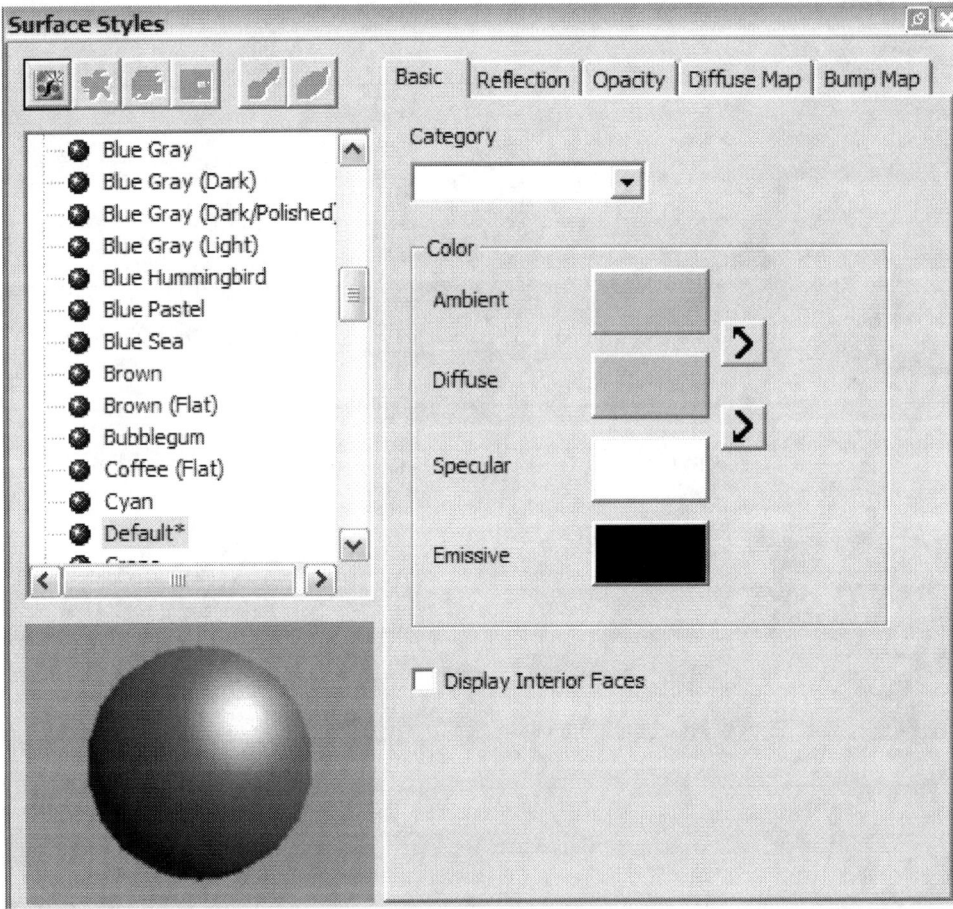

Inventor Studio comes with a library of styles already defined for your use. These are in addition to the standard colors and materials.

Rendering with Inventor Studio

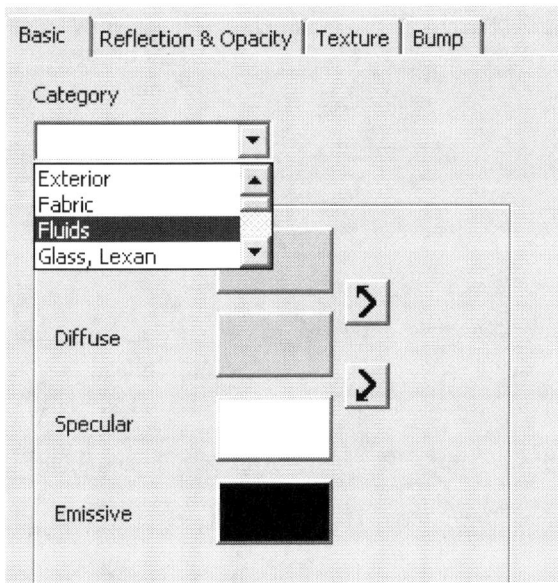

The Basic tab allows you to assign a new surface style to any one of four categories:

- Exterior
- Fabric
- Fluids
- Glass/Lexan

Ambient, Diffuse, Specular, and Emissive determines how the object's colors react to direct and indirect lighting.

Ambient	Controls the color of the light reflected from the object
Diffuse	Controls the face color in response to direct light
Emissive	Controls the color of any surface glow
Specular	Controls the color of the reflection from the light source

When this is enabled, the surface/texture is applied to the interior faces. This is not recommended unless you are using see-through materials as it is memory-intensive.

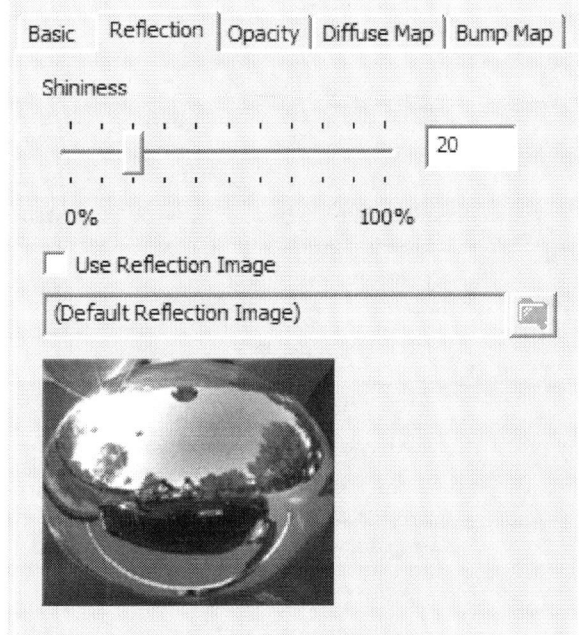

The Reflection tab defines how much light passes through the object and how much light is refracted.

Shininess	Determines how much of the specified specular color is reflected – 100% indicates 100% of the specular color is reflected.

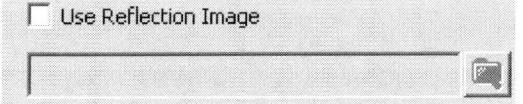

The **Use Reflection Image** option allows you to select an image map. An example might be an image that appears on a chrome bumper or on a windshield. You might want to map a sky as a reflection image onto a windshield or other cars on a chrome bumper.

20-3

Autodesk Inventor Fundamentals

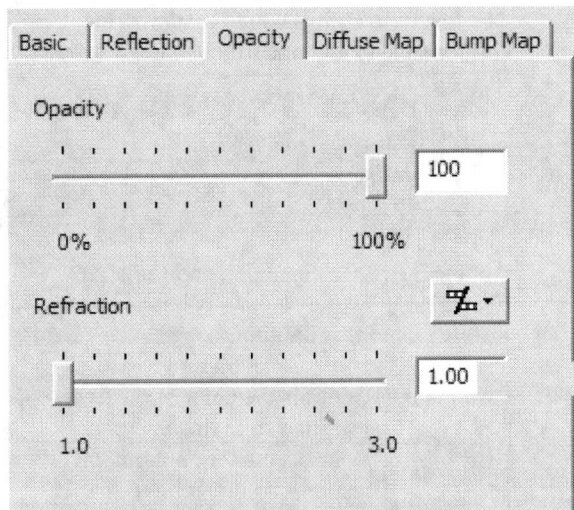

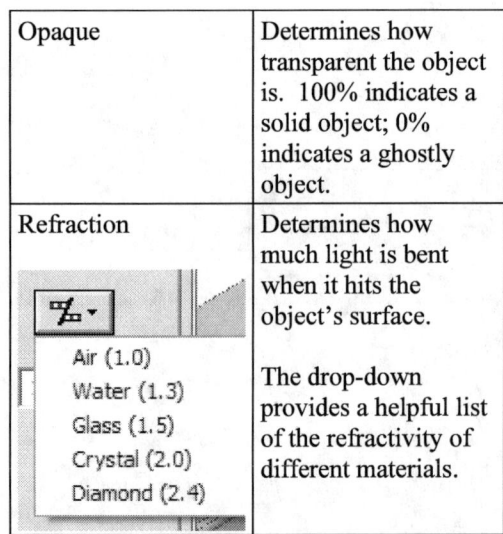

Opaque	Determines how transparent the object is. 100% indicates a solid object; 0% indicates a ghostly object.
Refraction	Determines how much light is bent when it hits the object's surface. The drop-down provides a helpful list of the refractivity of different materials.

The Diffuse Map tab is similar to defining a color or material texture.

The texture image file must be located in the same folder as the projects in order for it to be used. Any image file may be used.

Scale controls the scale of the image as it is mapped to the surface of the object where it is applied.

Rotation determines the direction of the mapping.

For example, for a wood surface you may wish to rotate the image so that the grain appears to follow a particular direction.

The rotation angle is based on the UCS of the selected plane.

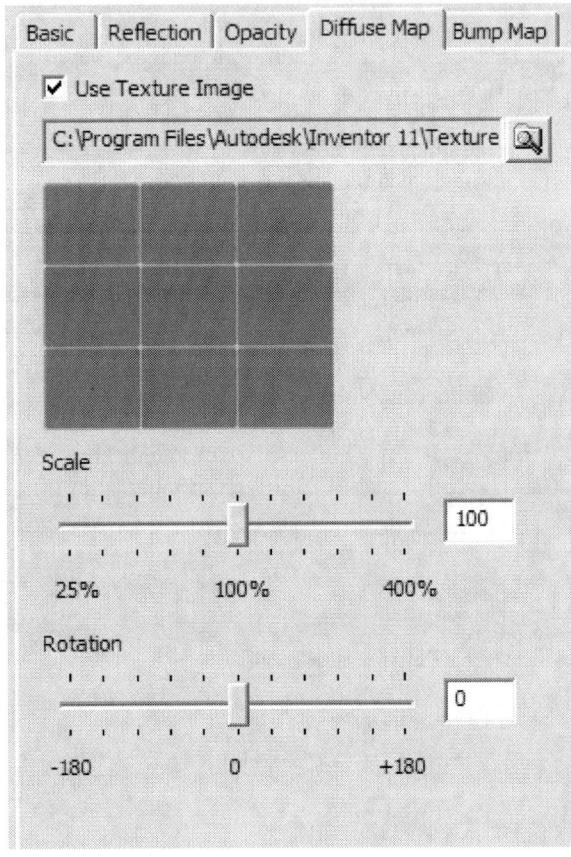

Rendering with Inventor Studio

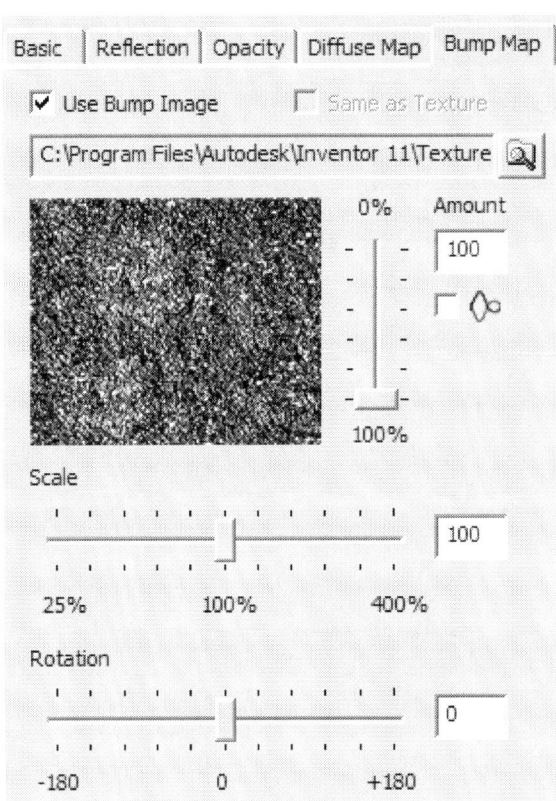

The Bump Map tab allows the user to add "noise" to the surface style to create a more 3D effect.

The user can set the material to be more "bumpy" using the same image as the texture or a different image.

The scale and rotation can be set to create a different effect.

☑ ◊	The Invert Effect button changes the bump pattern so it is inverted.

TIP: If you get an error stating that your Styles Library is corrupted, you can correct the error. To do this, locate the DesignData.exe file located in the Styles folder under *Program Files/Inventor 11/Styles*. Extract the file to the Styles folder located in your project. To extract, simply double click on the file name.

Autodesk Inventor Fundamentals

Exercise 20-1:
Creating a Surface Style

File: Ex17-3.ipt
Estimated Time: 15 minutes

1. Open *ex17-3.ipt*.

2. 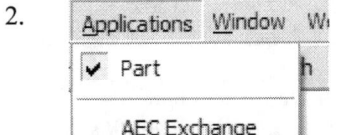 Go to **Applications→Inventor Studio**.

3. Select the **Surface Styles** tool.

4.  Expand the *Wood* folder.

    Locate the **Wood (Ash)** material.

    Select the **New Style** tool.

20-6

5.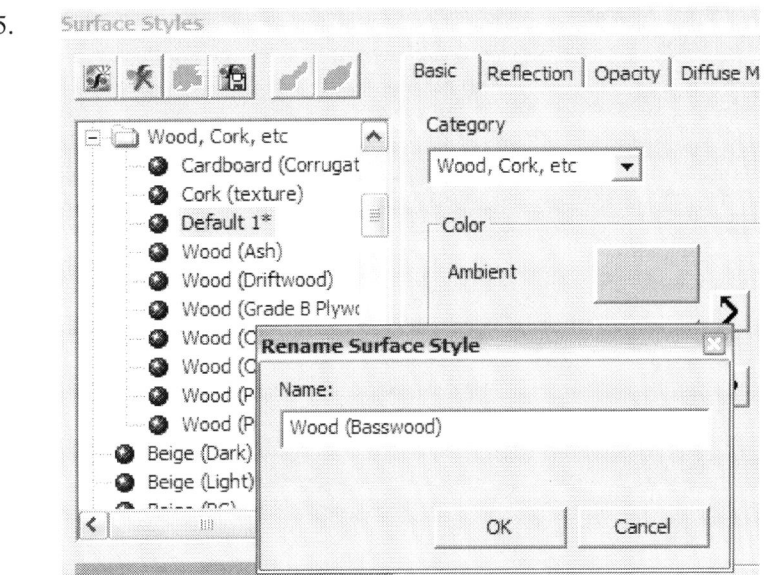

Set the **Category** to **Wood, Cork, etc**.

Right click and select **Rename Surface Style**.

Name the new style **Wood (Basswood)**.

Press **OK**.

6. Select the **Diffuse Map** tab.

7.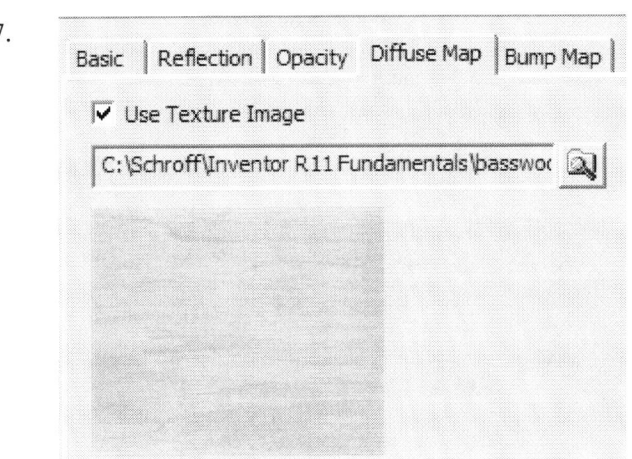

Enable **Use Texture Image**.

Browse for the *basswood.jpg* file (downloadable from the publisher's website).

8. Press **Save**.

Autodesk Inventor Fundamentals

9. In order for this new style to be available for all parts and assemblies, it needs to be saved to the Style Library.

Highlight the Basswood style.
Right click and select **Save to Style Library**.

10. Press **Yes**.

11. Press **Done**.

12. Basswood is now available as a material in the Style Library.

Select the **Basswood** material to apply to the part.

13. Select **Wood (Pine)** from the Style drop-down.

Note the direction of the grain.

14. Select the **Surface Styles** tool.

15. Select the **Diffuse Map** tab.

16. 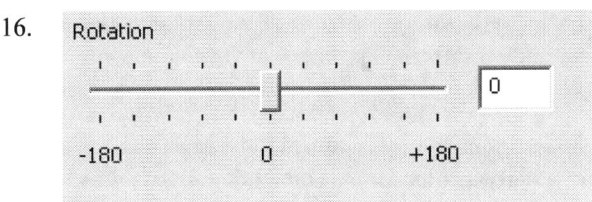 Set the **Rotation** to **0**.

    Press **Save** and **Done**.

Note the direction of the grain has changed.

17. Save as *ex20-1.ipt*.

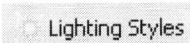

Lighting Styles

A Lighting Style basically saves a set of lights. You can create a new style and then add lights to it. In addition to defined lights, you may set the amount of ambient lighting for the scene.

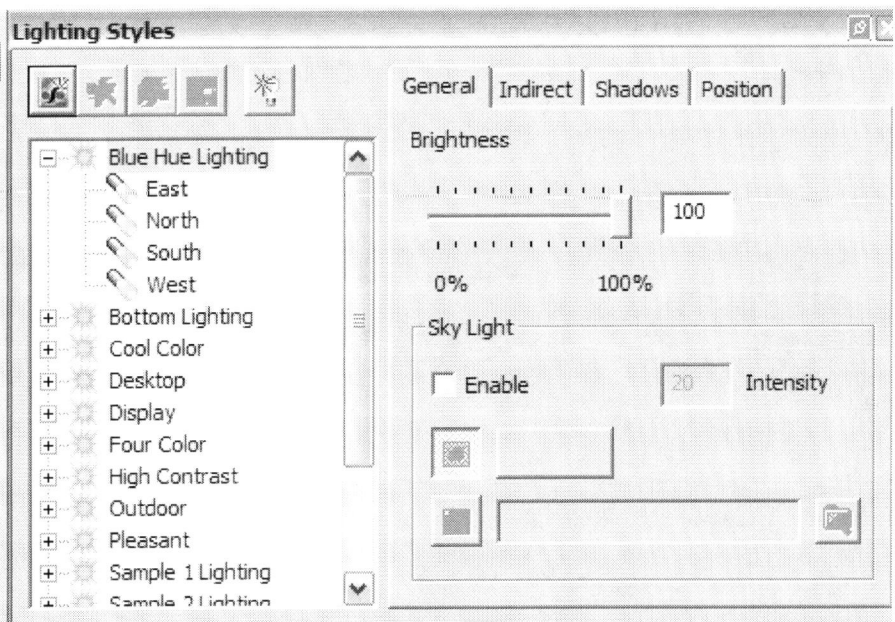

20-9

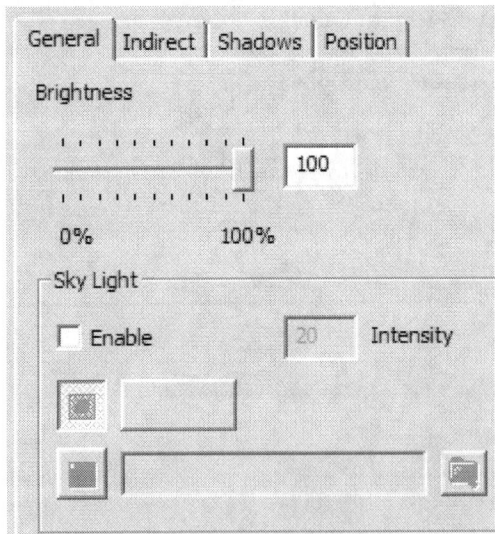

The slider control for brightness controls the intensity of all the lights, behaving like a universal dimmer switch.

The Sky Light controls ambient lighting. The user may specify a color for the ambient lighting. The user may also specify an image, in which case the image colors control the colors used by the Sky Light.

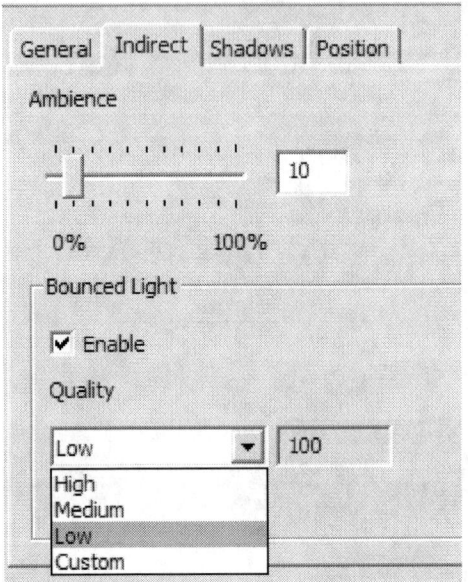

Ambience controls the amount of ambient lighting in the scene.

100% would be full daylight in an exterior setting. 0% would be a dark room.

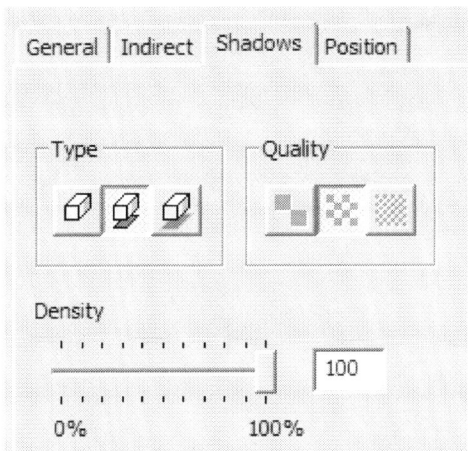

Shadows

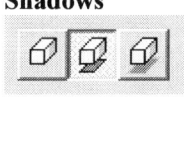

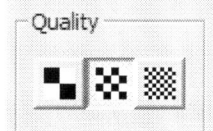

No Shadows – no shadows are shown
Hard Shadows – calculates shadows created by the light
Soft Shadows – more photo-realistic shadows, takes more calculation time
If Soft Shadows are enabled, you may set the Quality of the Shadow Calculation: low, medium, or high. High requires the most amount of time and provides the most photo-realism.

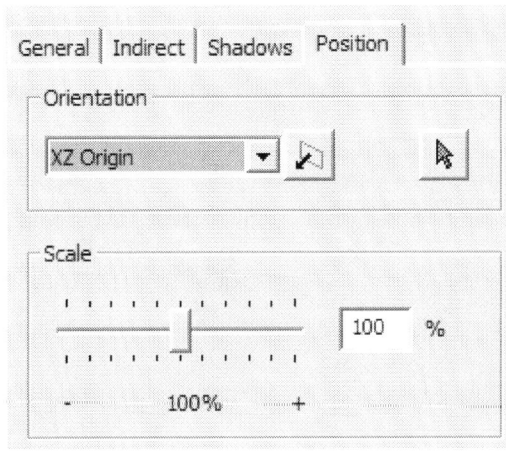

This tab provides a list of existing work planes that can be selected to orient the light styles.

The Flip Button flips the direction of the light 180 degrees.

The Select Button allows the user to select an alternate plane or face to be used for orienting the light style.

The Scale defines the relative scale of the lighting to the model.

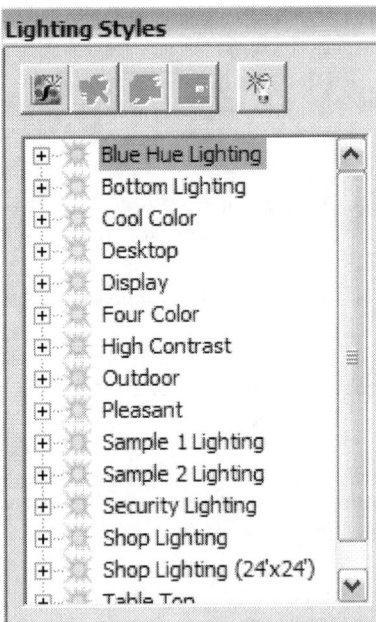 Inventor comes with several samples of lighting sets for your use.

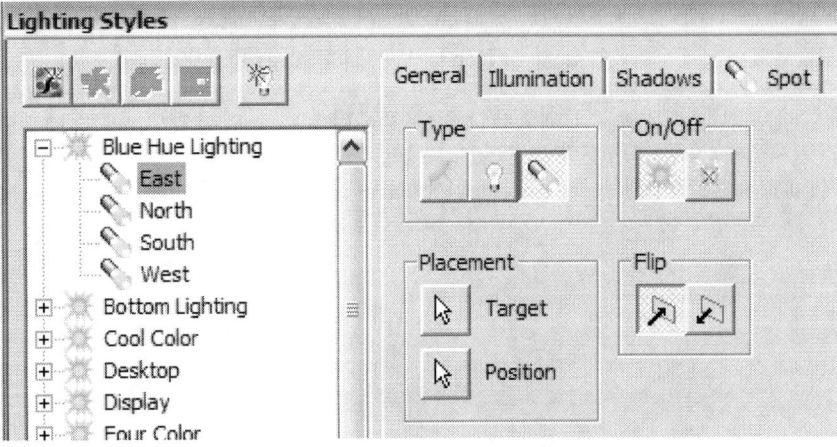 If you expand any of the lighting styles, you can see the lights defined for that style.

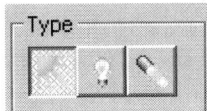 Lights can be defined as:
    Directional – a light source like the Sun.
    Point – a light source like a light bulb.
    Spot – a light source like a flashlight or spot light.

Exercise 20-2:
Creating a Lighting Style

File: Ex20-1.ipt
Estimated Time: 30 minutes

1. Open *ex20-1.ipt*.

2. 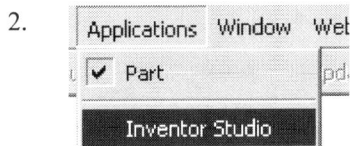 Go to **Applications→Inventor Studio**.

3. Select the **Lighting Styles** tool.

4. Highlight the **Blue Hue Lighting** style.
   Right click and select **Copy Lighting Style**.

5. Name the new style **Demo Lighting**.

6.  Locate the **Demo Lighting** style.
   Right click and set it to **Active**.

7. Close the dialog.

8.  Zoom out so you can see the lights.

   NOTE: If you double click your mouse's scroll wheel, your display will zoom to extents.

Autodesk Inventor Fundamentals

9. 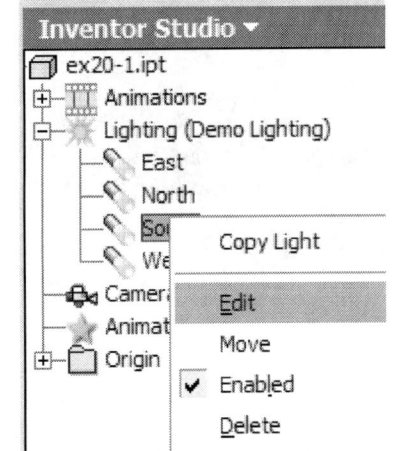 In the Browser, expand the **Lighting** category.

Note that it shows Demo Lighting as the active style.

Highlight each light and note that it highlights in the graphics window.

10. 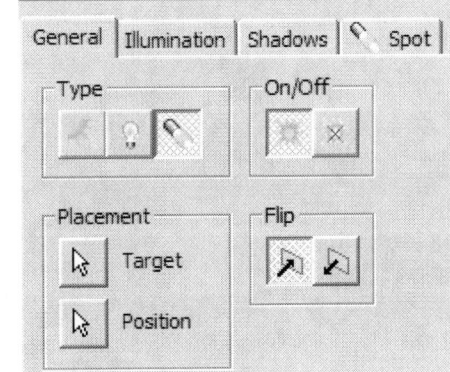 Highlight the **South** light in the Browser.

Right click and select **Edit**.

11. 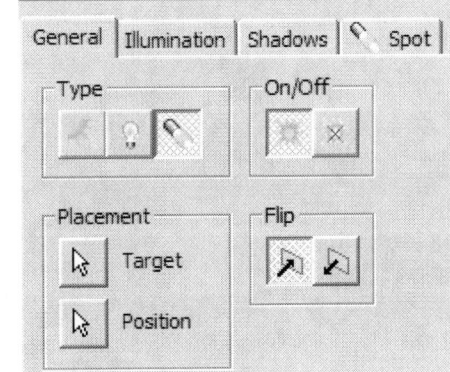 The light is set as a Spot light.
It is set to ON.

Select the **Flip** button and note that the direction of the spot changes.

Flip it back to the original setting.

12. 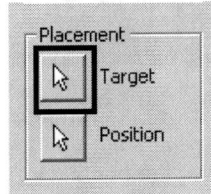 Select the **Target** button.

20-14

13.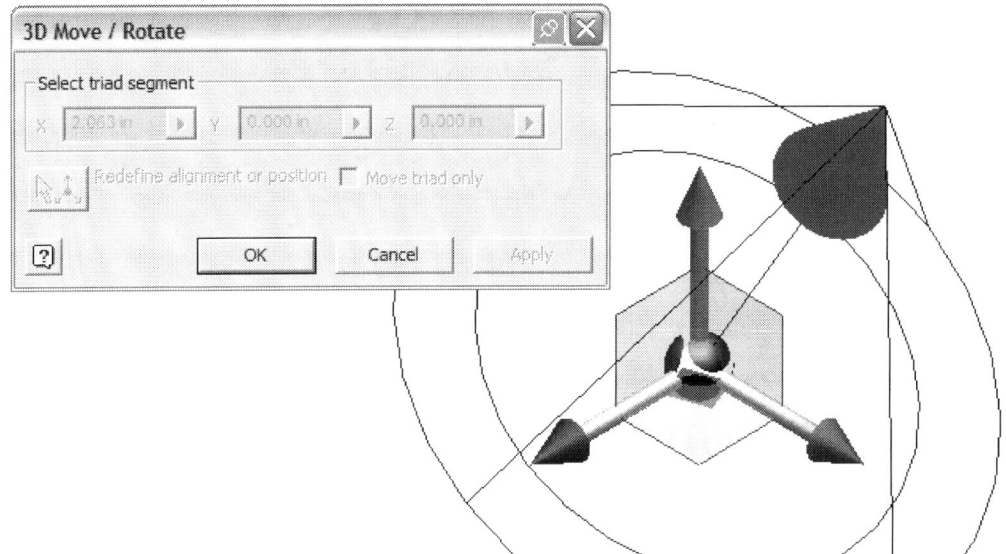

Click on the black box indicating the target area.
The 3D move/rotate triad and dialog will appear.
Shift the target to align with the part location by selecting an axis and moving the box.
Press **OK**.

14. 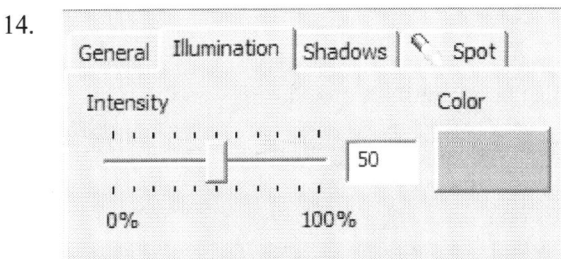   Select the **Illumination** tab.
Set the **Intensity** of the light to **50**.

15. 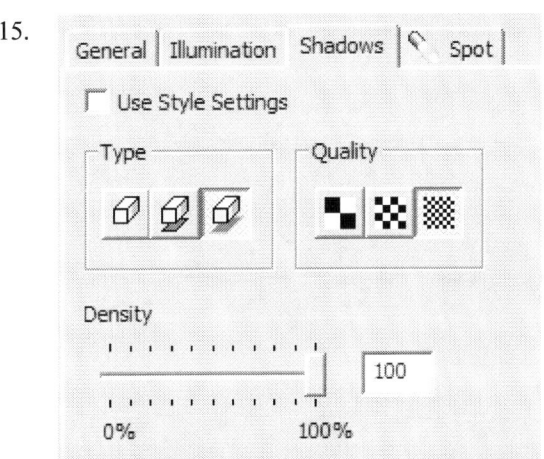   Select the **Shadows** tab.
Set the Shadows to **Soft Shadows**.

Set the Quality to **High**.

16. 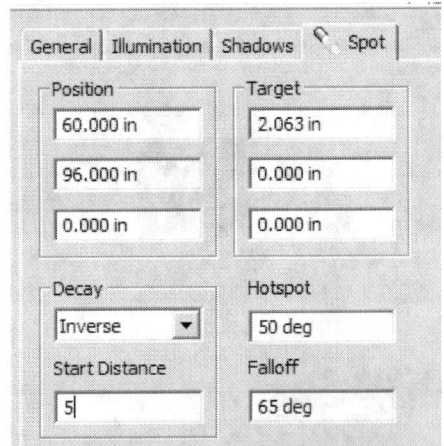 Select the **Spot** tab.

Set **Decay** to **Inverse**.

Set **Start Distance** to **5**.

Press **Save** and **Done** to close the dialog.

17. 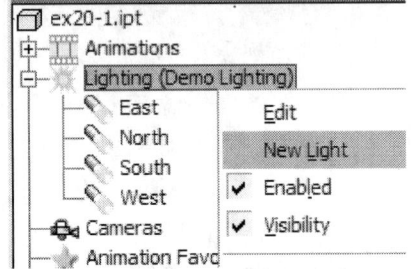 Highlight the lighting style in the Browser.
Right click and select **New Light**.

18. 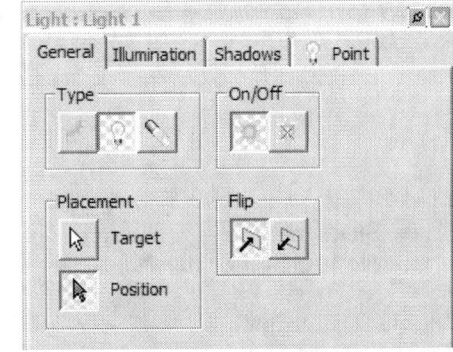 Select the **Point light** option.

Target will automatically be enabled.

Select the part.
Select a position above the part for the light.

19. 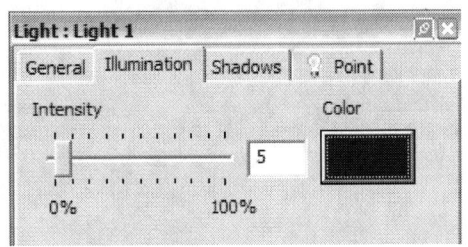 Select the **Illumination** tab.

Set the **Intensity** to **5**.
Set the **Color** to **Blue**.

20. 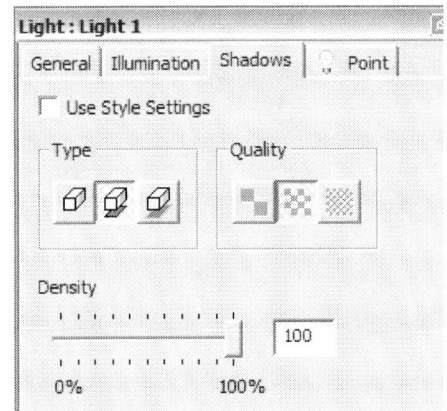 Select the **Shadows** tab.

Set the **Shadows** to **Sharp Shadows**.

21. 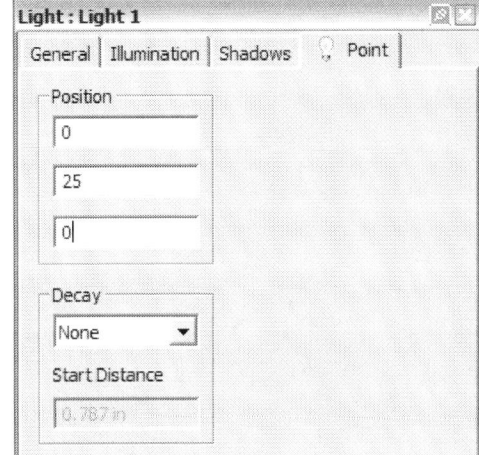 Select the **Point** tab.

Set the **X** position to **0**.
Set the **Y** position to **25**.
Set the **Z** position to **0**.

Note how the light position shifts.

Press **OK**.

22. 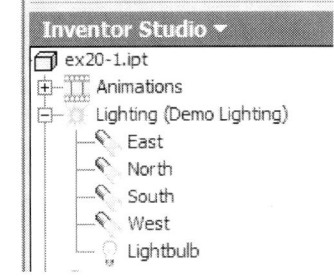 Rename the point light **Lightbulb** in the Browser.

NOTE: If you do not save to the Styles Library, you will not see your changes.

23.  Select the **Lighting Styles** tool.

24. 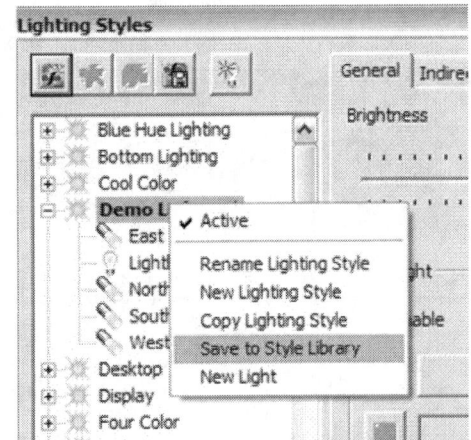 Highlight the Demo Lighting style.

    Right click and select **Save to Style Library**.

    Press **Done**.

25. 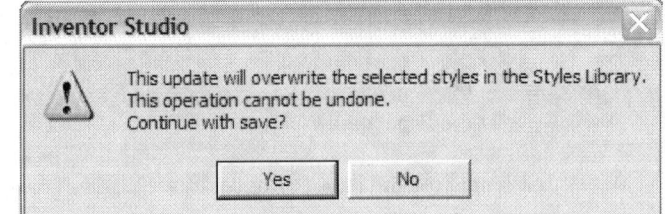 Press **Yes**.

26. Press **Done**.

27. Set **Applications** to **Part** mode.

28. Save as *ex20-2.ipt*.

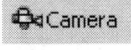

Camera

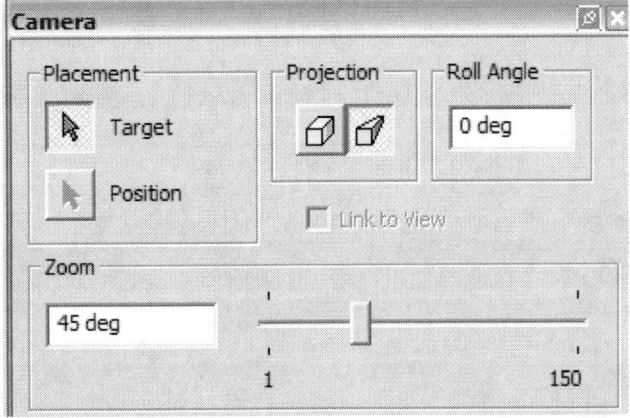 The Camera tool allows you to place a camera to define a viewpoint.

The Target indicates the area where the camera will be pointed.

The Position indicates the location of the camera.

The Roll Angle defines the angle of rotation, pointing the camera up or down along the Y axis. The value can be between -180 and +180 degrees.

The Zoom Angle determines the angle on the horizontal axis and can be between 1 and 150 degrees.

Exercise 20-3:
Creating a Camera

File: Ex20-2.ipt
Estimated Time: 10 minutes

1. Open *ex20-2.ipt*.

2.  Go to **Applications→Inventor Studio**.

3. 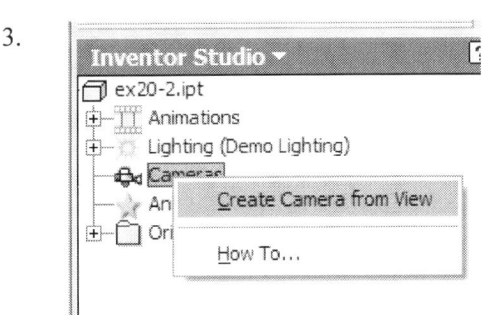 Highlight **Cameras** category in the Browser.
Right click and select **Create Camera from View**.

4.  Rename the view **Default Iso**.

5. Rotate the view slightly using the **3D Orbit** tool.

6. 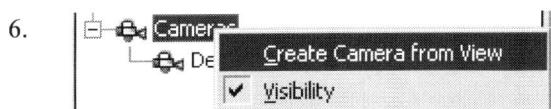 Highlight the **Cameras** category in the Browser.
Right click and select **Create Camera from View**.

7. 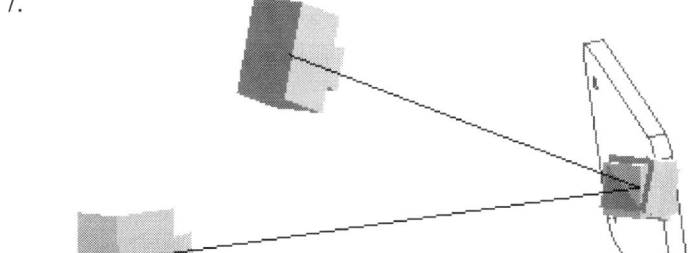 If you orbit the view, you will see both of the cameras you created.

8.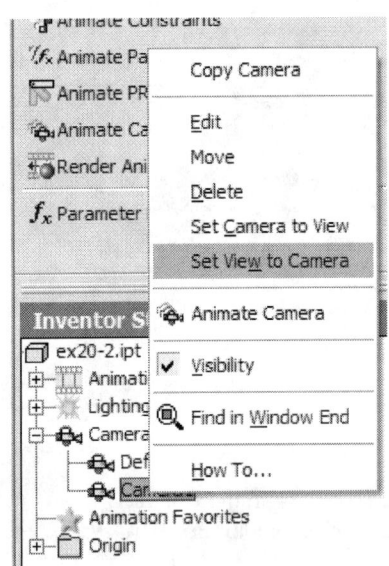

Highlight the **Default Iso** camera.

Right click and select **Set View to Camera**.

9. Note that the view returns to the previous view.

10. Return to Part mode.

11. Save as *ex20-3.ipt*.

Exercise 20-4:
Creating a Rendering

File: Arbor Press.iam (available in the Samples folder)
Estimated Time: 15 minutes

1. Select **Open**.

   Browse to the *Arbor Press* folder under Inventor 11.

2.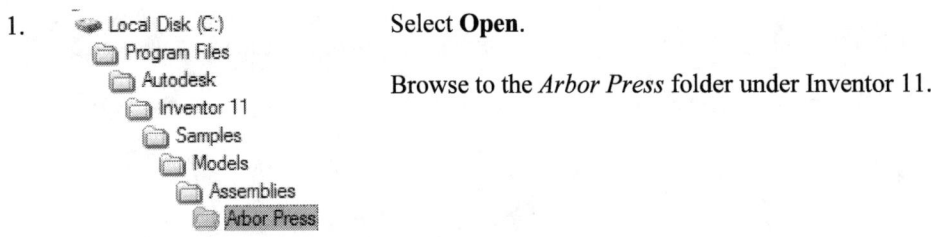

   Open the *Arbor_Press.iam* file.

3. Switch to **Inventor Studio** mode.

20-20

Rendering with Inventor Studio

4. **Render Image** — Select the **Render Image** tool.

4. Camera1 / (Current View) / Camera1 / Camera2 — Select **Camera1** from the **Camera** drop down.

5. Table Top / (Current Lighting) / Desktop / Four Color / Pleasant / Shop Lighting / Table Top — Select **Table Top** under **Lighting Style**.

6. XZ Reflective GP (Tan) / (Current Background) / Forest (Gradient) / XZ Reflective GP / XZ Reflective GP (Tan) — Select the **XZ Reflective GP (Tan)** from the **Scene Style**.

7. Render — Select the **Render** button.

8. 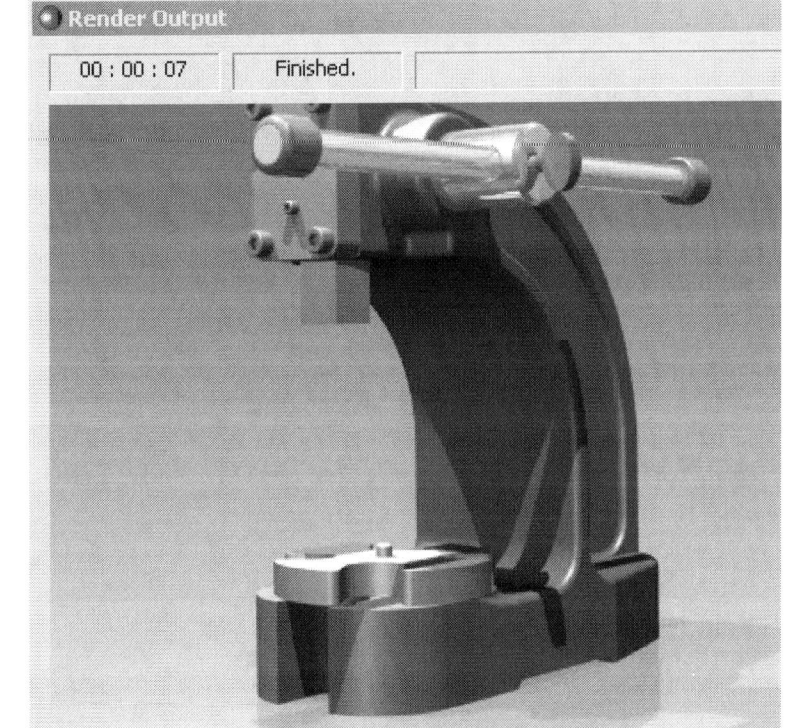 A rendering window appears and the model is rendered.

9. Close the rendering window.

10. Close the **Render** dialog.

20-21

Autodesk Inventor Fundamentals

We can copy lighting styles, surface styles, and scene styles from existing files to new files.

Exercise 20-5:
Copying Styles

File: Arbor Press.iam (available in the Samples folder)
Estimated Time: 15 minutes

1. 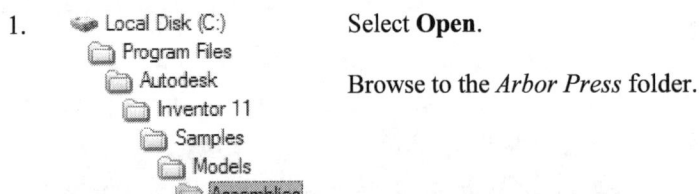 Select **Open**.

   Browse to the *Arbor Press* folder.

2. 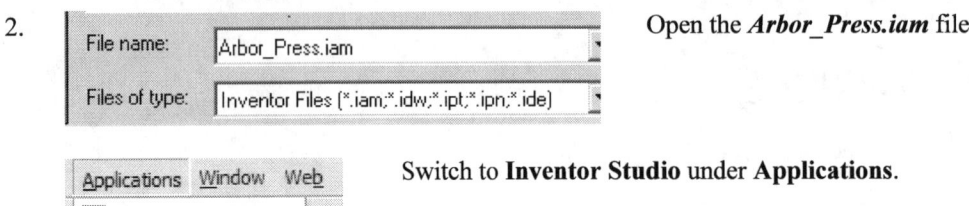 Open the *Arbor_Press.iam* file.

   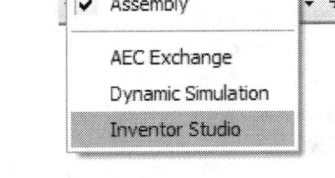 Switch to **Inventor Studio** under **Applications**.

3. ![Lighting Styles] Select the **Lighting Styles** tool.

4. ![Table Top] Highlight the **Table Top** Style.

5. ![save icon] Select **Save to Style Library**.

6. Repeat for the remaining Lighting Styles.

7. Press **Done** to close the dialog.

8. ![Scene Styles] Select **Scene Styles**.

9. ![Forest (Gradient)] Select **Forest (Gradient)**.

10. ![save icon] Select **Save to Style Library**.

11. Repeat for the remaining Lighting Styles.

12. Press **Done** to close the dialog.

13. The copied styles will now be available in your project.

Rendering with Inventor Studio

Exercise 20-6:
Creating a Rendering with Different Backgrounds

File: C-clamp.iam
Estimated Time: 30 minutes

1. Open *c-clamp.iam*.

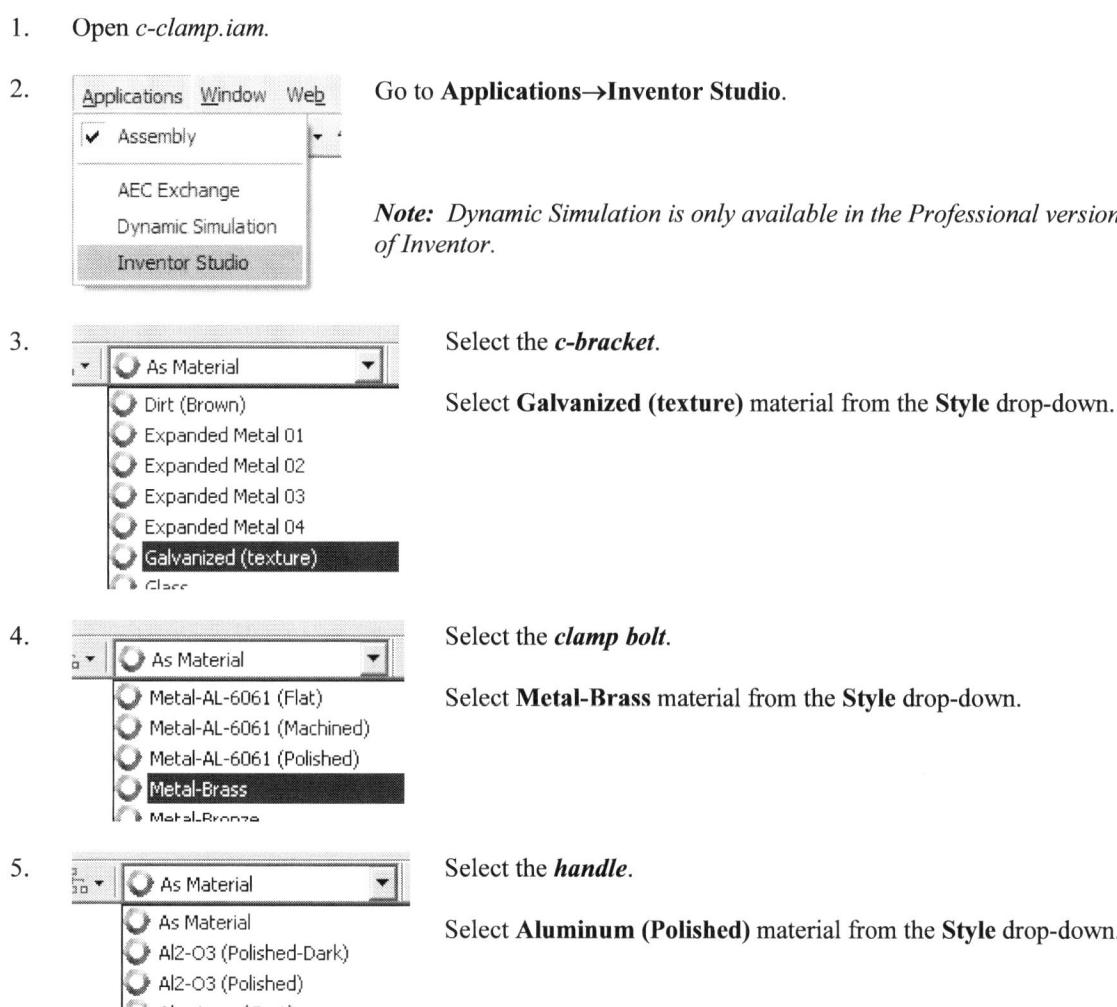

2. Go to **Applications→Inventor Studio**.

   Note: Dynamic Simulation is only available in the Professional version of Inventor.

3. Select the *c-bracket*.

   Select **Galvanized (texture)** material from the **Style** drop-down.

4. Select the *clamp bolt*.

   Select **Metal-Brass** material from the **Style** drop-down.

5. Select the *handle*.

   Select **Aluminum (Polished)** material from the **Style** drop-down.

6. Select the **Render Image** tool.

20-23

Autodesk Inventor Fundamentals

7.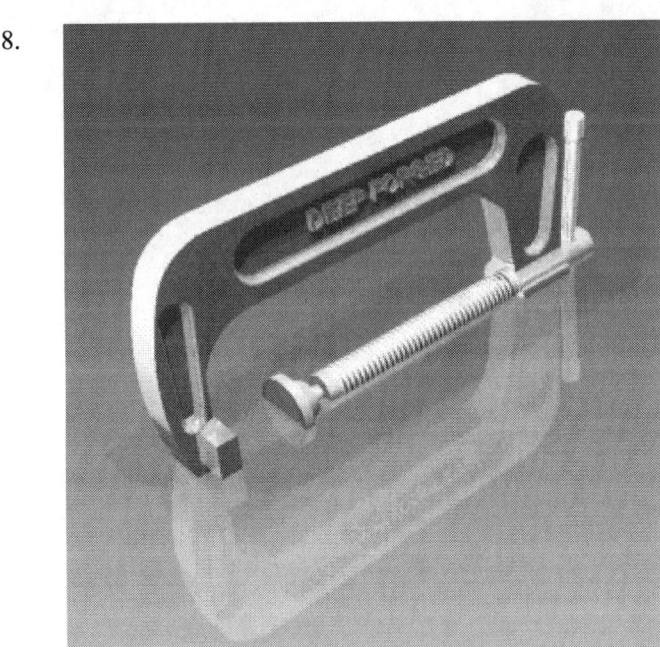

    Set the **Camera** to the **Current View**.
    Set the **Lighting Style** to **Table Top**.
    Set the **Scene Style** to **XZ Reflective GP (Tan)**.

    Press **Render**.

8. The model is rendered.
    You may notice that rendering has greatly improved in R11.

9. Close the Render window.

10. 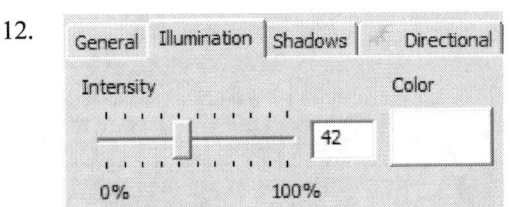 Select the **Lighting Styles**.

11. Highlight **Directional 1** under **Table Top**.

12. Select the **Illumination** tab.

    Set the **Intensity** to **42**.

20-24

13. 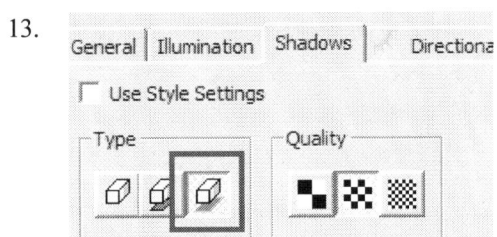 Select the **Shadows** tab.

    Set Shadows to **Soft Shadows**.

    Press **Done**.

14. 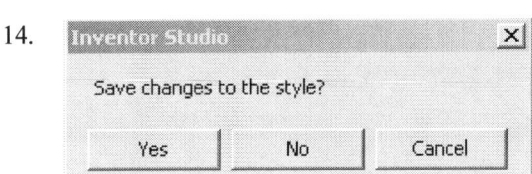 Press **Yes**.

15. Scene Styles   Select **Scene Styles**.

16. XZ Reflective GP (Tan)   Highlight **XZ Reflective GP (Tan)**.

17. 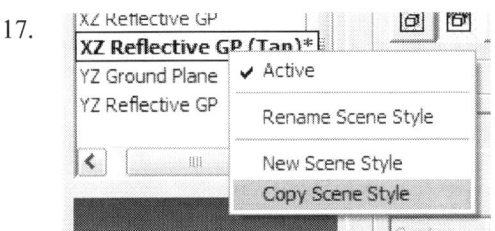 Right click and select **Copy Scene Style**.

18. 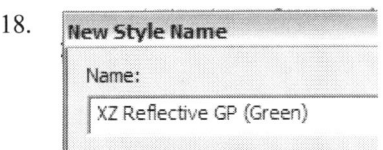 Change the style name to **XZ Reflective GP (Green)**.

19. 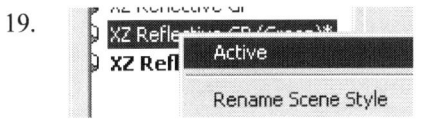 Highlight the new style.
    Right click and select **Active**.

20. Colors   Select the black square under **Colors**.

21. Define Custom Colors >>   Select **Define Custom Colors**.

20-25

Autodesk Inventor Fundamentals

22. Select a lighter color.

    Press **OK**.

23. Press **Save**.

24. Highlight **Forest (Gradient)**.

25. Select the **New Style** tool.

26. Name the new style **Sky**.

    Press **OK**.

27. Press the **Image Sphere** button under **Type**.

28. Select the *sky3.bmp* file.

    You can also use your own image file.
    Press **Open**.

29. Select the **Image** button.

30. Press **Save**.

31. Highlight **Sky**.
    Select **Save to Styles Library**.

    NOTE: Remember – if you don't save to the Styles Library, the style is only available in the local file.

32. Press **Done** to close the dialog.

33. Select the **Render Image** tool.

34. 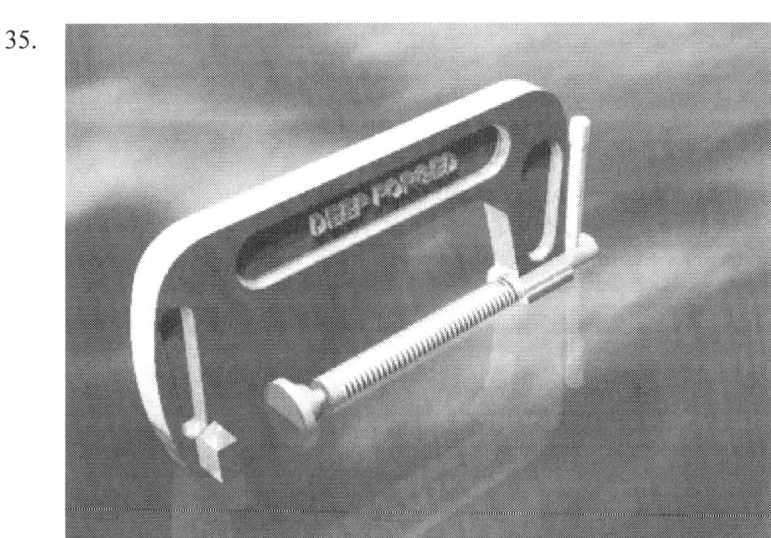 Set **Camera** to **Current View**.
    Set **Lighting Style** to **Table Top**.
    Set **Scene Style** to **Sky**.

    Press **Render**.

35. The image is rendered.

36. Close the two dialogs.

37. Save as *ex20-6.iam*.

 TIP: When you create an animation, it is similar to creating a positional representation. First, set up the constraint you want to use to define the motion. Then, suppress the constraint so it can be animated. The suppression has to be done in the Assembly environment.

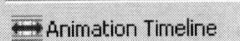

Animation Timeline

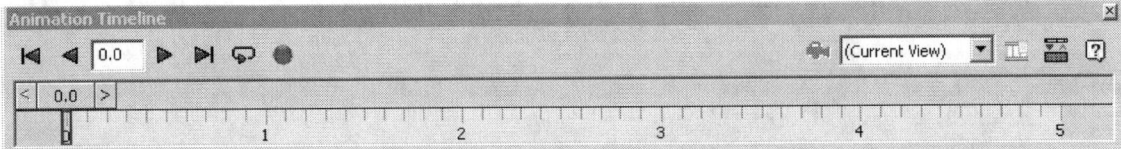

◄◄	Reset to start (Time 0)
◄	Play Animation in Reverse
1.0	Enter in a value to advance to a specific time frame.
►	Play Animation forward.
►►	Advance to last frame.
↻	Toggle Repeat – loop animation
●	Record animation
(Current View) / Front View / Iso View	Lists the camera views available
	Animation Options – sets the number of frames in the animation
	Expand Editor – allows you to see what occurs in each frame or frame set.

Animate Components

This tool is used to animate the position and rotation of parts in an assembly.

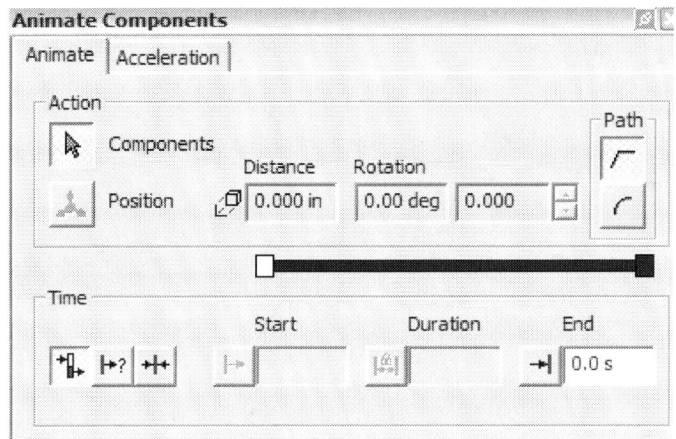

Components	Select one or two components in the graphics window or the Browser to be animated.
Position	Once the component(s) is selected, a triad will appear where you can adjust the movement of the component
1.688 in Distance	The distance selected when you adjust the motion of the component is displayed in the text box. You can manually edit the distance.
Path	The path may be sharp or smooth. If you enable smooth path, the program will provide a transitional curve between the start, middle, and end of the motion.
Time	Specify a start time, end time, and instance time.
	0.0s — Specify the time mark to begin the motion.
	3.0s — The total duration for the animation
	3.0s — The end time for the motion.

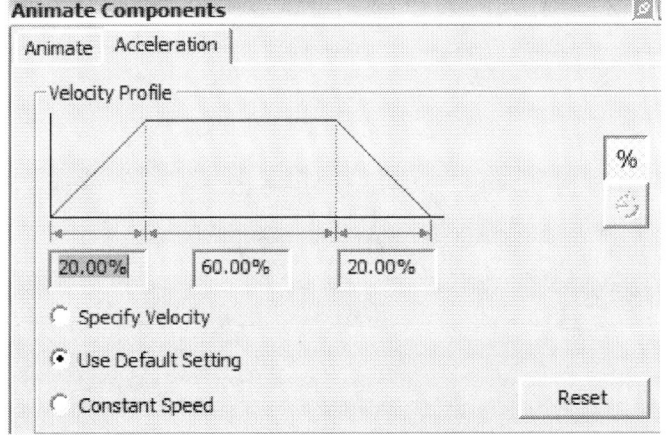

Acceleration controls how many frames are used for a complete transformation.

% defines a percentage for start, end, and middle of the transition.

The watch defines a set number of seconds or minutes.

20-29

Exercise 20-7:
Animate a Component

File: Ex20-6.iam
Estimated Time: 15 minutes

1. Open *ex20-6.iam*.

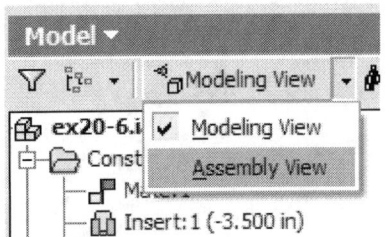

   Verify that you are in **Assembly View** and not in Modeling View.

2.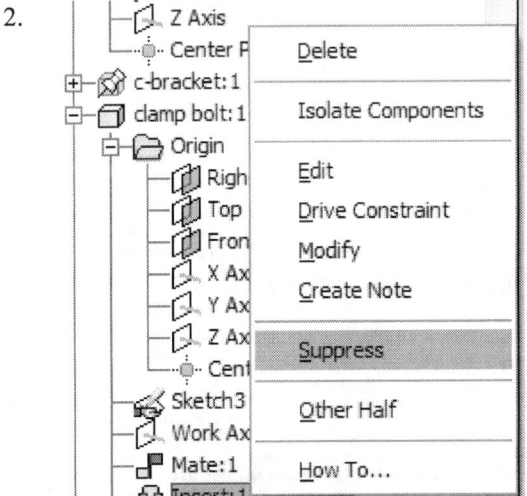

   Locate the **Insert Constraint** for the clamp bolt.

   Right click and select **Suppress**.

3.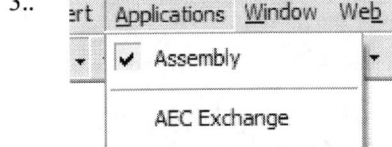

   Go to **Applications→Inventor Studio**.

4.

   Highlight the clamp bolt and handle in the Browser.

   Right click and select **Animate Components**.

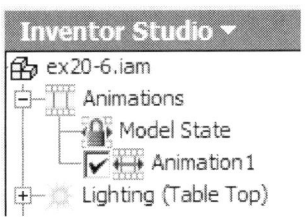
Note in the Browser, you should see an Animations folder. **Animation1** should be active.

5. Select the **Position** button.

6. Select the blue arrow tip and move the triad back slightly.

7.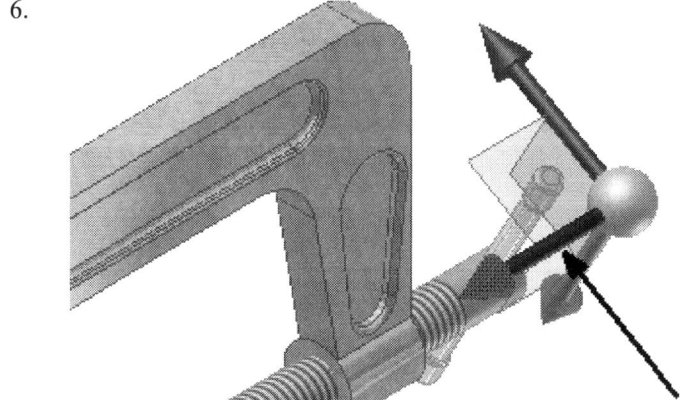

Select the blue axis and set the **Rotation** to **180 degrees**.
Press **OK**.

8. Modify the **Distance** value to **1.5**.

The degree value will automatically update.

9. Set the **Length** of animation to **5 seconds**.

Press **OK**.

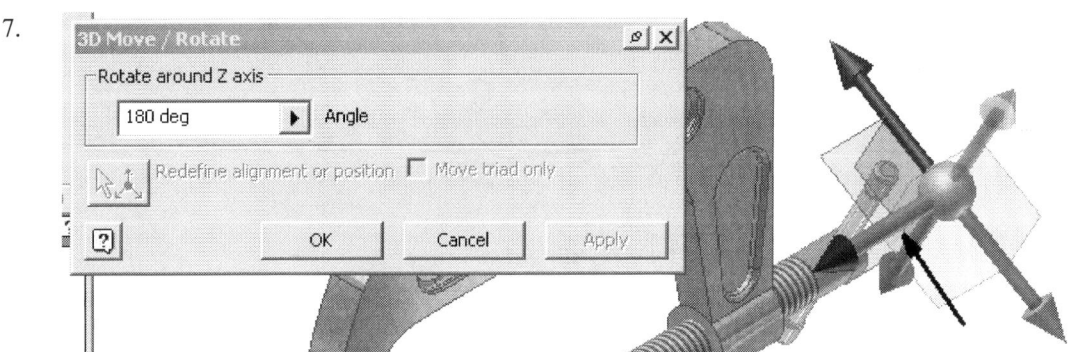

10. 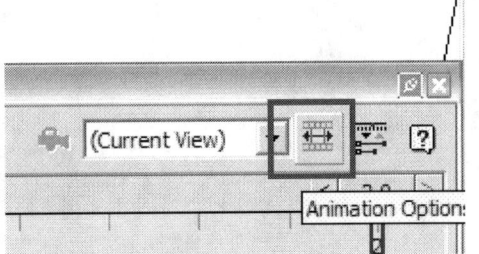 Select the **Animation Options** button on the timeline dialog.

11.  Set the **Length** time for the animation to **2.0 seconds**.
    Click **OK**.

12. Press **Go to Start**.

13. Press **Play**.

14. Return to Assembly mode.

15. Save as *ex20-7.iam*.

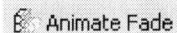

Animate Fade

Animate Fade is used to make one or more of the components in assembly appear transparent during an animation.

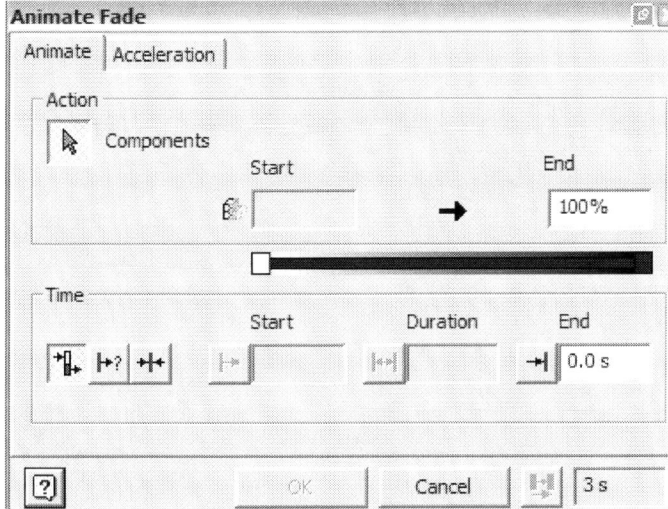

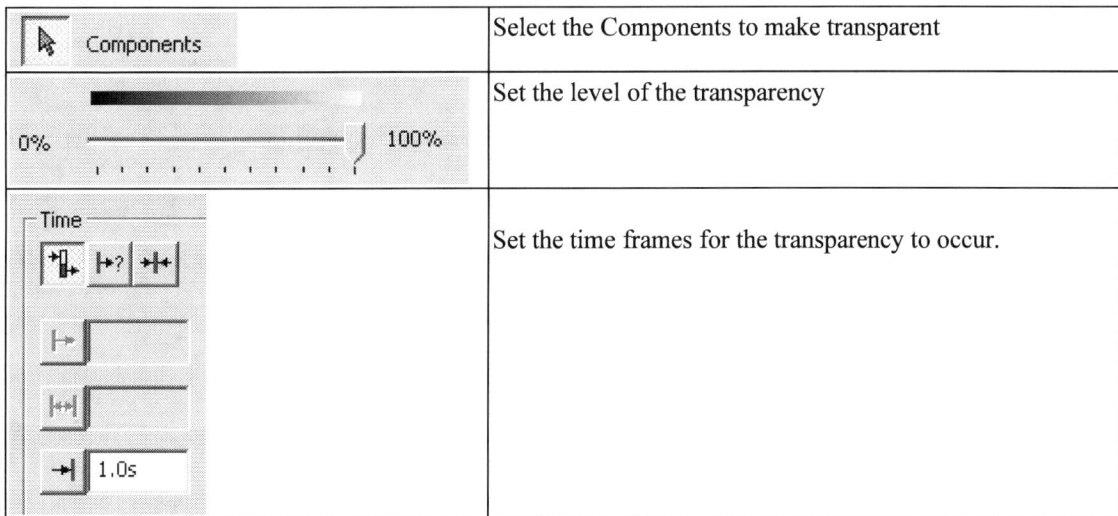

	Select the Components to make transparent
	Set the level of the transparency
	Set the time frames for the transparency to occur.

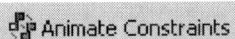

Animate Constraints

Animate Constraints allows you to suppress, enable and change the values of constraints.

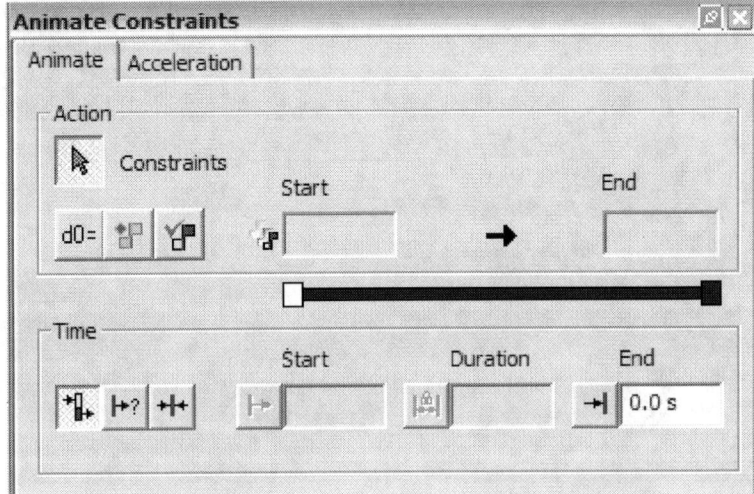

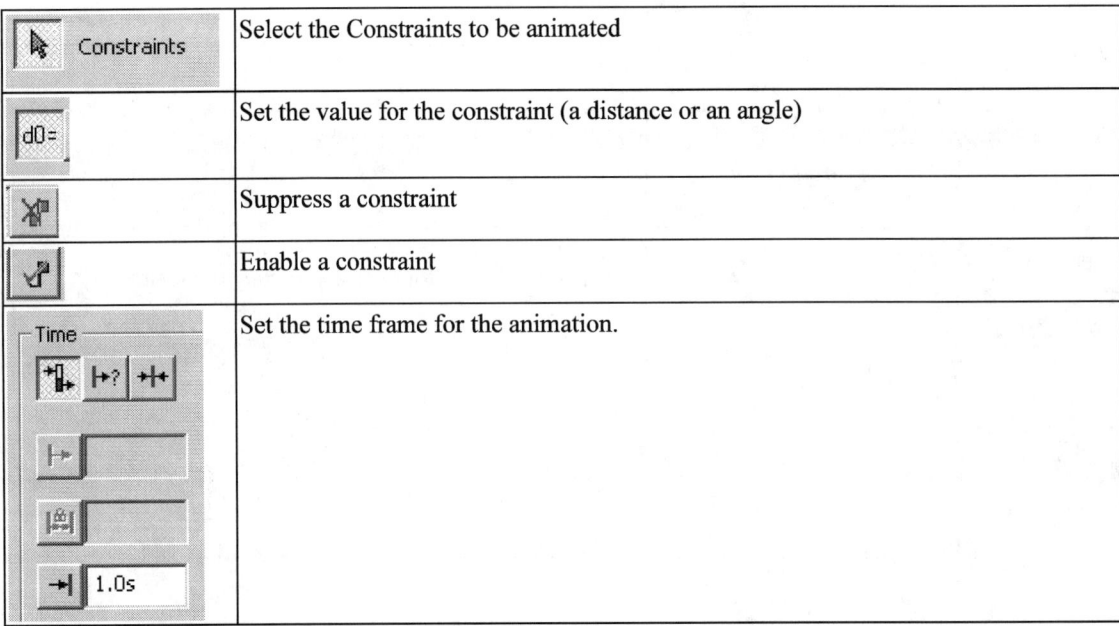

	Select the Constraints to be animated
	Set the value for the constraint (a distance or an angle)
	Suppress a constraint
	Enable a constraint
	Set the time frame for the animation.

Animate Camera

The Animate Camera tool is used to change the view display during an animation. Before you can animate a camera, you need to have at least one camera defined. You will then assign a view to that camera.

Exercise 20-8:
Animate a Camera

File: Ex17-8.iam
Estimated Time: 30 minutes

1. 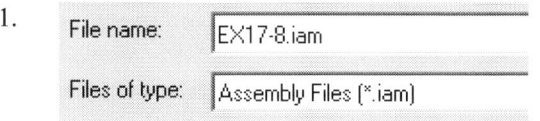 Open *ex17-8.iam*.

2.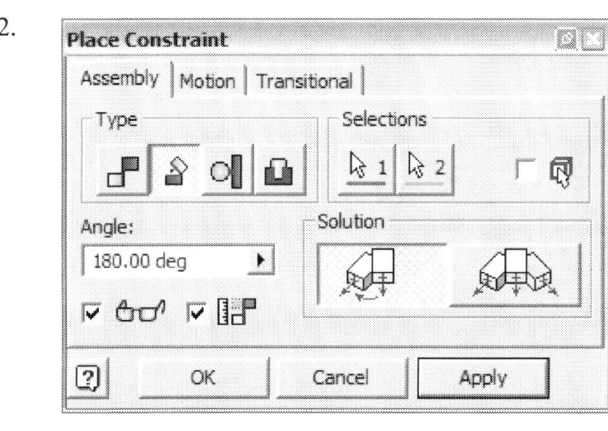

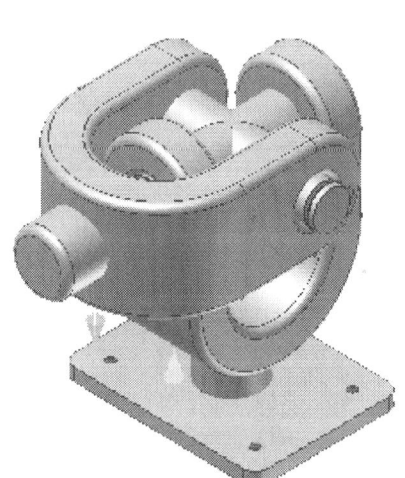

Add an angle constraint between the face of the top link and the top face of the plate.

3. 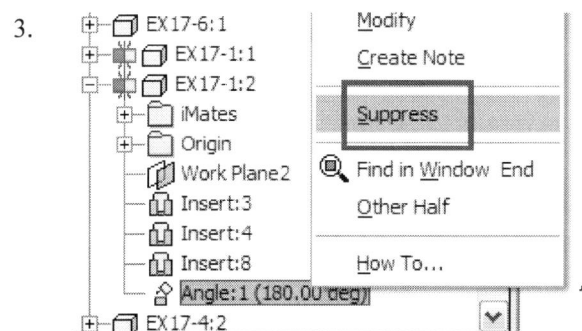 Highlight the angle constraint.

Right click and select **Suppress**.

4. 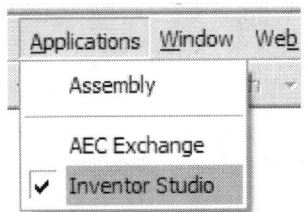 Go to **Applications→Inventor Studio**.

5. 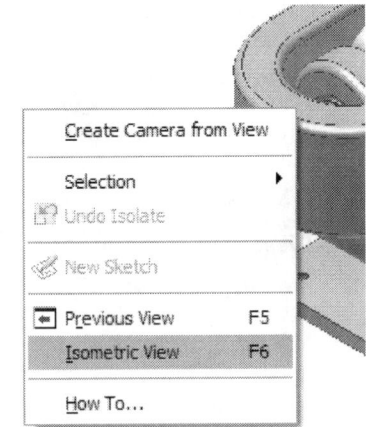 Right click in the graphics window and switch to an Isometric View.

6. 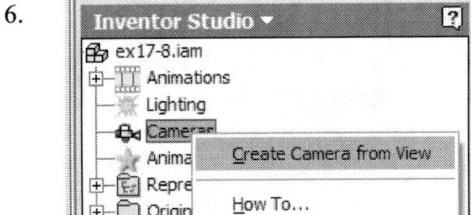 Highlight the **Cameras** category.
Right click and select **Create Camera from View**.

7. Rename **Iso Camera**.

8. Select the **Animation Timeline** tool.

9.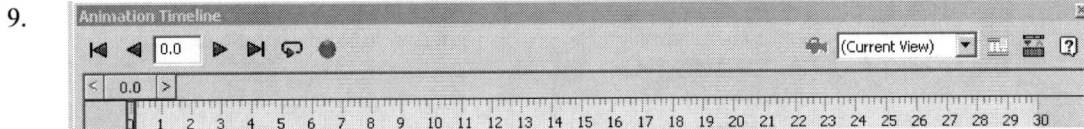

A toolbar with the frames in seconds appears.

10. Select the **Options** button located in the upper right of the Animation Timeline.

Rendering with Inventor Studio

11. 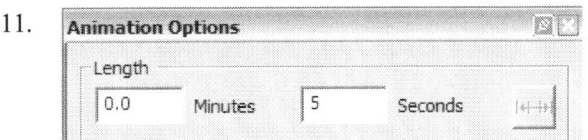 Set the **Length** of the animation to **5 seconds**. Press **OK**.

12. 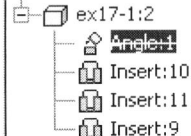 Click the **Animate Constraints** tool.

13. The **Select Constraints** button is enabled.

    Select the angle constraint you just defined and suppressed.

14.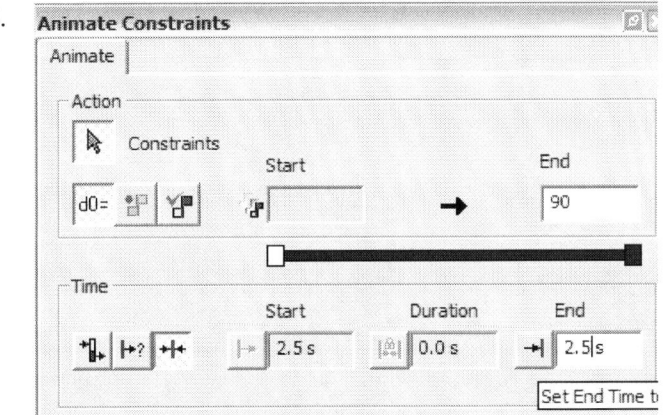

    Click the **Enable Constraint** button.

    Click the **Specify Constraint Value** button.
    Then enter a value of **90**.

    Set the end time to **2.5**s.

    Press **OK**.

15. 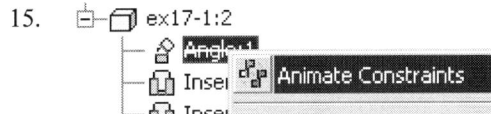 Highlight the **Angle** constraint in the Browser.

    Right click and select **Animate Constraints**.

16.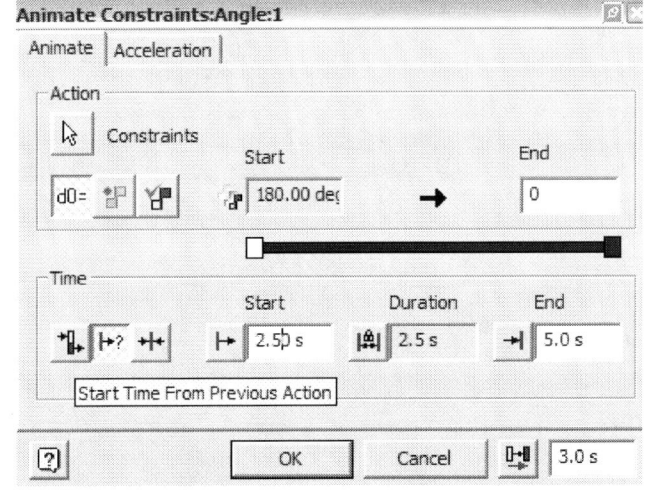

    Click the **Enable Constraint** button.

    Click the **Specify Constraint Value** button.
    Then enter an **End** value of **0**.

    Set the **Start** time to **2.5**.
    Set the **End** time to **5.0**s.

    Press **OK**.

20-37

17. 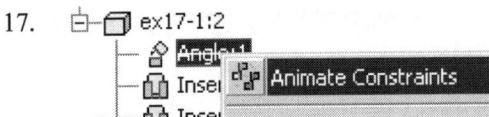 Highlight the **Angle** constraint in the Browser.

    Right click and select **Animate Constraints**.

18. 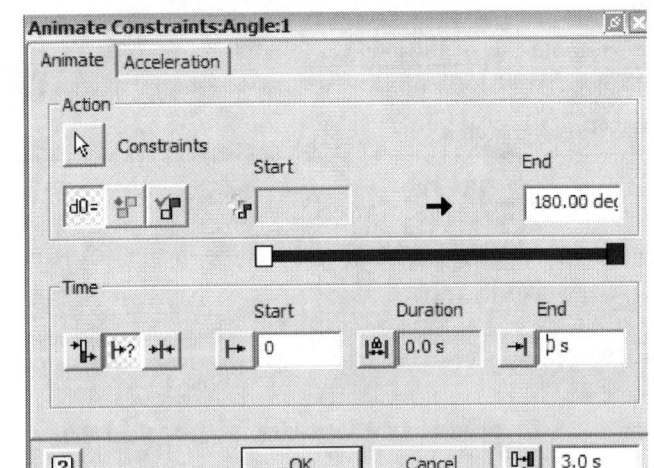 Click the **Enable Constraint** button.

    Click the **Specify Constraint** button. Then enter a value of **180**.

    Set the **Start** time to **0.0**.
    Set the **End** time to **0.0s**.

    Press **OK**.

19. Select the **Go to Start** button.

20. Select the **Play** button.

21. Select the **Go to Start** button.

22. Select **Iso Camera** from the camera drop-down on the Animation timeline.

23. 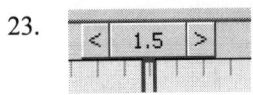 Move the time frame to **1.5s**.

20-38

Rendering with Inventor Studio

24. Enable the **Common View** tool.

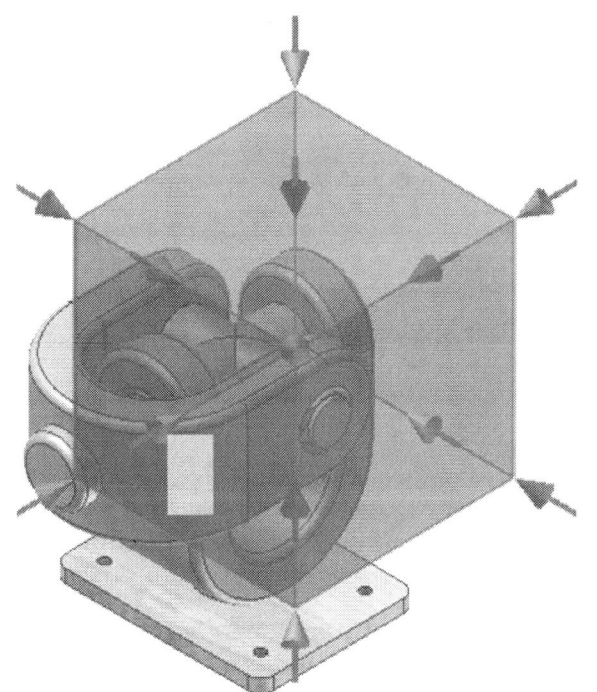

25. Select the front view.

26. 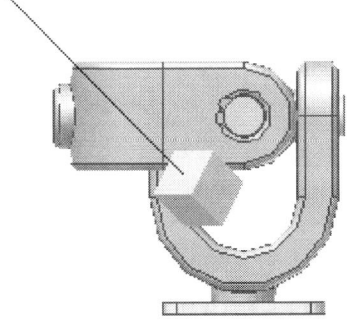 Click on the **Add Camera Action** button.

27. 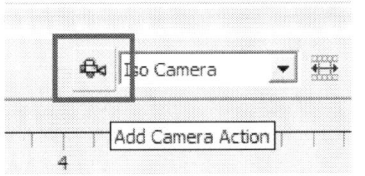 Highlight the Cameras category in the Browser. Right click and disable **Visibility**.

28. 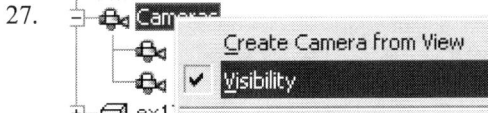 Select the **Go to Start** button.

29. Select the **Play** button.

20-39

30. Select the **Expand Action Editor**.

31.

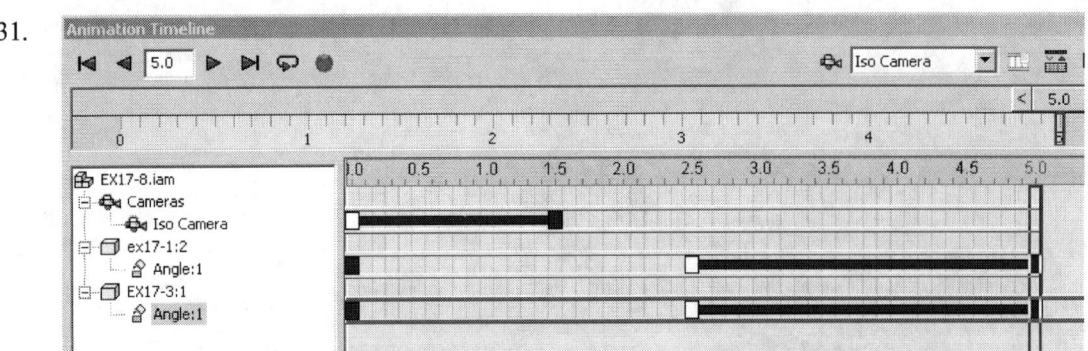

You can see the Camera view you defined and that it is active until the 1.5s mark. You see the two angle constraints that you animated.

32. Save as *ex20-8.iam*.

Render Tools

Surface Styles	Applies materials to parts.
Lighting Styles	Creates lighting sets consisting of one or more lights.
Scene Styles	Sets the background of a scene.
Camera	Creates named views.
Render Image	Creates a rendered image that can be saved to a file.
View Last Image	Brings up the previous render window.
Animation Timeline	Brings up a timeline for an animation.
Animate Components	Controls the movement of components.
Animate Fade	Controls the visibility of components
Animate Constraints	Controls the value of constraints as well as enabling and disabling constraints.
Animate Parameters	Controls the value of parameters
Animate PR	Uses positional representations as key frames in an animation
Animate Camera	Controls the display of a part or an assembly
Render Animation	Creates an avi file of a rendered part or assembly.
f_x Parameter Favorites	Lists the parameters that are used in the animation.

Review Questions

1. Each surface style has a one-to-one relationship with the color style of the same name. Thus, creating or editing a surface style, creates or edits the corresponding color style.

    A. True
    B. False

2. Inventor Studio is best used to create and animate exploded views.

    A. True
    B. False

3. Animate Fade is used to:

    A. Change the color of a component
    B. Change the lighting of a component
    C. Change the visibility of a component
    D. Change the camera view

4. Animate Parameters controls a change in:

    A. dimensions
    B. user parameters
    C. global parameters
    D. design parameters

5. You can only record one camera at a time.

    A. True
    B. False

6. The Inventor Studio tools are accessed through:

    A. The Tools Menu
    B. The Applications Menu
    C. The File Menu
    D. The View Menu

7. The Animation Timeline can be set to different animation time lengths.

    A. True
    B. False

8. Create Camera From View:

    A. Creates a camera definition based on the current view display
    B. Creates a view based on a camera created by the user
    C. Creates a Camera from a Pre-Set View
    D. Creates a Camera and the user must select a view

ANSWERS: 1) A; 2) B – that is a better use for Presentation files; 3) C; 4) B; 5) A; 6) B; 7) A; 8) A

Index

2D Sketch	3-4	Break Link with Base Part	5-107
3D Fillet	8-5	Break-out View	12-31, 12-34
3D Indicator	3-19, 19-10	Broken View	12-31, 14-35
3DRotate	1-11, 4-4	Browser Bar	3-2

A

Activate	12-15
Activate Contact Solver	17-31
Active Standard	14-12
Adaptive	16-14
Add Path	3-18
Add Text Parameter	12-54
All Around	8-9
Ambience	20-9
Analyze Faces	4-8
Angle Constraint	16-26
Animate	19-14
Animate Camera	20-35
Animate Components	20-29
Animate Constraints	20-34, 20-37
Animate Fade	20-33
Animation Timeline	20-28
Annotation Scale	3-19
Application Options	2-1, 3-44, 6-37, 13-4, 20-2
Arc	6-2
Assembly Tools	16-47
Auto Balloon	13-52
Auto Dimension	6-15
Auto-Resize	2-7
Automated Centerlines	13-27
Auxiliary View	12-12, 12-15
Axes	3-35

B

Background	3-24
Balloon	14-20
Balloon All	13-50, 20-7
Base View	12-1, 12-5, 12-32, 13-24, 20-2, 20-3
Baseline Dimension	13-13, 14-38
Bend	9-22
Bend Radius	9-3
Bend Relief	9-3

C

Camera	20-18
Cartesian Coordinate System	2-1
Caterpillar	13-69
Center Point	1-6
Centerline	5-6, 6-7, 13-29
Centerline Bisector	13-33
Center Mark	13-25, 14-16, 14-37
Centered Pattern	13-33, 13-34
Chamfer	5-67, 6-2
Circle	6-2
Circular Pattern	5-83, 5-101 6-14
Clearance	5-10
Close Loop	6-8
Coil	5-52
Coincident Constraint	1-6
Colors	14-4
Color Options	3-26
Color Scheme	3-24
Column Chooser	13-59
Column Width	13-59
Common	14-14
Common View	1-11, 4-5
Component Opacity	3-40, 4-7
Component Visibility	16-35
Constraint, assembly	16-26
Constraints, sketch	6-22
Contact Set	17-30
Content Center	3-21, 5-87, 16-14
Contour Flange	9-8
Coordinate System Indicator	3-35, 6-38
Copy	12-15
Corner Chamfer	9-24
Corner Round	9-23
Corner Seam	9-20
Cosmetic Weld	8-8
Create Axis	2-5
Create Camera From View	20-19
Create Component	16-1, 16-9, 18-18

Autodesk Inventor Fundamentals

Create Text	6-29
Create View	19-2
Create Welding Symbol	8-7, 8-8
Cursor Cues	6-39
Custom iProperties	3-10, 12-61
Custom Filters	14-20
Customize	3-1, 3-3, 3-42
Cut	9-11

D

Datum Identifier Symbol	13-46
Datum Target	13-46, 14-23
Decal	5-80, 6-35
Defer Update	3-39
Define New Border	12-81
Define New Symbol	12-69, 14-15
Define New Title Block	12-48
Degrees of Freedom	16-31
Delete Leader	12-73
Delete Member	13-15
Delete Tweaks	19-19
Delta Input	7-7
Derived Component	5-106
Design Assistant	13-60
Design Data	3-21
Design View	12-2
Details	3-22
Detail View	12-27, 12-29
Dimension, General	1-4
Dimension Properties	9-29
Dimension Styles	13-3, 14-9, 14-13, 14-25
Display	4-6, 14-26
Display Options	3-27
Document Settings	7-8, 12-10, 12-64, 14-2
Draft View	12-40
Drawing Annotation	12-72, 13-1, 13-75
Drawing Borders	12-78
Drawing Options	3-33
Drawing Organizer	14-34
Drawing Resources	12-36
Drive Constraint	18-35
Dynamic Zoom	4-4

E

Edit Arrowhead	13-39
Edit Balloon	13-53
Edit Break	12-33
Edit Column Properties	13-63
Edit Coordinate System	5-59, 6-35
Edit Definition	12-76
Edit Dimension	6-14
Edit Feature	11-10, 18-15
Edit Field Text	12-76
Edit Hole	5-14
Edit Hole Note	13-15
Edit Parts List	13-59
Edit Select Filters	14-20
Edit Sheet	12-45, 14-8
Edit Sketch	6-19
Edit Solid	7-1
Edit Surface Texture	13-28
Edit Tag	13-55, 13-67
Edit Trail	19-14
Edit View	12-25, 20-4
Edit View Label	12-24
Edit Welding Symbol	13-31
Ellipse	6-2
Emboss	5-72, 5-75, 18-15
End Treatment	13-70
Expand Action Editor	20-40
Expert Mode	1-8
Export Parts List	13-59
Extend	6-18
Extend/Contract Body	7-6
Extract iFeature	5-96, 9-29
Extrude	1-1, 1-9, 5-2

F

Face	9-5, 10-2
Face Draft	5-70
Feature Control Frame	13-42
Feature Identifier Symbol	13-44
Feature Properties	15-2
Features toolbar	5-1, 5-115
File Extensions	1-3
File Options	3-23
Fill/Hatch Sketch Region	12-49
Fillet	5-61, 6-2, 11-9, 17-7
Fillet Welds	8-6, 8-10
Finish Sketch	1-7
Finish Solid Edit	7-7
Flange	9-13, 10-3
Flat Pattern	9-4, 10-8
Flip Axis	6-36
Fold	9-17
Font Size	3-21
Format Column	13-60

G

General Dimension	1-4, 6-15, 13-2
General Options	3-21
Grid Lines	3-35
Groove Welds	8-10
Grounded	16-5

H

Hardware	3-30
Hatch	14-15
Hem	9-15
Hide Dimensions	12-55
Hide Origin Indicator	13-19
Hide Trails	19-13
Hidden Edge	1-12
Hole	5-7, 5-12, 9-23, 10-8
Hole Chart	13-66
Hole Data	5-20
Hole Options	5-10
Hole Size	5-10
Hole Threads	5-12
Hole/Thread Notes	13-23, 14-31
Hole Types	5-9

I

Idrop	5-92
IFeature Options	3-42
Import Destination Options	6-26
Import Points	6-37
Insert AutoCAD File	6-24
Insert Constraint	16-27, 17-29
Insert Custom Part	13-55
Insert Drawing Border	12-79
Insert iFeature	5-98, 11-4, 18-6
Insert Image	5-77, 6-32, 12-52
Insert Sheet	12-78
Insert Sketched Symbol	14-42
Interactive Contact	3-41
Inventor Content	16-15
Inventor Studio	20-1
iProperties	3-6, 12-61, 14-38
Isometric View	1-8

L

Launchpad	1-2
Leader Text	13-50
Lighting Styles	20-9
Line	1-3, 6-1
Loft	5-32
Look At	4-5

M

Management	3-42
Mate Constraint	16-26
Measure Distance	6-27
Minimum Remnant	9-3
Mirror	6-4, 6-8
Mirror Component	16-23
Mirror feature	5-86, 18-11
Motion Constraints	16-28
Move	6-18
Move Component	16-12, 16-37
Move Face	5-70, 7-2
Multi-User Options	3-13

N

New File	1-2
New Sheet	12-17, 12-45
New Sketch	2-3
Notebook Options	3-37

O

Object Defaults	14-13
Occurrence Properties	16-40
Offset	6-14
Old Versions	3-24
Open file	2-1, 4-1
Options dialog	3-20
Ordinate Dimension	13-18, 14-37
Ordinate Dimension Set	13-18
Orthographic Camera	4-6
Overlay View	12-38

P

Pan	1-10, 4-4
Panel Bar	3-2
Parallel View	2-2
Parameters	6-23
Part Features	3-4
Part Options	2-2, 3-40

Parts List	13-55, 13-46, 14-21
Pattern Component	16-22
Pattern tools	5-82
Perspective Camera	4-6
Physical iProperties	3-12
Place Component	16-3
Place Constraint	16-25
Point, Hole Center	2-10, 5-12, 6-4
Polygon	6-4
Place Component	16-3, 16-15
Play animation	19-10
Plug welds	8-10
Precise View Rotation	19-15
Prefix/Suffix	14-27
Presentation Tools	19-22
Project Cut Edges	6-24
Project Editor	13-3
Project Flat Pattern	6-24
Project Geometry	1-6, 6-23
Project iProperties	3-7
Projected View	12-11, 14-36
Projects	3-8, 3-15
Projects folder	3-21
Promote	5-106
Prompts	3-31
Properties	15-1
Properties dialog	3-6
Property Field	12-54, 12-69
Publish Feature	5-101
Punch Tool	9-25, 9-33
Purge Styles	14-2

R

Rails	5-31
Realtime Zoom	1-10
Rectangle	6-2
Rectangular Pattern	5-79, 5-100, 6-13, 10-7, 11-6
Redefine Feature	2-6, 2-11
Redefine Isometric	11-2
Redefine Sketch	6-6
Redo	4-3
Reflections	3-26
Relative Orientation	7-7
Relative Origin	7-6
Render Image	20-21
Replace All	16-36
Replace Component/Replace All	16-36
Retrieve Dimensions	13-73
Return	4-3, 7-5, 16-11
Revision Table	13-70
Revision Tag	13-71
Revolve	5-4, 5-5, 17-4
Rib	5-28, 11-8
Rotate	6-19, 6-35
Rotate Component	16-37
Rotate View	1-10

S

Save Copy As	2-8, 16-19
Save iProperties	3-11
Save Options	3-22
Save Sketched Symbol	12-71, 14-16
Save Styles	13-10
Save Title Block	12-60, 14-11
Scene Styles	20-21
Search	5-92
Section View	12-17, 12-20
Section View, Assembly	16-34
Select	4-3
Select Other	3-19
Sequence View	19-12
Set Camera	19-13
Set View to Camera	20-20
Shaded	1-12
Shadow	4-7
Shaded	1-12
Shadows	20-11
Share Sketch	5-29
Shared Project	3-19
Sheet	12-7, 14-18
Sheet Format	12-46
Sheet Margins	12-80
Sheet Metal Tools	9-35
Sheet Properties	12-55
Shell	5-22, 5-57
Show/Delete Constraints	6-23
Show Dimensions	6-10, 9-27
Show Startup	3-19
Sketch	1-3, 2-3, 4-3
Sketch Dimensions	1-4
Sketch Options	3-38
Sketch Overlay	12-68
Sketched Symbols	12-69
Sketch Tools	6-40
Slot Weld	8-10
Snap to Grid	3-35
Solids	7-1
Solids Editing Tools	7-8
Spline	6-1
Split	5-72
Spot Weld	8-9
Standard Toolbar	4-10

Startup Dialog	3-19	**V**	
Status Properties	3-8		
Summary Properties	3-3	Versions	3-19
Surface Texture	14-17	View Catalog	5-80
Surface Texture Symbol	13-36	View Justification	3-31
Standard Toolbar	4-9	View Name	12-54
Standards	12-7, 14-1, 14-12	Visibility	2-7, 5-22, 13-68
Status Bar	3-2	Volo View Express	6-25
Status iProperties	3-9		
Styles	9-1, 10-2	**W**	
Styles Editor	13-5, 14-1		
Summary	3-7	Weld	8-5, 8-11, 17-25
Surface Styles	20-2		
Sweep	5-46, 17-3	Weld Symbol	13-39, 14-18
Symbols	12-69, 13-73	Weldment	8-4, 12-12
Symbol Visibility	8-8	Wire frame	1-12
		Work Axis	2-8, 2-10, 5-105
T		Work Plane	2-1, 2-5, 5-105
		Work Points	2-12, 5-105
Table Layout	13-58		
Tangent Constraint	16-26	**Z**	
Team Web	3-21		
Templates folder	3-21	Zoom All	4-3
Terminator	14-28	Zoom Selected	4-4
Text	5-58, 6-29, 6-31, 12-53, 13-48	Zoom Window	4-3
Text Appearance	3-19		
Text Box	14-9		
Text, Dimension	14-27		
Text Styles	14-33		
Texture	15-4, 20-4		
Threads, Hole	5-9		
Thread tool	5-58, 18-22		
Thumbnails	5-91, 12-52		
Title block	12-37		
Toggle Precise UI	7-6		
Tolerance	14-28		
Toolbar	13-1		
Trim	6-18		
Tweak Components	19-5		
Tweak View	19-19		
Tweak Visibility	19-16		

U

Undo	4-3
Undo file	3-19
Update	4-3, 7-5
Use Texture Image	20-7
Username	3-19

About the Author

Elise Moss has worked for the past twenty years as a mechanical designer in Silicon Valley, primarily creating sheet metal designs. She has written articles for Autodesk's Toplines magazine and AUGI's PaperSpace. She is President of Moss Designs, creating custom applications and designs for corporate clients. She has taught at DeAnza College and Evergreen Valley College. She holds a BSME from San Jose State University.

She has been married more than twenty-five years to Ari Stassart, a computer scientist. They have two sons. Benjamin is an electrical engineer at a Fortune 500 firm in Silicon Valley. Daniel is a project manager for a local construction firm.

She can be contacted via email at elise_moss@mossdesigns.com

More information about the author and her work can be found on her website at www.mossdesigns.com.